Advanced and Emerging Technologies in Radiation Oncology Physics

Series in Medical Physics and Biomedical Engineering

Series Editors: John G. Webster, E. Russell Ritenour, Slavik Tabakov,
and Kwan-Hoong Ng

Recent books in the series:

Advanced and Emerging Technologies in Radiation Oncology Physics
Siyong Kim and John Wong (Eds)

A Guide to Outcome Modeling In Radiotherapy and Oncology: Listening to the Data
Issam El Naqa (Ed)

Advanced MR Neuroimaging: From Theory to Clinical Practice
Ioannis Tsougos

Quantitative MRI of the Brain: Principles of Physical Measurement, Second edition
Mara Cercignani, Nicholas G. Dowell, and Paul S. Tofts (Eds)

A Brief Survey of Quantitative EEG
Kaushik Majumdar

Handbook of X-ray Imaging: Physics and Technology
Paolo Russo (Ed)

Graphics Processing Unit-Based High Performance Computing in Radiation Therapy
Xun Jia and Steve B. Jiang (Eds)

Targeted Muscle Reinnervation: A Neural Interface for Artificial Limbs
Todd A. Kuiken, Aimee E. Schultz Feuser, and Ann K. Barlow (Eds)

Emerging Technologies in Brachytherapy
William Y. Song, Kari Tanderup, and Bradley Pieters (Eds)

Environmental Radioactivity and Emergency Preparedness
Mats Isaksson and Christopher L. Rääf

The Practice of Internal Dosimetry in Nuclear Medicine
Michael G. Stabin

Radiation Protection in Medical Imaging and Radiation Oncology
Richard J. Vetter and Magdalena S. Stoeva (Eds)

Statistical Computing in Nuclear Imaging
Arkadiusz Sitek

The Physiological Measurement Handbook
John G. Webster (Ed)

Radiosensitizers and Radiochemotherapy in the Treatment of Cancer
Shirley Lehnert

Advanced and Emerging Technologies in Radiation Oncology Physics

Edited by

Siyong Kim

John Wong

CRC Press
Taylor & Francis Group
Boca Raton London New York

CRC Press is an imprint of the
Taylor & Francis Group, an **informa** business

CRC Press
Taylor & Francis Group
6000 Broken Sound Parkway NW, Suite 300
Boca Raton, FL 33487-2742

First issued in paperback 2020

© 2018 by Taylor & Francis Group, LLC
CRC Press is an imprint of Taylor & Francis Group, an Informa business

No claim to original U.S. Government works

ISBN 13: 978-0-367-57154-2 (pbk)
ISBN-13: 978-1-4987-2004-5 (hbk)

**Visit the Taylor & Francis Web site at
http://www.taylorandfrancis.com**

**and the CRC Press Web site at
http://www.crcpress.com**

Contents

Preface

THESE DAYS, IT IS not unusual to frequently realize how fast technological change does occur. For example, even a law that the doubling of computer processing speed happens every 18 months—known as Moore's Law—does exist in the area of computer science. The field of radiation oncology physics is not an exception and has recently experienced significant technological developments. Such rapid change is expected to be continued even with higher speed. As commonly agreed, medical physics is the driving force in adapting new technologies in radiation therapy. Thus, it is desirable for physicists to be continuously up-to-date in technical aspects, and this book describes advanced and emerging technologies in radiation oncology physics. The main intention of the book is to help medical physicists get proactively prepared for advanced and emerging technologies so that such technologies, when become available for their clinic, can be implemented properly and efficiently to maximize the benefit patients would get from them. In addition, this book is expected to provide important information to both students and researchers that would help them timely find their research topics and directions.

In alignment with the main intention, chapters of this book has been grouped in five based on their topic. Brachytherapy is an important part of radiation therapy but was intentionally excluded in this book.

1. *Topic I—Imaging*: New technologies in imaging physics that are under early-test or have great potential for radiation therapy are mainly described under this topic. To cover the wide range of imaging modalities (e.g., CT, MR, and PET) and related biological modeling, a total of four chapters are allocated under this topic.

2. *Topic II—Treatment planning*: Enhancing computer power is a continuing subject of research and recent architectural advances such as GPU-based computing and cloud-based computing can bring significant benefit for radiation therapy. This topic deals with technological advances in both computer infrastructures and treatment planning algorithms. There are two chapters under this topic.

3. *Topic III—Treatment delivery*: Techniques for setup/target localization in radiation therapy is a unique area and its importance is getting bigger and bigger as more precise delivery within short period time becomes popular. Reviews on both technical improvements in the current systems and new methods in setup/localization are included. When a magnetic resonance imaging unit is integrated

into a treatment unit, it can enhance intra-fraction target monitoring as well as target localization. There are several groups working on such machine, and both the current status and emerging technologies related to those units are described. Interest in charged particle therapy is growing and there is huge effort for improving charged particle therapy system (e.g., super-conducting magnet technology for minimizing unit size), and such developments are described. There are a fair amount of interesting technologies that may not fit into conventional categories but have potential for being useful. One of chapters under this topic describes several of them. Obviously, this is one of major topics and contains four chapters.

4. *Topic IV—Dosimetry, QA, and safety*: Dosimetry is an essential area of radiation therapy, and every new dosimetry technology can make significant impact on routine practice of clinical physics. However, many clinical physicists are not familiar with those methods unless they are directly subject to using such systems. A chapter under this topic describes new developments in dosimetry materials, devices, and systems. Both safety and QA are important subjects in radiation therapy and huge effort is continuously being given on how to enhance them by the radiation therapy societies (e.g., adopting FMEA tool for QA and establishing web-based medical incident reporting systems). Such subjects are dealt with under this topic. Two chapters are included.

5. *Topic V—Informatics*: The importance of informatics in medicine is rapidly growing. One of the outstanding problems in radiotherapy related informatics is how to integrate radiation oncology information into overall medical informatics system. So-called big data is another interesting area, and appropriate utilization of big data has great potential. However, most clinical physicists haven't paid much attention to it. Such issues are described in two chapters assigned under this topic.

There are a total of fourteen chapters in this book. Even though significant effort was made by every author to introduce as many technologies as possible it was practically impossible to handle every technology from every subject. Instead, focus was given on technologies considered either feasibility already demonstrated or heavily impactful when realized. Regarding many other technologies not mentioned in this book, it is the hope of editors that readers would be able to find at least a clue how to get the necessary information through this book.

This book is available in both paper and electronic form. Although there are many colored figures in the e-book version, all of figures are in greyscale in the paper-book version. However, color versions of many figures are available on the CRC Press website (https://www.crcpress.com/9781498720045) and hard-cover book buyers can download them at no additional charge.

Siyong Kim and John Wong

Acknowledgments

THE EDITORS WOULD LIKE to cordially thank all the authors who have made an enormous effort to provide a book chapter that well describes both state-of-the-art and emerging technologies in radiation therapy.

The editors also would like to extend their thanks for the support of the Taylor & Francis team, particularly Francesca McGowan and Rebecca Davies who kept providing administrative help through the whole process of the project.

Lastly, the editors would like to further extend their thanks to their families for the endless support, encouragement, and patience.

About the Series

THE *SERIES IN MEDICAL Physics and Biomedical Engineering* describes the applications of physical sciences, engineering, and mathematics in medicine and clinical research. The series seeks (but is not restricted to) publications in the following topics:

- Artificial organs
- Assistive technology
- Bioinformatics
- Bioinstrumentation
- Biomaterials
- Biomechanics
- Biomedical engineering
- Clinical engineering
- Imaging
- Implants
- Medical computing and mathematics
- Medical/surgical devices
- Patient monitoring
- Physiological measurement
- Prosthetics
- Radiation protection, health physics, and dosimetry
- Regulatory issues
- Rehabilitation engineering
- Sports medicine
- Systems physiology

- Telemedicine

- Tissue engineering

- Treatment

The *Series in Medical Physics and Biomedical Engineering* is an international series that meets the need for up-to-date texts in this rapidly developing field. Books in the series range in level from introductory graduate textbooks and practical handbooks to more advanced expositions of current research.

The *Series in Medical Physics and Biomedical Engineering* is the official book series of the International Organization for Medical Physics.

THE INTERNATIONAL ORGANIZATION FOR MEDICAL PHYSICS

The International Organization for Medical Physics (IOMP) represents over 18,000 medical physicists worldwide and has a membership of 80 national and 6 regional organizations, together with a number of corporate members. Individual medical physicists of all national member organizations are also automatically members.

The mission of IOMP is to advance medical physics practice worldwide by disseminating scientific and technical information, fostering the educational and professional development of medical physics and promoting the highest quality medical physics services for patients.

A World Congress on Medical Physics and Biomedical Engineering is held every three years in cooperation with International Federation for Medical and Biological Engineering (IFMBE) and International Union for Physics and Engineering Sciences in Medicine (IUPESM). A regionally based international conference, the International Congress of Medical Physics (ICMP) is held between world congresses. IOMP also sponsors international conferences, workshops and courses.

The IOMP has several programs to assist medical physicists in developing countries. The joint IOMP Library Programs supports 75 active libraries in 43 developing countries, and the Used Equipment Programs coordinates equipment donations. The Travel Assistance Programs provides a limited number of grants to enable physicists to attend the world congresses.

IOMP co-sponsors the *Journal of Applied Clinical Medical Physics*. The IOMP publishes, twice a year, an electronic bulletin, *Medical Physics World*. IOMP also publishes e-Zine, an electronic news letter about six times a year. IOMP has an agreement with Taylor & Francis for the publication of the *Medical Physics and Biomedical Engineering* series of textbooks. IOMP members receive a discount.

IOMP collaborates with international organizations, such as the World Health Organizations (WHO), the International Atomic Energy Agency (IAEA) and other international professional bodies such as the International Radiation Protection Association (IRPA) and the International Commission on Radiological Protection (ICRP), to promote the development of medical physics and the safe use of radiation and medical devices.

Guidance on education, training and professional development of medical physicists is issued by IOMP, which is collaborating with other professional organizations in development of a professional certification system for medical physicists that can be implemented on a global basis.

The IOMP website (www.iomp.org) contains information on all the activities of the IOMP, policy statements 1 and 2 and the 'IOMP: Review and Way Forward' which outlines all the activities of IOMP and plans for the future.

Editors

Dr. Siyong Kim is Professor and Director of Clinical Physics, Department of Radiation Oncology, Virginia Commonwealth University (VCU), Richmond, Virginia. VCU provides a full range of clinical services at the VCU Medical Center and its six satellite facilities. In addition, three medical physics education programs are offered: Medical Physics Graduate Program; Medical Physics Residency Program; and Medical Physics Certificate Program. Since joining VCU in 2013, he has been the Director of Medical Residency Program as well. He received a master's degree in nuclear engineering from Seoul National University, Seoul, Korea, in 1986. He entered the University of Florida, Gainesville, Florida, to study medical physics in 1993 and earned his PhD with a thesis on Modeling of a Multileaf Collimator (MLC) in 1997. Through the PhD project, he developed a software module of MLC for a commercial treatment planning system, introduced a more accurate equivalent square field formula for head scatter, and established a general solution for head scatter calculation especially in intensity-modulated radiation therapy (IMRT). Dr. Kim has performed research in the area of dose uncertainty estimation, patient motion management, and image guidance method, resulting in over 60 peer-reviewed journal articles and over 150 conference abstracts. He also has mentored students both nationally and internationally, served as either a member of editorial boards or reviewer for over a dozen scientific journals, published 12 book chapters, and published 1 report book. He has been actively involved in professional service including multiple American Association of Physicists in Medicine (AAPM) and IMPCB (International Medical Physic Certification Board) committees.

Dr. John Wong is Professor and Director of Division of Medical Physics in the Department of Radiation Oncology and Molecular Radiation Sciences at Johns Hopkins University School of Medicine.

Dr. Wong oversees the physics and dosimetry services of the department. Dr. Wong is the primary or contributing author of over 170 peer-reviewed scientific publications and 20 book chapters. He has been a principal investigator or co-investigator on 20 research initiatives funded by public agencies and industries. He is a co-inventor of the Active Breathing Coordinator, flat panel Cone-Beam CT, and the Small Animal Radiation Research Platform (SARRP) that have been commercialized as radiation therapy products for the clinical and research community.

Dr. Wong is a Fellow of the American Association of Medical Physicists, the recipient of the George Edelstyn Medal from the Royal College of Radiology, United Kingdom in 2001, and the awardee of the Edith Quimby Lifetime Achievement Award of the American Association of Physicists in Medicine (AAPM) in 2017. His current research focus is on molecular optical imaging for pre-clinical radiation research, robotic ultrasound imaging for Image-guided radiation therapy (IGRT), and informatics infra-structure for data sharing in radiation oncology. He is committed to advancing cancer treatment through education, research, and collaboration.

Contributors

Stephen R. Bowen
Radiation Oncology and Radiology
University of Washington
Seattle, Washington

Stephen Boyd
Electrical Engineering
Stanford University
Stanford, California

Theodore L. DeWeese
Radiation Oncology and Molecular
    Radiation Sciences
Johns Hopkins University
Baltimore, Maryland

Toshifumi Gabata
Radiology
Kanazawa University
Kanazawa, Japan

Matthias Guckenberger
Radiation Oncology
University of Zürich
Zürich, Switzerland

Kristi R. G. Hendrickson
Radiation Oncology
University of Washington
Seattle, Washington

David Hoffman
Radiation Medicine and Applied Sciences
University of California San Diego
La Jolla, California

Geoffrey Hugo
Radiation Oncology
Washington University
St. Louis, Missouri

Robert Jeraj
Medical Physics
University of Wisconsin
Madison, Wisconsin

Steve Jiang
Radiation Oncology
University of Texas Southwestern
Dallas, Texas

Hosang Jin
Radiation Oncology
University of Oklahoma
Oklahoma City, Oklahoma

Daniel Johnson
Radiation Oncology
University of Oklahoma
Oklahoma City, Oklahoma

Alan M. Kalet
Radiation Oncology
University of Washington
Seattle, Washington

Siyong Kim
Radiation Oncology
Virginia Commonwealth University
Richmond, Virginia

Taeho Kim
Radiation Oncology
Virginia Commonwealth University
Richmond, Virginia

John P. Kirkpatrick
Radiation Oncology
Duke University
Durham, North Carolina

Satoshi Kobayashi
Radiology
Kanazawa University
Kanazawa, Japan

Choonik Lee
Radiation Oncology
University of Michigan
Ann Arbor, Michigan

Ruijiang Li
Radiation Oncology
Stanford University
Stanford, California

Lawrence B. Marks
Radiation Oncology
University of North Carolina
Chapel Hill, North Carolina

Kosuke Matsubara
Quantum Medical Technology
Kanazawa University
Kanazawa, Japan

Todd McNutt
Radiation Oncology
Johns Hopkins University
Baltimore, Maryland

Vitali Moiseenko
Radiation Medicine and Applied Sciences
University of California San Diego
La Jolla, California

Yang-Kyun Park
Radiation Oncology
University of Texas Southwestern
Dallas, Texas

Eric Paulson
Radiation Oncology
Medical College of Wisconsin
Milwaukee, Wisconsin

Julian Perks
Radiation Oncology
University of California Davis
Sacramento, California

Mark H. Phillips
Radiation Oncology
University of Washington
Seattle, Washington

Harry Quon
Radiation Oncology and Molecular
    Radiation Sciences
Johns Hopkins University
Baltimore, Maryland

Lei Ren
Radiation Oncology
Duke University
Durham, North Carolina

Wade P. Smith
Radiation Oncology
University of Washington
Seattle, Washington

Jao Jang Su
Physics
University of Maryland
College Park, Maryland

Jinsoo Uh
Radiation Oncology
St. Jude Children's Research Hospital
Memphis, Tennessee

Baris Ungun
Radiation Oncology
Stanford University
Stanford, California

Elisabeth Weiss
Radiation Oncology
Virginia Commonwealth University
Richmond, Virginia

John Wong
Radiation Oncology
Johns Hopkins University
Baltimore, Maryland

Lei Xing
Radiation Oncology
Stanford University
Stanford, California

Yulong Yan
Radiation Oncology
University of Texas Southwestern
Dallas, Texas

Yinyu Ye
Management Science and Engineering
Stanford University
Stanford, California

Seonghwan Yee
Radiology
University of Colorado
Aurora, Colorado

Fang-Fang Yin
Radiation Oncology
Duke University
Durham, North Carolina

Alicia Yingling
Radiation Oncology
University of Texas Southwestern
Dallas, Texas

Masoud Zarepisheh
Medical Physics
Memorial Sloan Kettering Cancer Center
New York, New York

CHAPTER 1

Recent Advances in Computed Tomography

Choonik Lee, Satoshi Kobayashi, Kosuke Matsubara, and Toshifumi Gabata

CONTENTS

1.1 INTRODUCTION

Since the first patient brain scan was performed in 1971, computed tomography (CT) has gained unmatched popularity in the medical imaging field. It is performed for close to 70 million cases in the United States alone in 2006 (Beckmann, 2006; Mettler et al., 2008, 2009; Schauer and Linton, 2009). The increased speed of the scan, especially with the multislice detector, has accelerated the adaptation of CT. In external beam radiation therapy, CT imaging has almost eliminated the need for the conventional simulator. The volumetric anatomical information that CT provides enables three-dimensional treatment planning, in which normal tissues can be spared while maximizing the dose to target volumes. Digitally reconstructed beam's-eye views from CT also provide critical anatomical information for treatment planning and daily patient setup (McShan et al., 1990).

CT images provide fast and robust morphological information but, more important, they also provide relative quantitative information of different tissues. For megavoltage photon and electron radiation therapy, the CT number-to-electron density or physical density relations have been used for accurate dose calculation within patients (Kijewski and Bjärngard, 1978; Parker et al., 1979; Cassell et al., 1981; McShan, 1987; Schneider et al., 2000). This CT number–to–density conversion is a required step due to the difference in the energies of diagnostic X-rays (typically around 100 kVp), which are far less than the energies of therapeutic beams that conventional medical linear accelerators produce.

With the advancement of the sixth generation multi-slice helical CT, it was possible to capture temporal displacements within the patient (Low et al., 2003; Vedam et al., 2003; Keall et al., 2004). This so-called four-dimensional CT (4DCT) enables the assessment of organ motion, especially for organ motion caused by respiration. Target volume encompassing the full extent of motion could be defined and used for radiation therapy treatment planning optimization (Underberg et al., 2004), further reducing the dose to nearby normal tissues.

There is no question that current CT technology provides invaluable morphological and radiological information that is essential to radiation therapy treatment calculation and image guidance; however, CT suffers from its own limitations compared to other imaging modalities, such as magnetic resonance imaging (MRI). Compared to MRI, which can provide excellent soft-tissue contrast without the use of ionization radiation, conventional CT has limited soft-tissue contrast and can impart non-negligible radiation exposure to patients. Its poorer contrast has been shown to result in less robust prostate contouring than MRI (Roach et al., 1996; Rasch et al., 1999). The inherent radiation exposure and its potential risk of secondary cancer is of concern (Brenner and Hall, 2007; Berrington de González et al., 2009; Mettler et al., 2009; Schauer and Linton, 2009; Pearce et al., 2012). Practice changes have been recommended for lowering the overall imaging dose as much as possible, especially for the pediatric patient group (Kalra et al., 2004; Goske et al., 2008, 2012).

In this chapter, we will discuss recent development in the field of multispectral CT, or dual-energy CT, which can provide improved soft-tissue contrast without additional radiation exposure, along with other recent exciting developments in CT.

1.2 DUAL-ENERGY CT

1.2.1 Background and Physics of Dual-Energy CT

It has been recognized since the inception of CT that dual-energy acquisition could improve tissue characterization by taking advantage of the energy dependency of photoelectric effects (Chiro et al., 1979; Millner et al., 1979), but dual-energy CT (DECT) has not been clinically implemented until recently because the technological requirements to achieve near-simultaneous acquisition were not available (Johnson et al., 2011; Coupal et al., 2014).

In typical clinical practice, X-ray tube voltages used in CT scanning range between 70 and 140 kVp. Due to the probabilistic nature of the bremsstrahlung production in the rotating anode target, the resulting X-rays have a broad spectrum ranging between 30 and 140 keV, depending on X-ray tube design. In this energy range, Compton scattering and photoelectric effects play dominant roles as the X-rays interact with human tissues. Soft tissues with low atomic numbers are not affected much with tube voltage selections because their K-shell binding energy values are not much different and are relatively low. For tissues with higher atomic numbers, however, the resulting image contrast is affected heavily by the selection of tube voltage because the contributions from photoelectric effects are accentuated by steep power functions of X-ray energy and atomic number. With dual-energy CT, it is possible to decompose the inherent contrast in attenuation coefficients as a linear combination of distinctive Compton scattering and photoelectric components (Saba et al., 2015).

For example, in the conventional CT energy range, calcium and iodine may result in similar attenuation. At a lower energy range, however, the attenuation of iodine is higher than that of calcium, thus making it possible to differentiate the two materials (Figure 1.1)

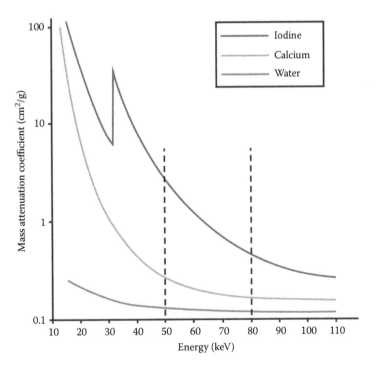

FIGURE 1.1 Mass-attenuation coefficients for iodine (blue), calcium (green), and water (red) on CT images obtained at two different energies (vertical dashed lines) shows that these materials can be characterized by comparing their attenuation at the lower energy with that at the higher energy. When dual-energy images reconstructed for 50 and 80 keV are compared, iodine demonstrates a greater decrease in attenuation than calcium does at the higher energy, whereas the attenuation of water remains more or less constant. (From Kaza, R.K. et al., *Radiographics*, 32, 353–369, 2012. With permission.)

(Kalender et al.. 1986; Kaza et al., 2012). This feature can be used to create virtual noncontrast images (contrast subtraction) along with the usual contrast-enhanced images from a single acquisition of DECT (Brodoefel et al., 2009; Toepker et al., 2012). This implies that overall patient exposure from DECT could be lower than the conventional contrast-enhanced CT, which typically involves two or more scans with and without contrast. This ability of material decomposition can enable the diagnosis of various conditions, including cancerous tissues, where conventional CT would fail to provide enough conspicuousness (Coursey et al., 2010).

The settings of 80 and 140 kVp are commonly used in DECTs because they provide enough difference and less overlap between the spectra produced with standard X-ray tube (Johnson, 2012). Newer generation DECTs, second-generation DECTs, may have a pair of 100 and 140 kVp sources where the spectral overlap is limited by applying additional filtering to 140 kVp source, resulting in better decomposition of materials (Primak et al., 2009).

Different approaches have been attempted by different investigators to achieve optimal decomposition. These approaches can be categorized as follows: (1) dual-source, dual-detector CT scanner (DSCT); (2) single-source, fast energy switching CT scanner; (3) single-source, dual-layer detector CT scanner; and (4) single-source with split-filter CT scanner. Figure 1.2 illustrates three different approaches.

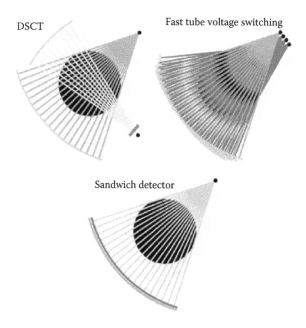

FIGURE 1.2 Different hardware approaches to dual-energy CT (DECT) imaging. In particular, three implementations are available: dual-source CT (DSCT), fast kilovoltage (kV) switching, and sandwich (dual-layer) detector techniques. (From Simons, D. et al., *Eur. Radiol.*, 24, 930–939, 2014. With permission.)

1.2.2 Implementation of Dual-Energy CT

1.2.2.1 Dual-Source, Dual-Detector CT

Dual-source, dual detector CT (DSDT) scanners employ two independent X-ray source and detector pairs attached on a single CT gantry. The two pairs scan at the same anatomical level in the z-direction (craniocaudal direction), but their projection angles are offset by 95°. Each source-detector array pair can be optimized independently for spectrum and tube current to maximize the spectral separation while minimizing the patient dose (Schenzle et al., 2010). Due to the physically different angular projection and nonidentical detector sizes, however, the material decomposition process can be processed only in the image-domain (instead of projection domain), which may result in inferior image quality compared to fast energy-switching DECT, which is explained in the next section. Also, the smaller second source-detector pair limits the field of view (FOV) of 33 cm, which may limit oncologic application where a large FOV is critical (1) to accommodate various immobilization devices and (2) to consider the external body contour for accurate dose calculation.

1.2.2.2 Single-Source, Fast Energy Switching

In this approach, the two voltage settings, typically 80 and 140 kVp, of a single X-ray tube are alternated rapidly during a single axial or helical scan. The Gemstone Spectral Imaging (GSI) mode of the Discovery CT750 HD scanner (GE Healthcare, Milwaukee, Wisconsin, United States) can switch between high kVp and low kVp rapidly (at intervals of 0.5 ms or less) in adjacent projections. This enables near-simultaneous acquisition of two energies, resulting in almost no temporal displacement between the two acquisitions (Zhang et al., 2011). Because the individual projections are created from nearly the same anatomical location and projection angle, projection-level (sinogram domain instead of image domain) material decomposition can be performed (Kaza et al., 2012). The sequential and alternating acquisition is also inherently less prone to signal contamination compared to the DSCT system (Ginat and Gupta, 2014). Due to the lower output of the 80 kVp (higher attenuation via the photoelectric effect), a unequal exposure time is assigned to the two voltages (about 40% to 140 kVp and 60% to 80 kVp) to obtain equivalent image quality between the two scans (Figure 1.3) (Li et al., 2011).

Single-source, fast energy switching CT systems can achieve the conventional FOV of 50 cm compared to the smaller dual-source CT system, as previously mentioned before. However, only one physical X-ray tube for two energies makes it difficult to optimize the filtration of each spectrum.

1.2.2.3 Single-Source, Dual-Layer Detector CT

The single-source, dual-layer detector CT utilizes the polychromatic spectrum of a single X-ray source as measured by a special energy-resolving multilayer detector system (Figure 1.4).

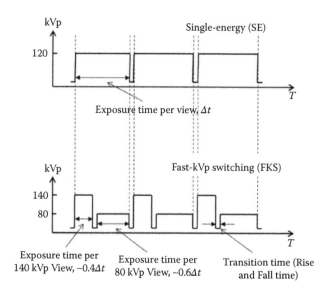

FIGURE 1.3 The exposure time of each 140 or 80 kVp view in fast-kVp switching is roughly only 40% or 60%, respectively, of that of each single-energy view. (From Li, B. et al., *Med. Phys.*, 38, 2595–2601, 2011. With permission.)

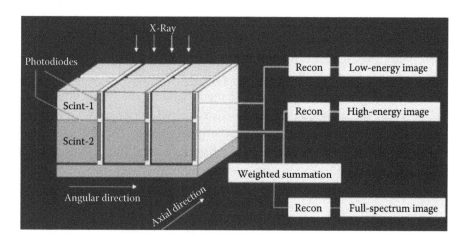

FIGURE 1.4 The dual-layer detection system (only a few detector elements are shown). The photodiodes are parallel to the X-ray direction, attached to the sides of the two types of scintillator elements. (From Johnson, T. et al., *Dual Energy CT in Clinical Practice*, Springer, Berlin, Germany, 2011. With permission.)

The two layers of scintillating materials are sandwiched together in a way that, under high voltage settings (120 or 140 kVp), the top layer will absorb about 50% of the lower-energy spectrum, while the bottom layer absorbs the higher-energy spectrum (Vlassenbroek, 2011; Gabbai et al., 2015). This approach has an advantage over the others because it inherently eliminates any temporal or spatial displacement. There is also no sacrifice in FOV.

The material decomposition can be processed at the sinogram domain, and it is also free from cross-scatter problems of the dual-source (dual-detector) technique.

This technique has been implemented by Philips in its IQon Spectral CT (Philips Healthcare, Eindhoven, The Netherlands). However, the spectral separation is not as good as other dual-energy approaches with a distinct, two-X-ray source because the top layer of the dual-layer detector may also absorb higher-energy photons (Saba et al., 2015). Furthermore, the detector system is optimized for a higher tube potential of 120 or 140 kVp, which is a disadvantage for reducing the radiation dose.

1.2.2.4 Single-Source, Split-Filter CT Scanner

This approach is the most recent development. It utilizes two different prefilter materials (Au and Sn) within the X-ray tube housing to split the beam along the patient's longitudinal direction. In the Twin Beam mode of Somatom Definition Edge (Siemens Healthcare, Forchheim, Germany), one half of the filter is made of tin and aluminum, and very thin gold and aluminum makes up the other half of the filtration (Saba et al., 2015). The energy separation may not be as good as it is with the dual-source CT, but it does not suffer from the reduced scan area and provides dual-energy data for the full FOV of 50 cm for both energy groups.

1.2.3 Implication of Dual-Energy CT in Radiation Therapy Dose Calculation

The material decomposition ability of dual-energy CT has shown potential to improve radiation therapy dose calculations. Conventional CT has long suffered from challenges such as beam hardening artifacts, metal artifacts, and good but not perfect correlation with the stopping power ratio. The last is especially critical for proton radiation therapy dose calculation because the pencil beam dose calculation algorithm relies on the stopping power ratio of the medium to water to calculate proton range. It has been shown that DECT can be used more reliably to estimate tissue-specific stopping power ratios compared to conventional CT for various types of tissues (Yang et al., 2010; van Elmpt et al., 2016).

With improved soft-tissue contrast, DECT has shown a great potential for better tumor detection and characterization while reducing the total scan time and patient dose (Simons et al., 2014; De Cecco et al., 2015).

1.3 ADVANCES IN CT RECONSTRUCTION ALGORITHM

As noted earlier, conventional CT imaging is limited in resolving metal artifacts, which poses significant challenges in tumor identification and accurate dose calculation. Recent developments in the CT reconstruction algorithm enabeled much improved metal artifact reduction (MAR). More advanced algorithms, once considerd too computation-intensive, are now clinically available to achieve equivalent image quality while delievering significantly less doses to the patient.

1.3.1 Metal Artifact Reduction Techniques

The presence of high-density meterial implants, such as hip prostheses, dental fillings, spinal fixations, and vascular coils, can cause significant CT image artifacts. Increased beam hardening effect (absorption of low energy X-rays) and photon starvation (lack of photon transmission and detection) result in bright streak and dark void artifacts in the reconstructed images, which not only obscure clinical diagnosis but also may lead to inaccurate Hounsfield units (HU) for radiaton therapy dose calculation (Reft et al., 2003; Bamberg et al., 2011; Spadea et al., 2014).

Various MAR techniques have been proposed to alleviate these issues. The corrupted and lost projection information could be corrected by linear interpolation of adjacent projection data (Kalender et al., 1987; Mahnken et al., 2003) or iterative reconstructions (IRs) (Wang et al., 1996, 2000). The interpolation method can eliminate streaking due to photon starvation and beam hardening, but it may introduce new artifacts, especially at metal and bone interfaces (Prell et al., 2009).

Recently, MAR techniques have been introduced for commercial CT scanners, such as the orthopedic metal artifacts reduction (O-MAR) algorithm (Philips Healthcare, Cleveland, Ohio). The O-MAR algorithm identifies metal and tissue pixels out of the initlally reconstructed images and creates corresponding sinograms via forward projection. The metal sinogram data is then subtracted from the original sinogram before they are used to reconstruct a corrected image via back projection. The algorithm iteratively updates the corrupted projection data by using the corrected data until it reaches a user-defined optimization threshold. Investigators have found significant artifact reductions, but only a slight improvement in accuracy of the calculated dosimetry for patients with metal hip implants (Li et al., 2012). The O-MAR is a purely algorithm-based approach and can be implemented for an existing CT scanner without hardware modification (Figure 1.5).

As introduced in the previous sections, DECT can simulate virtual monoenergetic CT images. This feature could substantially reduce metal artefacts (Bamberg et al., 2011). The users of GE's HD750 Discovery CT system (GE Healthcare, Milwaukee, Wisconsin) may use its monochromatic GSI technique to generate synthesized virtual monochromatic images along with its MAR tool (Figure 1.6) (Lee et al., 2012; Pessis et al., 2013; Huang et al., 2015). After comparing O-MAR, GSI, and GSI with MAR techniques in phantom and clinical cases, Huang et al. found that no single method outperformed the others, suggesting that the reduction techniques should be used with caution and a good understanding of their weaknesses.

1.3.2 Iterative Reconstruction Technique

With the subsequent radation exposure to patients that comes with the increased use of CT for various diagnostic examinations, concern for the risk of secondary malignancies, such as leukemia and brain cancer, has also grown (Hall and Brenner, 2008, 2012; Pearce et al., 2012). With the conventional Filtered Back Projection (FBP) reconstruction algorithm, patient radiation exposure can be reduced most effectively by optimizing

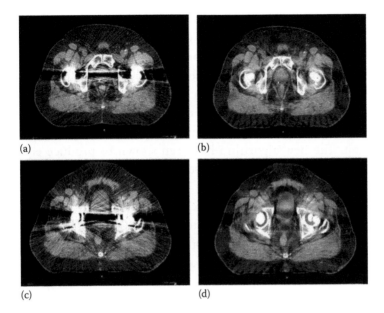

FIGURE 1.5 Metal artifact reduction for two different axial planes on a patient case with bilateral metal hip implants. The image display window width is 400 HU, and the window level is 800 HU. (a) Before O-MAR correction. (b) After O-MAR correction. (c) Before O-MAR correction. (d) After O-MAR correction. (From Li, H. et al., *Med. Phys.*, 39, 7507–7517, 2012. With permission.)

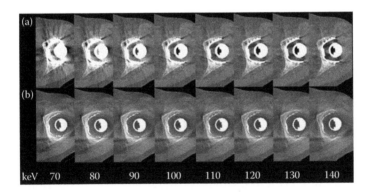

FIGURE 1.6 A 63-year-old woman with total hip arthroplasty of the left hip: Acetabular cup and femoral stem with stainless steel and femoral head with alumina ceramic. CT images were reconstructed with different keV values: 70, 80, 90, 100, 110, 120, 130, and 140 keV. Reconstructions with different DFOVs without MARs (a) and with MARs (b) showed the changes with different keV energy levels. (From Lee, Y.H. et al., *Eur. Radiol.*, 22, 1331–1340, 2012. With permission.)

the hardware itself and the acquisition parameters, such as tube current (Kalra et al., 2004). However, when tube current is substantially reduced and/or patients are obese, the FBP algorithm produces substandard image quality for diagnostic purpose (Hara et al., 2009; Leipsic et al., 2010; Moscariello et al., 2011; Singh et al., 2011), limiting the options for further dose reduction.

Another class of image reconstruction algorithms known as IR algorithms has recently been introduced commercially for CT. This class of reconstruction algorithms was originally introduced with the development of CT as the algebraic reconstruction technique (ART), but it did not gain much popularity becasuse of its heavy computational burden and nonconvergence with realistic noisy data (Herman et al., 1973; Hounsfield, 1973). With recent advances in computing power and algorithms, all major CT manufacturers currently provide various forms of the IR algorithm to aid low-dose image acquisition. The new algorithm has been shown to produce equivalent diagonistic quality images with significantly lower tube currents than the conventional FBP algorithm (Hara et al., 2009).

In the IR algorithm, each iteration typically performs both forward and back projections of an intermediate image volume to identify and model the noise component of the projection image. The normal FBP of the measured projection data was initially used to reconstruct an estimated model of the subject. During forward projection, X-ray interactions are modeled based on the source and detector geometry, resulting in a set of simulated projections. This set of simulated projections is compared against the actual measurements collected by the CT detectors. The iterations continue, and the intermediate stage of the patient image is updated by alternating back and forward projections until the discrepancy between the actual projection and the simulated projection is within certain statistical criteria (Hsieh et al., 2013; Willemink et al., 2013). The ideal result of the IR is to generate CT images without the identified image artifacts and noises.

Each commercial solution of IR has a slight variation to achieve a model-based reconstructed image at low dose. Adaptive Statistical Iterative Reconstruction (ASIR) and Model-Based Iterative Reconstructions (MBIR); both from GE Healthcare, (Waukesha, Wisconsin, the United States) are IR techniques where images are reconstructed via iteration of full forward and back projection reconstructions, as described in Figure 1.7a. Iterative Reconstruction in Image Space (IRIS); Siemens Medical Solutions, Forchheim, Germany iterates in the image domain alone (Figure 1.7b), whereas Sinogram-Affirmed Itertive Reconstructinos (SAFIRE); (Siemens Medical Solutions, Forchheim, Germany), Adaptive Iterative Dose Reduction 3D (AIDR 3D), (Toshiba Medical Systems, Tokyo, Japan), and iDose (Philips Healthcare, Best, The Netherlands) algorithms iterate in both the raw data domain and image domain (Figure 1.7c) (Willemink et al., 2013).

Singh et al. demonstrated how ASIR algorithm can achieve clinically acceptable signal-to-noise ratios with significantly less tube current than the conventional filtered back projection algorithm (Figure 1.8). In radiation oncology, this advantagous low-dose reconstruction algorithm can benefit radiation oncology patients who undergo multiple CT scans for motion assessment or treatment response evaluation for possible adaptive plan modifications.

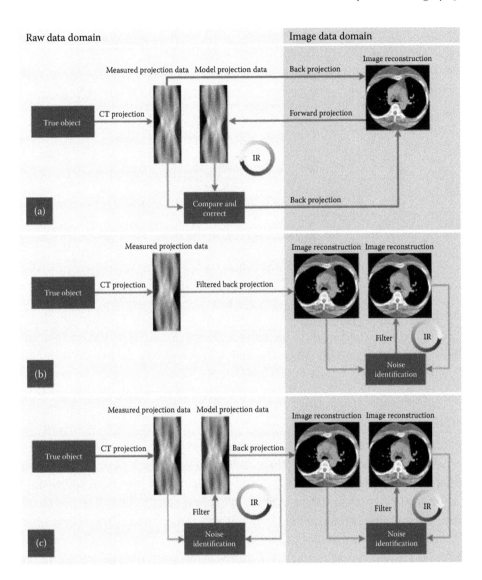

FIGURE 1.7 Overview of commercially available iterative reconstruction techniques. Full iterative reconstruction consists of forward and backward reconstruction steps (A-ASIR and MBIR). Less advanced iterative reconstruction algorithms iterate in the image domain alone (B-IRIS) or in both the image and the raw data domain (C-AIDR 3D, iDose4, and SAFIRE). IR Iterative reconstruction; IRIS Iterative Reconstruction in Image Space, Siemens Medical Solutions; AIDR 3D Adaptive Iterative Dose Reduction, Toshiba Medical Systems; ASIR Adaptive Statistical Iterative Reconstruction, GE Healthcare; iDose4, Philips Healthcare; SAFIRE Sinogram Affirmed Iterative Reconstruction, Siemens Medical Solutions; MBIR Model-Based Iterative Reconstruction, GE Healthcare. (From Willemink, M.J. et al., *Eur. Radiol.*, 23, 1623–1631, 2013. With permission.)

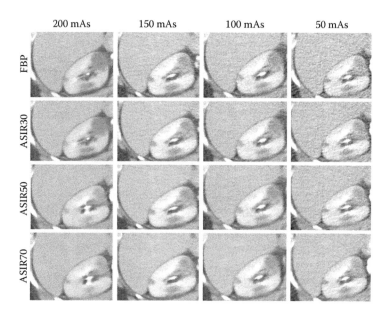

FIGURE 1.8 Transverse abdominal CT images of multiple hypo-intense renal lesion reconstructed with FBP and three levels of ASIR (30%, 50%, and 70%) at four tube current–time products (200, 150, 100, and 50 mAs). (From Singh, S. et al., *Radiology*, 259, 565–573, 2011. With permission.)

1.4 MOBILE CT

The recent development of mobile CT scanners allows more versatility in terms of the point of service. In conventional settings, patients in need of CT examinations have to be transferred and moved onto the CT scanner table, which may pose additional risk of injury for critically ill patients (Papson et al., 2007; Peace et al., 2010). It is also challenging to provide CT examination for patients undergoing surgery unless the operating room is specially equipped with a stationary CT scanner.

Mobile CT, also known as portable CT, can provide point-of-care volumetric imaging with reasonably good image quality. Several mobile CT scanners are available commercially, including the CereTom and BodyTom (NeuroLogica Corp, Danvers, Massachusetts, the United States), the xCAT (Xoran Technologies, Ann Arbor, Michigan, the United States), and the Airo (BrainLab, Inc., Feldkirchen, Germany) systems.

The BodyTom is a battery-powered full-body 32-slice CT scanner with an 85 cm diameter bore (60 cm FOV), whereas the CereTom has a smaller bore size of 32 cm to provide higher-quality images (Figure 1.9). The patient is immobilized and the gantry travels over the patient while acquiring tomographic images in both systems. The xCAT system is a mobile cone beam CT (CBCT) scanner that is mainly used intraoperatively in otolaryngologic procedures (Figure 1.10). The CBCT system acquires volumetric images, with the only moving part being the rotating gantry. The Airo CT scanner is similar to BodyTom system, but it provides much wider clearance with a bore size of 107 cm and a 50 cm FOV diameter. The Airo 32-slice helical scanner can navigate through tight spaces with its small footprint of 1.5 m^2.

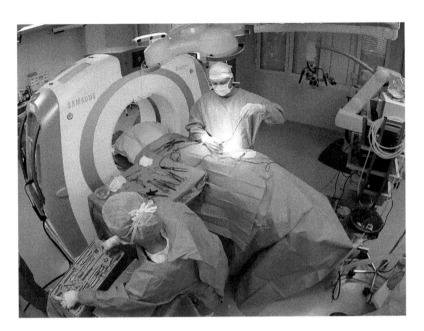

FIGURE 1.9 BodyTom® by NeuroLogica-Samsung is a battery-powered portable multi-slice mobile CT. (With permission.)

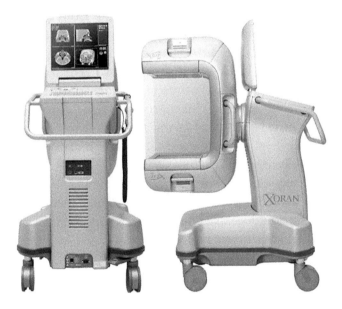

FIGURE 1.10 Xcat® ENT by Xoran Technologies is a mobile cone-beam CT unit that is specialized for otolaryngology surgery. (With permission.)

Although the image quality and FOV may not be as good as that of the conventional diagnostic CT scanners (Rumboldt et al., 2009), these mobile CT scanners will find their applications in brachytherapy procedures such as interstitial implants where the three-dimensional volumetric imaging provides improved accuracy of the implant and dose delivery.

1.5 PHASE-CONTRAST CT

Another breakthrough in CT in terms of soft-tissue contrast may come from phase-contrast CT. Even though conventional CT provides excellent cross-sectional anatomy, it still suffers from poor soft-tissue contrast when used without an added contrast enhancement agent. This is due to the fundamentally low differential X-ray attenuations among soft tissues. However, in phase-contrast imaging (PCI), in addition to X-ray absorption, the changes of X-ray wavefront at the interfaces of different soft tissues are detected and used to derive significantly enhanced soft-tissue contrast. Even tissues with similar X-ray attenuation properties may show significant differences in their refractive indices to enable high-contrast soft-tissue imaging. However, the detection of this subtle change in phases is more technically challenging than it is with conventional CT imaging.

Various approaches have been explored over the years to take advantage of the phase-shifting effects of X-ray interactions. Recent review papers typically classify them into five main categories (Bravin et al., 2012; Coan et al., 2013; Diemoz et al., 2012): (1) the interferometric methods (Bonse and Hart, 1965; Momose et al., 1996), (2) the propagation-based imaging (PBI) methods (Snigirev et al., 1995), 3) the analyzer-based imaging (ABI) methods (Förster et al., 1980; Davis et al., 1995), (4) the grating interferometric (Weitkamp et al., 2005), and (5) the grating noninterferometric methods (Olivo and Speller, 2007).

Initial effort with crystal-based interferometer relied on high-intensity monochromatic X-rays, which required large scale synchrotron facilities. However, grating-based phase contrast imaging allows the use of conventional polychromatic X-ray sourses (Pfeiffer et al., 2006; Tapfer et al., 2011). Breast tissue has been one of the early sites of this unique imaging modality, resulting in higher soft-tissue contrast and better spatial resolution than with conventional X-ray imaging in differentiating tumor versus grandular tissues. Recent developments have achieved extremely high resolution three-dimentional images of biological samples, with a pixel resolution of 7.4 μm and enough soft-tissue contrast to permit visualization of cellular-level compartmental structure information within soft tissues (McDonald et al., 2009; Beheshti et al., 2013).

This promising development may enable early screening and staging of cancerous tissue with higher accuracy and with lower radiation dose than conventional X-ray examinations. In addition, the spacially detailed morphological information should lead to better understanding of radiation treatment response.

1.6 CONCLUSION

Recent advancements in CT has enabled better or equivalent image quality while significantly reducing image artificats and patient radiation exposure. The DECT is commercially available from multiple vendors that utilize various hardware and software approaches to take advantage of the energy dependence of X-ray interactions within patient anatomy. The oncologic application of DECT is still early, but it is a very active area of research with great potential of improving cancer diagnosis and staging. The FOV of DECT is not yet wide enough to accommodate various immobilization devices of the radiation oncology patients. The smaller and portable mobile CT scanners can significantly improve image quality for brachytherapy cases, such as prostate or gyncological implants for which ultrasound or two-dimensional imaging is still widely used. Phase-contrast CT imaging is preclinical yet a very exciting new technology that can provide extremely high resolution and high soft-tissue contrast compared to conventional CT imaging. With increasing emphasis being given to personalized medicine in radiation oncology, these new improvements and discoveries will help improve the diagnosis of cancer, monitoring of treatment responses, and adaptive optimization of radiation treatments.

REFERENCES

Bamberg, F., A. Dierks, K. Nikolaou, et al. 2011. Metal artifact reduction by dual energy computed tomography using monoenergetic extrapolation. *Eur Radiol* 21 (7):1424–1429.

Beckmann, E. C. 2006. CT scanning: The early days. *Br J Radiol* 79 (937):5–8.

Beheshti, A., B. R. Pinzer, J. T. McDonald, M. Stampanoni, and L. Hlatky. 2013. Early tumor development captured through nondestructive, high resolution differential phase contrast X-ray imaging. *Radiat Res* 180 (5):448–454.

Berrington de González, A., M. Mahesh, K. P. Kim, et al. 2009. Projected cancer risks from computed tomographic scans performed in the United States in 2007. *Arch Intern Med* 169 (22):2071–2077.

Bonse, U, and M Hart. 1965. An x ray interferometer with long separated interfering beam paths. *Appl Phys Lett* 7 (4):99–100.

Bravin, A., P. Coan, and P. Suortti. 2012. X-ray phase-contrast imaging: from pre-clinical applications towards clinics. *Phys Med Biol* 58 (1):R1.

Brenner, D. J., and E. J. Hall. 2007. Computed tomography: An increasing source of radiation exposure. *N Engl J Med* 357 (22):2277–2284.

Brodoefel, H., C. Burgstahler, M. Heuschmid, et al. 2009. Accuracy of dual-source CT in the characterisation of non-calcified plaque: Use of a colour-coded analysis compared with virtual histology intravascular ultrasound. *Br J Radiol* 82 (982):805–812.

Cassell, K. J., P. A. Hobday, and R. P. Parker. 1981. The implementation of a generalized batho inhomogeneity correction for radiotherapy planning with direct use of CT numbers. *Phys Med Biol* 26 (5):825–833.

Chiro, G. D., R. A. Brooks, R. M. Kessler, et al. 1979. Tissue signatures with dual-energy computed tomography. *Radiology* 131 (2):521–523.

Coan, P., A. Bravin, and G. Tromba. 2013. Phase-contrast x-ray imaging of the breast: Recent developments towards clinics. *J Phys D Appl Phys* 46 (49):494007.

Coupal, T. M., P. I. Mallinson, P. McLaughlin, et al. 2014. Peering through the glare: Using dual-energy CT to overcome the problem of metal artefacts in bone radiology. *Skeletal Radiol* 43 (5):567–575.

Coursey, C. A., R. C. Nelson, D. T. Boll, et al. 2010. Dual-energy multidetector CT: How does it work, what can it tell us, and when can we use it in abdominopelvic imaging? *Radiographics* 30 (4):1037–1055.

Davis, T. J., T. E. Gureyev, D. Gao, A. W. Stevenson, and S. W. Wilkins. 1995. X-ray image contrast from a simple phase object. *Phys Rev Lett* 74 (16):3173.

De Cecco, C. N., A. Laghi, U. J. Schoepf, and F. G. Meinel. 2015. *Dual Energy CT in Oncology*. Springer International Publishing, Switzerland.

Diemoz, P. C., A. Bravin, and P. Coan. 2012. Theoretical comparison of three X-ray phase-contrast imaging techniques: Propagation-based imaging, analyzer-based imaging and grating interferometry. *Opt Express* 20 (3):2789–2805.

Förster, E., K. Goetz, and P. Zaumseil. 1980. Double crystal diffractometry for the characterization of targets for laser fusion experiments. *Kristall und Technik* 15 (8):937–945.

Gabbai, M., I. Leichter, S. Mahgerefteh, and J. Sosna. 2015. Spectral material characterization with dual-energy CT: Comparison of commercial and investigative technologies in phantoms. *Acta Radiol* 56 (8):960–969.

Ginat, D. T., and R. Gupta. 2014. Advances in computed tomography imaging technology. *Annu Rev Biomed Eng* 16:431–453.

Goske, M. J., K. E. Applegate, J. Boylan, et al. 2008. The Image Gently campaign: Working together to change practice. *AJR Am J Roentgenol* 190 (2):273–274.

Goske, M. J., K. E. Applegate, D. Bulas, et al. 2012. Image Gently 5 years later: What goals remain to be accomplished in radiation protection for children? *AJR Am J Roentgenol* 199 (3):477–479.

Hall, E. J., and D. J. Brenner. 2008. Cancer risks from diagnostic radiology. *Br J Radiol* 81 (965):362–378.

Hall, E. J., and D. J. Brenner. 2012. Cancer risks from diagnostic radiology: The impact of new epidemiological data. *Br J Radiol* 85 (1020):e1316–e1317.

Hara, A. K., R. G. Paden, A. C. Silva, et al. 2009. Iterative reconstruction technique for reducing body radiation dose at CT: Feasibility study. *AJR Am J Roentgenol* 193 (3):764–771.

Herman, G. T., A. Lent, and S. W. Rowland. 1973. Art-mathematics and applications: Report on mathematical foundations and on applicability to real data of algebraic reconstruction techniques. *J Theor Biol* 42 (1):1–32.

Hounsfield, G. N. 1973. Computerized transverse axial scanning (tomography). 1. Description of system. *Br J Radiol* 46 (552):1016–1022.

Hsieh, J., B. Nett, Z. Yu, et al. 2013. Recent advances in CT image reconstruction. *Curr Radiol Rep* 1 (1):39–51.

Huang, J. Y., J. R. Kerns, J. L. Nute, et al. 2015. An evaluation of three commercially available metal artifact reduction methods for CT imaging. *Phys Med Biol* 60 (3):1047.

Johnson, T. R. 2012. Dual-energy CT: General principles. *AJR Am J Roentgenol* 199 (5 Suppl):S3–S8.

Johnson, T., C. Fink, S. O. Schoenberg, and M. Reiser. 2011. *Dual Energy CT in Clinical Practice*. Berlin, Germany: Springer.

Kalender, W. A., R. Hebel, and J. Ebersberger. 1987. Reduction of CT artifacts caused by metallic implants. *Radiology* 164 (2):576–577.

Kalender, W. A., W. H. Perman, J. R. Vetter, and E. Klotz. 1986. Evaluation of a prototype dual-energy computed tomographic apparatus. I. Phantom studies. *Med Phys* 13 (3):334–339.

Kalra, M. K., M. M. Maher, T. L. Toth, et al. 2004. Strategies for CT radiation dose optimization. *Radiology* 230 (3):619–628.

Kaza, R. K., J. F. Platt, R. H. Cohan, et al. 2012. Dual-energy CT with single- and dual-source scanners: Current applications in evaluating the genitourinary tract. *Radiographics* 32 (2):353–369.

Keall, P. J., G. Starkschall, H. Shukla, et al. 2004. Acquiring 4D thoracic CT scans using a multislice helical method. *Phys Med Biol* 49 (10):2053–2067.

Kijewski, P. K., and B. E. Bjärngard. 1978. The use of computed tomography data for radiotherapy dose calculations. *Int J Radiat Oncol Biol Phys* 4 (5–6):429–435.

Lee, Y. H., K. Kyu Park, H.-T. Song, S. Kim, and J.-S. Suh. 2012. Metal artefact reduction in gemstone spectral imaging dual-energy CT with and without metal artifact reduction software. *Eur Radiol* 22 (6):1331–1340.

Leipsic, J., G. Nguyen, J. Brown, D. Sin, and J. R Mayo. 2010. A prospective evaluation of dose reduction and image quality in chest CT using adaptive statistical iterative reconstruction. *Am J Roentgenol* 195 (5):1095–1099.

Li, B., G. Yadava, and J. Hsieh. 2011. Quantification of head and body CTDI(VOL) of dual-energy x-ray CT with fast-kVp switching. *Med Phys* 38 (5):2595–2601.

Li, H., C. Noel, H. Chen, et al. 2012. Clinical evaluation of a commercial orthopedic metal artifact reduction tool for CT simulations in radiation therapy. *Med Phys* 39 (12):7507–7517.

Low, D. A., M. Nystrom, E. Kalinin, et al. 2003. A method for the reconstruction of four-dimensional synchronized CT scans acquired during free breathing. *Med Phys* 30 (6):1254–1263.

Mahnken, A. H., R. Raupach, J. E. Wildberger, et al. 2003. A new algorithm for metal artifact reduction in computed tomography: In vitro and in vivo evaluation after total hip replacement. *Invest Radiol* 38 (12):769–775.

McDonald, Samuel Alan, Federica Marone, Christoph Hintermüller, et al. 2009. Advanced phase-contrast imaging using a grating interferometer. *J Synchrotron Radiat* 16 (4):562–572.

McShan, D. 1987. 3-D treatment planning: II. Integration of gray scale images and solid surface graphics. Ninth International Conference on the Use of Computers in Radiation Therapy, The Hague, The Netherlands.

McShan, D. L., B. A. Fraass, and A. S. Lichter. 1990. Full integration of the beams eye view concept into computerized treatment planning. *Int J Radiat Oncol Biol Phys* 18 (6):1485–1494.

Mettler, F. A., Jr., M. Bhargavan, K. Faulkner, et al. 2009. Radiologic and nuclear medicine studies in the United States and worldwide: Frequency, radiation dose, and comparison with other radiation sources—1950–2007. *Radiology* 253 (2):520–531.

Mettler, F. A., Jr., B. R. Thomadsen, M. Bhargavan, et al. 2008. Medical radiation exposure in the U.S. in 2006: Preliminary results. *Health Phys* 95 (5):502–507.

Millner, M. R., W. D. McDavid, R. G. Waggener, et al. 1979. Extraction of information from CT scans at different energies. *Med Phys* 6 (1):70–71.

Momose, A., T. Takeda, Y. Itai, and K. Hirano. 1996. Phase-contrast X-ray computed tomography for observing biological soft tissues. *Nat Med* 2 (4):473–475.

Moscariello, A., R. A. Takx, U. J. Schoepf, et al. 2011. Coronary CT angiography: Image quality, diagnostic accuracy, and potential for radiation dose reduction using a novel iterative image reconstruction technique-comparison with traditional filtered back projection. *Eur Radiol* 21 (10):2130–2138.

Olivo, A. and R. Speller. 2007. A coded-aperture technique allowing x-ray phase contrast imaging with conventional sources. *Appl Phys Lett* 91 (7):074106.

Papson, J. P., K. L. Russell, and D. M. Taylor. 2007. Unexpected events during the intrahospital transport of critically ill patients. *Acad Emerg Med* 14 (6):574–577.

Parker, R. P., P. A. Hobday, and K. J. Cassell. 1979. The direct use of CT numbers in radiotherapy dosage calculations for inhomegeneous media. *Phys Med Biol* 24 (4):802–809.

Peace, K., E. M. Wilensky, S. Frangos, et al. 2010. The use of a portable head CT scanner in the intensive care unit. *J Neurosci Nurs* 42 (2):109–116.

Pearce, M. S., J. A. Salotti, M. P. Little, et al. 2012. Radiation exposure from CT scans in childhood and subsequent risk of leukaemia and brain tumours: A retrospective cohort study. *Lancet* 380 (9840):499–505.

Pessis, E., R. Campagna, J. M. Sverzut, et al. 2013. Virtual monochromatic spectral imaging with fast kilovoltage switching: Reduction of metal artifacts at CT. *Radiographics* 33 (2):573–583.

Pfeiffer, F., T. Weitkamp, O. Bunk, and C. David. 2006. Phase retrieval and differential phase-contrast imaging with low-brilliance X-ray sources. *Nat Phys* 2 (4):258–261.

Prell, D., Y. Kyriakou, M. Beister, and W. A. Kalender. 2009. A novel forward projection-based metal artifact reduction method for flat-detector computed tomography. *Phys Med Biol* 54 (21):6575–6591.

Primak, A. N., J. C. Ramirez Giraldo, X. Liu, L. Yu, and C. H. McCollough. 2009. Improved dual-energy material discrimination for dual-source CT by means of additional spectral filtration. *Med Phys* 36 (4):1359–1369.

Rasch, C., I. Barillot, P. Remeijer, et al. 1999. Definition of the prostate in CT and MRI: A multi-observer study. *Int J Radiat Oncol Biol Phys* 43 (1):57–66.

Reft, Chester, Rodica Alecu, Indra J. Das, et al. 2003. Dosimetric considerations for patients with HIP prostheses undergoing pelvic irradiation. Report of the AAPM Radiation Therapy Committee Task Group 63. *Med Phys* 30 (6):1162–1182.

Roach, M., P. Faillace-Akazawa, C. Malfatti, J. Holland, and H. Hricak. 1996. Prostate volumes defined by magnetic resonance imaging and computerized tomographic scans for three-dimensional conformal radiotherapy. *Int J Radiat Oncol Biol Phys* 35 (5):1011–1018.

Rumboldt, Z., W. Huda, and J. W. All. 2009. Review of portable CT with assessment of a dedicated head CT scanner. *AJNR Am J Neuroradiol* 30 (9):1630–1636.

Saba, L., M. Porcu, B. Schmidt, and T. Flohr. 2015. Dual energy CT: Basic principles. In *Dual Energy CT in Oncology*, 1–20. Springer International Publishing, Switzerland.

Schauer, D. A., and O. W. Linton. 2009. NCRP report No. 160, ionizing radiation exposure of the population of the United States, medical exposure—are we doing less with more, and is there a role for health physicists? *Health phys* 97 (1):1–5.

Schenzle, J. C., W. H. Sommer, K. Neumaier, et al. 2010. Dual energy CT of the chest: How about the dose? *Invest Radiol* 45 (6):347–353.

Schneider, W., T. Bortfeld, and W. Schlegel. 2000. Correlation between CT numbers and tissue parameters needed for Monte Carlo simulations of clinical dose distributions. *Phys Med Biol* 45 (2):459–478.

Simons, D., M. Kachelriess, and H. P. Schlemmer. 2014. Recent developments of dual-energy CT in oncology. *Eur Radiol* 24 (4):930–939.

Singh, S., M. K. Kalra, M. D. Gilman, et al. 2011. Adaptive statistical iterative reconstruction technique for radiation dose reduction in chest CT: A pilot study. *Radiology* 259 (2):565–573.

Snigirev, A., I. Snigireva, V. Kohn, S. Kuznetsov, and I. Schelokov. 1995. On the possibilities of x-ray phase contrast microimaging by coherent high-energy synchrotron radiation. *Rev Sci Instrum* 66 (12):5486–5492.

Spadea, M. F., J. M. Verburg, G. Baroni, and J. Seco. 2014. The impact of low-Z and high-Z metal implants in IMRT: A Monte Carlo study of dose inaccuracies in commercial dose algorithms. *Med Phys* 41 (1):011702.

Tapfer, A., M. Bech, B. Pauwels, et al. 2011. Development of a prototype gantry system for preclinical x-ray phase-contrast computed tomography. *Med Phys* 38 (11):5910–5915.

Toepker, M., T. Moritz, B. Krauss, et al. 2012. Virtual non-contrast in second-generation, dual-energy computed tomography: Reliability of attenuation values. *Eur J Radiol* 81 (3):E398–E405.

Underberg, R. W., F. J. Lagerwaard, J. P. Cuijpers, et al. 2004. Four-dimensional CT scans for treatment planning in stereotactic radiotherapy for stage I lung cancer. *Int J Radiat Oncol Biol Phys* 60 (4):1283–1290.

van Elmpt, W., G. Landry, M. Das, and F. Verhaegen. 2016. Dual energy CT in radiotherapy: Current applications and future outlook. *Radiother Oncol*.

Vedam, S. S., P. J. Keall, V. R. Kini, et al. 2003. Acquiring a four-dimensional computed tomography dataset using an external respiratory signal. *Phys Med Biol* 48 (1):45–62.

Vlassenbroek, A. 2011. Dual layer CT. In *Dual Energy CT in Clinical Practice*, 21–34. Springer International Publishing, Switzerland.

Wang, G., D. L. Snyder, J. A. O'Sullivan, and M. W. Vannier. 1996. Iterative deblurring for CT metal artifact reduction. *IEEE Trans Med Imaging* 15 (5):657–664.

Wang, G., T. Frei, and M. W. Vannier. 2000. Fast iterative algorithm for metal artifact reduction in X-ray CT. *Acad Radiol* 7 (8):607–614.

Weitkamp, T., A, Diaz, C. David, et al. 2005. X-ray phase imaging with a grating interferometer. *Opt Express* 13 (16):6296–6304.

Willemink, M. J., P. A. de Jong, T. Leiner, et al. 2013. Iterative reconstruction techniques for computed tomography Part 1: Technical principles. *Eur Radiol* 23 (6):1623–1631.

Yang, M., G. Virshup, J. Clayton, et al. 2010. Theoretical variance analysis of single- and dual-energy computed tomography methods for calculating proton stopping power ratios of biological tissues. *Phys Med Biol* 55 (5):1343–1362.

Zhang, D., X. Li, and B. Liu. 2011. Objective characterization of GE discovery CT750 HD scanner: Gemstone spectral imaging mode. *Med Phys* 38 (3):1178–1188.

CHAPTER 2

Advances in Magnetic Resonance Imaging for Radiation Oncology

Jinsoo Uh and Seonghwan Yee

CONTENTS

2.1 INTRODUCTION—CHALLENGES AND OPPORTUNITIES IN MAGNETIC RESONANCE IMAGING FOR RADIATION ONCOLOGY

Magnetic resonance imaging (MRI) is a nonionizing imaging technique that is based on the interaction of the magnetic moment of the nuclei with external electromagnetic fields. Since its conception in the 1970s (Lauterbur, 1973), MRI has become the medical imaging modality of choice in a wide range of clinical practices because of its superior diagnostic capability. The relatively high spatial resolution and excellent soft-tissue contrasts that are powered by a plethora of imaging protocols, namely, pulse sequences, provide tailored visualization of anatomical structures and tissue abnormalities. Functional MR images weighted by blood perfusion or diffusion of extracellular water molecules can also provide physiological and metabolic insights into disease pathology.

The interest in MRI in radiation oncology has increased exponentially over the years. If successfully implemented, MRI can offer unprecedented improvements in all stages of radiotherapy, including planning, delivery, verification, and adaptation. The conventional use of MRI for radiotherapy has been limited to assisting computed tomography (CT)-based target volume definition (Chung et al., 2004; Voroney et al., 2006). The MR images in this procedure are often adopted from those acquired for diagnostic purposes, which are not necessarily optimized for treatment planning, particularly in terms of image coverage and patient positioning. As such, the potential error in registration between MRI and CT is a limiting factor, specifically when the patient position during MRI is not consistent with that of the treatment. Because recent advances in radiotherapy have been directed at highly conformal techniques such as intensity-modulated radiation therapy (IMRT), volumetric modulated arc therapy (VMAT), and proton therapy, the need for MR images acquired at the treatment position is becoming increasingly important. Therefore, many radiation oncology departments now have their own MRI scanners for acquiring MR images more relevant to treatment planning (Karlsson et al., 2009). Major MRI vendors currently supply scanners dedicated to radiation oncology. These scanners are typically equipped with an indexed flat patient tabletop, solutions for radiofrequency (RF) receiver coils accommodating immobilization devices, and a laser beam bridge. These scanners produce anatomical and physiological MR images in the treatment position, which are favorable for spatial registration to the CT images for radiotherapy planning. Furthermore, they have the potential to simulate a treatment without the need to acquire the planning CT (Kapanen et al., 2013), which provides several benefits, such as eliminating the positioning errors between MR and CT scans, reducing the exposure to ionizing radiation, and decreasing the imaging costs.

On the other side of these exciting promises remain several challenges in the use of MRI for radiotherapy. The most critical issues are to ensure geometric accuracy and to minimize image artifacts. The spatial information of MRI inherently relies on the homogeneity of the static magnetic field and the linearity of the dynamic magnetic field gradient. However, it is not always feasible to maintain the desired conditions of the magnetic fields because of limitations in the hardware and, more important, patient-induced heterogeneities in the fields. In comparison with CT, additional care needs to be taken in MR images to discern various image artifacts that originate from multiple factors associated with hardware, imaging protocol, and imaging subject. The key limitation in MRI-only simulation without CT is the inherent lack of electron density information. A related problem is the unavailability of radiograph-like reference images during treatment, unless MR images are acquired for patient setup verification as in the integrated MRI teletherapy systems (Fallone et al., 2009; Raaymakers et al., 2009; Mutic, 2012). Another relatively minor concern of the radiation oncology-dedicated MR scanners is that the receiver coil is configured differently from that of conventional diagnostic scanners: additional spaces between the coil and the patient are often introduced with a reduced number of coil elements to accommodate immobilization devices, which can result in compromised signal strength or uniformity. Other challenges include poor bone imaging, higher costs than for CT, and the lack of standardization of MR protocols.

In this chapter, we provide insights into the current and future roles of MRI in radiation oncology. Because Chapter 8 is dedicated to MRI for treatment delivery, this chapter focuses on planning and assessment of radiation treatment. Sections 2.2–2.4 deal with the developments in MRI acquisition and post-processing methods used for treatment planning. Strategies to overcome the aforementioned challenges as well as recent technical advances are also introduced. Sections 2.5–2.8 review the various MR imaging modalities that have been studied to evaluate tumor responses and integrity of nontargeted tissues. The possibilities in adaptive image-guided radiotherapy by using these MRI modalities are also discussed. Because the application of MRI is relatively new in radiation oncology, several techniques have not yet been fully established for routine clinical practices. These emerging MRI techniques and well-known features will also be covered to highlight the potential impact of these techniques.

For those who are not familiar with MRI terminology, a glossary focusing on those mentioned in this chapter is provided in the following:

Chemical Shift: The shift of resonance frequency of a given nucleus due to the molecule-specific magnetic field surrounding the nucleus. The chemical shift of hydrogen nuclei in fat from those in water (approximately 3.5 ppm) is the most commonly referred chemical shift in clinical MRI.

Frequency/Phase Encoding: One of the ways of encoding spatial information in an MRI signal. A magnetic field gradient is temporarily applied so that the resonance frequency or phase of a given nuclei (typical hydrogen) has a spatial dependency.

K-Space: Mathematical space in reciprocal relation to real space in terms of Fourier transformation. A spatially resolved MRI signal is typically acquired in the k-space. Then, the k-space data are converted to images by Fourier transformation. Due to the symmetric nature (complex conjugate symmetry) of a k-space, it is possible to construct the image using only the partially filled k-space, which can be used to reduce the scan time at the cost of image quality.

Magnetic Susceptibility: Measure of the ability of a substance to induce magnetization in response to the external magnetic field. Most of the biological materials either reduce (diamagnetic) or enhance (paramagnetic) the external magnetic field. In contrast, ferromagnetic materials retain strong magnetization even in the absence of the external field. Distinct magnetic susceptibilities between substances (e.g., tissue versus metallic insert) can cause heterogeneity in the magnetic fields and, in turn, image artifacts.

Pulse Sequence: A collection of serial events generated by the scanner to collect MRI signals. More specifically, a pulse sequence defines the order and the time of pulsed events of transmission of RF radiation, the switching of magnetic field gradient, and the timing of signal reception during the MRI scan. Rich variations of MRI features, including image resolution and contrast, are achieved by the corresponding design of pulse sequence.

Radio Frequency Pulse: Electromagnetic radiation transmitted to the imaging subject during the MRI scan for a short time period (typically in order of milliseconds). The carrier frequency of the electromagnetic radiation is typically matched to the resonance frequency of the MR-active nucleus (42.58 MHz/T for a proton, ^{1}H) in order to excite or perturb the equilibrium magnetization so that the excited magnetization can be detected by the receiver coil.

Receiver Coil: The antenna for receiving RF signals from the imaging subject. In general, the receiver coil is designed for the specific anatomical region. The intensity of the observed signal decreases sharply as the receiver coil is distanced from the subject being imaged.

Shimming: The process of improving the homogeneity of the static magnetic field. Shimming can be performed either by inserting a ferromagnetic material around the magnet (passive shimming), or by adjusting currents in the shim coils (active shimming). Subject-specific active shimming is performed at the time of imaging.

Specific Absorption Rate (SAR): Amount of energy deposited in the imaging subject by the RF pulse. The typical unit of SAR is W/kg. SAR is higher for pulse sequences requiring more intensive use of RF pulses. As such, SAR needs controlled to be carefully for patient safety.

Static Magnetic Field/Magnetic Field Gradient: The static magnetic field is the primary magnetic field in an MRI scanner, which is presumably maintained constant by superconducting coils. Contemporary clinical scanners are typically operated with the static magnetic field of 1.5 or 3.0 T. In contrast to the static field, the spatially varying magnetic field gradient is induced only for a brief period time by a separate gradient coil. The typical magnitude of a magnetic field gradient is on the order of tens of mT/m. Because the frequency and phase encoding rely on the magnetic field gradient, ensuring the spatial linearity of the field is critical for geometrical accuracy in MRI.

2.2 SUPPRESSION OF ARTIFACTS AND DISTORTIONS IN MR IMAGES

One of the challenges in the application of MRI for radiotherapy is to minimize artifacts and distortions associated with specific patient, imaging parameters, and system hardware. This section reviews common issues in MR image quality and strategies to suppress them.

Involuntary or voluntary motion in patients causes ghosting or blurring artifacts in MR images along the phase-encoding direction, regardless of the actual direction of the motion. Respiratory motion causes multiple surfaces of the skin or overlap of the anatomical structures in abdominal images. A thorax MR image often shows blurring in the heart and vessels as a result of cardiac motion. Eyeball movement during a scan often creates a series of ghosting of eyeballs. A similar artifact can be seen when blood flows into imaging slices.

Voluntary motion at a large scale would be of less concern if it could be controlled by immobilization devices, as in radiation treatment. This procedure requires MR-safe immobilization devices and preferably the use of a wide-bore (70 cm) system. Note that not all plastic materials are MR-safe because materials including carbon fibers may carry charges and may heat up or interfere with RF signals during the scan (Devic, 2012; Liney and Moerland, 2014). Figure 2.1 demonstrates the configurations of receiver coils designed for MRI-based planning of radiotherapy. To position patients in the same way as in the treatment with immobilization devices, flexible coils of circular shape have been used (Figure 2.1a) instead of the standard phase-array head coil for head imaging. Potential trade-offs of this modification in coil configuration include compromised image uniformity and signal-to-noise ratio (SNR); however, these can be compensated for in part by the advances in postprocessing. In contrast to typical diagnostic imaging, the anterior receiver coil is mounted on a bridge for abdominal imaging (Figure 2.1b) to avoid the alteration of body shape due to compression against the abdomen.

Another approach to suppress the effect of motion is to use special imaging techniques. For instance, the blood signal flowing into imaging slices can be suppressed by a spatial-saturation RF pulse or gradient-moment nulling (Bernstein et al., 2004). A widely used technique to suppress bulk motion is periodically rotated overlapping parallel lines with enhanced reconstruction (PROPELLER; also known as BLADE or multivane) MRI (Pipe et al., 2014). PROPELLER MRI acquires k-space data in radially directed strips to mitigate the motion effect. Caveats in this technique are a longer scan time and artifacts associated with the unconventional acquisition such as streaking. Simply repeating and averaging the same scan can also mitigate the motion effect over multiple acquisitions, especially when the motion is random, but the scan time needs to be increased significantly to make the averaging effective. Fast imaging techniques are preferably used to suppress motion-induced artifacts. Examples for abdominal and pelvic imaging are balanced steady-state free precession (bSSFP); fast low-angle shot (FLASH); and half-Fourier acquisition single-shot turbo spin echo (HASTE), which gives T1/T2-, T1-, and

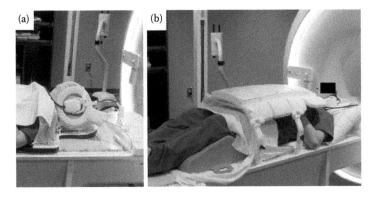

FIGURE 2.1 Configurations of receiver coils to accommodate immobilization devices for MRI-based planning of radiation therapy. (a) Flexible head coil mounted on a thermoplastic mask for head imaging. (b) Anterior receiver coil mounted on a bridge.

T2-weighted image contrasts, respectively. These techniques can acquire one image slice in 200–400 ms, which can help mitigate respiratory or peristaltic motions. Fast imaging can be combined with respiratory gating or triggering using external sensor or internal navigator RF pulse to acquire images at a selected window of respiratory phase. The most critical challenge in this approach is for the patient to maintain regular breathing because the images at the same respiratory phase are not necessarily coherent if breathing pattern changes temporally. Alternatively, fast imaging can be assisted by breath-holding during imaging, but the typical length of breath-hold (15–20 s) in conventional techniques is not feasible for all patients. An interesting alternative technique is a quasi-breath-hold that utilizes audiovisual guidance (Park et al., 2011; Kim et al., 2014a, b). In this technique, the patients are instructed to continuously follow a programmed breathing pattern in which the length of either the end-inhalation or end-exhalation is slightly elongated (e.g., 2 s) than normal breathing. The imaging data are acquired during the quasi-breath-hold to minimize the motion effect, while the convenience of the patient is accommodated by the relatively short length of breath-hold.

Other than motion, the difference in magnetic susceptibility is another major source of degraded MR image quality. Substances responsible for susceptibility differences in clinical imaging include air cavity, hard bone, implants, and artificial prosthetics in patients. Fixation devices attached to patients can also sometimes cause susceptibility differences. These substances alter the homogeneity of the static magnetic field, which is the key prerequisite in MRI, resulting in geometric distortion of the acquired images. A clinically feasible solution to suppress these effects is to use higher-order shimming on the magnetic field. Image protocols based on spin echo, as opposed to gradient echo, are preferable to partly recover the signals decayed by the inhomogeneous magnetic field. Also, a higher value of receiver bandwidth is helpful because a larger difference in susceptibility translates to a smaller spatial difference as the bandwidth increases. However, a higher receiver bandwidth requires a stronger magnetic field gradient and a shorter sampling time, which in turn can increase noise, the significant absorption rate (SAR), and the risk of peripheral nerve stimulation (PNS). Susceptibility-induced image distortions are often pronounced in diffusion- or perfusion-weighted functional MR images acquired by single-shot echo planar imaging (EPI). Because single-shot EPI acquires the entire image slice after a single RF excitation, the effect of an inhomogeneous magnetic field becomes more severe than for other conventional anatomical imaging techniques. Characterization and correction of susceptibility-induced distortions have been discussed extensively in the literature (Morgan et al., 2004; Reinsberg et al., 2005; Baldwin et al., 2009; Stanescu et al., 2012; Wang et al., 2013). Inhomogeneity of the magnetic field can be measured by phase images with different echo times, which can aid distortion correction. Acquisition of two sets of identical images with the opposite distortions has been investigated to cancel out the distortion effect.

The difference in resonance frequency (approximately 3.5 ppm) and MR relaxation times between water and fat can lead to another type of patient-induced MRI artifact.

Examples of such artifacts include signal reduction in gradient echo images and spatial displacement of the fat signal (Schick et al., 1998; Dietrich et al., 2008; Hernando et al., 2011). In addition, the off-resonant fat signal can cause dark bands in images acquired by fast imaging techniques that require steady-state conditions. Widely used techniques to reduce such artifacts include suppressing the fat signal by applying a frequency-selective saturation RF pulse (Frahm et al., 1985; Merchant et al., 1992; Halligan et al., 1998) or employing the inversion of magnetization (Bydder et al., 1985; Del Grande et al., 2014). Alternatively, fat saturation can be achieved by selectively exciting the water signal (Rosen et al., 1984; Redpath and Wayte, 1993; Thomasson et al., 1996). When fat suppression is not feasible because of increased scan time, the artifact can be minimized by using a higher receiver bandwidth at the cost of increased SAR and reduced SNR.

The sources of system-induced image distortions include magnetic field inhomogeneity and gradient nonlinearity. The effect of magnetic field inhomogeneity is similar to that of susceptibility difference, but system-induced inhomogeneity is relatively minimal and tends to spread broadly rather than being localized. The major contribution to system-induced distortion is nonlinearity in the magnetic-gradient field caused by limitations in the gradient coil, which is more pronounced with a larger field of view (FOV). System-induced distortions can be mitigated by higher-order shimming and vendor-provided correction algorithms, which are readily available for most commercial MRI scanners. However, it is highly recommended that geometric accuracy be ensured by using a geometric phantom (Doran et al., 2005; Baldwin et al., 2009) during quality assurance.

Other common artifacts attributed to suboptimal imaging parameters include the fold-over artifact, which is also known as wraparound or aliasing artifact. This artifact is seen at both sides in the phase-encoding direction when the FOV is too small. The fold-over artifact can occur on all four sides in an image slice if phase encoding is used for two dimensions in the three-dimensional (3D) imaging. The simplest way to remove this artifact is to increase the FOV or the number of encoding steps in the phase-encoding direction. A side benefit of this change is increased SNR. On the other hand, the trade-offs are increased scan time or decreased spatial resolution.

2.3 MR-BASED SIMULATION

Conventionally, the use of MRI in treatment planning generally involves spatial registration of MRI to patient-specific CT so that target volumes, organs at risk, and dosimetrically relevant structures delineated on MRI are transferred to CT. This is followed by dose calculation based on calibration from the CT number to electron density. Recently, considerable attention has been paid to developing MRI-based treatment planning that does not require patient-specific CT (Kapanen et al., 2013). The critical challenge in this endeavor is to provide electron density for dose calculation, which is not readily available in MRI. Several approaches have been explored to generate

CT-like images (also known as pseudo-CT or synthetic CT) from MRI for deriving the information uniquely provided by CT. Such approaches were also developed in the context of positron emission tomography (PET)/MRI, for which the attenuation correction is not readily available without an additional CT scan. A relatively simple approach is to segment MRI into distinct structures that have similar CT numbers (e.g., bony structure, fat, water, and air) and assign a corresponding bulk CT value to each structure (Lee et al., 2003; Stanescu et al., 2008; Jonsson et al., 2010; Lambert et al., 2011; Hoogcarspel et al., 2014). A key challenge in this approach is to distinguish bony structures from air, which is not feasible in conventional MR images because of fast MR relaxation in solid tissues. Furthermore, human intervention is often needed when autosegmentation is not feasible. The partial volume effect among segments and inhomogeneous density within the bone are additional confounding factors, but they may be overcome by segmentwise conversion of MRI intensity to CT number (Korhonen et al., 2014).

An advanced approach for deriving pseudo-CT is voxelwise estimation of CT intensity by using multiple MR images, including ultra-short echo time (UTE) that can detect bony structures. A notable technique is to use multiple UTE images acquired with various echo times and flip angles (Johansson et al., 2011, 2012; Rank et al., 2013). The model parameters are estimated by Gaussian mixture regression from a preexisting training set of MR and CT data. Then, they are used to generate pseudo-CT from MR images. Another similar technique uses the Dixon method in addition to UTE to separate fat and water volumes and probabilistically classifies tissues by fuzzy c-means clustering (Hsu et al., 2013). The voxelwise approach performs considerably better than the segmentation method (Edmund et al., 2014). The need for extra MR images such as UTE and fat–water images may be a limitation, although this can be mitigated by, for example, generating multiple image contrasts by a small number of scans using advanced imaging protocols (Yee, 2016). The image quality of current UTE protocols is, unfortunately, not sufficient for accurate pseudo-CT. For instance, blood vessels appear dark in UTE and cannot be differentiated from bony structures (Hsu et al., 2013). Methods to improve UTE images for better discrimination of air from bone are currently being investigated (Hsu et al., 2015).

Another alternative approach to generate pseudo-CT is to deform one or more atlas CT images to patient-specific MRI (Dowling et al., 2012; Uh et al., 2014; Sjolund et al., 2015). Deformation mapping is typically performed by nonlinear spatial registration between patient-specific MRI and atlas CT (Schreibmann et al., 2010) or, more commonly, an MRI conjugated with the atlas CT. The biggest challenge in this approach is the uncertainty in atlas deformation, particularly for anatomies with atypical shapes. One way to reduce the uncertainty in atlas deformation is to create a virtual atlas image by combining many deformed images (Dowling et al., 2012). Otherwise, multiple deformed CT images can be generated from multiple atlases to be "synthesized," as illustrated in Figure 2.2 (Uh et al., 2014). The selection of the atlases and their synthesis are important aspects in this approach. As opposed to a simple arithmetic average, increasing the number of atlases and using an advanced pattern recognition process (Hofmann et al., 2008) can improve robustness, but the gain appeared marginal considering the computational cost (Figure 2.3) (Uh et al., 2014). In a conceptually similar

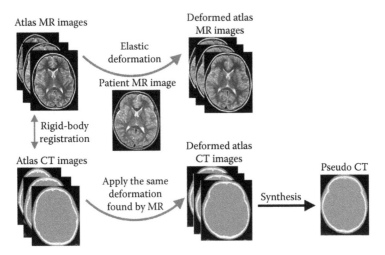

FIGURE 2.2 The process for constructing a pseudo-CT from atlas images. The deformations for atlas MR images were applied to the corresponding atlas CT images to generate multiple deformed atlas images. Then they were combined to produce a pseudo-CT by the arithmetic mean process or the pattern recognition Gaussian process. MR, magnetic resonance; CT, computed tomography.

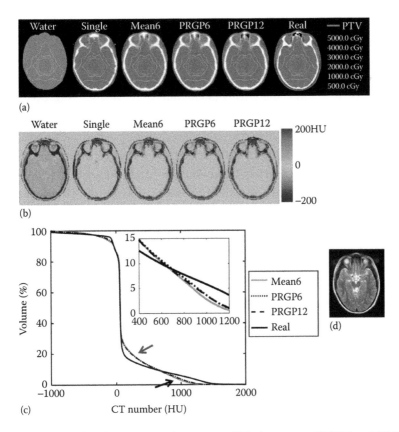

FIGURE 2.3 The blue and red arrows → The gray and black arrows; PRGP6 and PRGP12 (green and blue lines) → PRGP6 and PRGP12 (dotted and dashed lines); real CT (red line) → real CT (black line); MEAN6 (black line) → MEAN6 (gray line).

method, an electron density value of a voxel was found from combined information of the MR image intensity and geometry in a probabilistic Bayesian framework, which produced a more accurate pseudo-CT than that based on either intensity or geometry information only (Gudur et al., 2014). Another patch-based approach using nonlocal means prediction can generate a high-quality pseudo-CT by linear registration only without a cumbersome deformation process (Andreasen et al., 2015). An extension of atlas-based approaches is conceivable with additional atlases of UTE images. However, because this comes with the pros and cons associated with UTE, the trade-off between robustness and cost needs to be evaluated carefully.

2.4 FOUR-DIMENSIONAL MRI

Four-dimensional (4D) imaging refers to the acquisition of temporally resolved multiple volumes of images. Recently, 4D MRI has been explored to assess patient-specific respiratory-induced motion for advanced treatment planning. 4D MRI has several potential advantages over 4D CT, such as soft-tissue contrast, no need for ionizing radiation, and flexibility in image orientation. In principle, 4D imaging can be achieved by simply repeating the scan of a 3D volume if the scanning time per volume is sufficiently short (e.g., a few milliseconds). However, such rapid imaging is not feasible currently without severe compromise in image resolution and quality (Blackall et al., 2006; Dinkel et al., 2009). Consequently, present 4D MRI techniques are mostly based on retrospective sorting of two-dimensional (2D) images (Remmert et al., 2007) or k-space segments (Liu et al., 2015) by respiratory phases. Compared with the sorting of 2D images, the sorting of k-space segments provides opportunities for higher spatiotemporal resolution and flexibility in acquisition schemes, but it is vulnerable to aliasing artifacts when the temporal coherency of the sorted segments is not sufficiently precise (Liu et al., 2015; Stemkens et al., 2015).

A typical 2D image-based retrospective 4D MRI protocol uses multi-slice fast imaging sequences (see Section 2.2). Because the primary direction of respiratory motion is along the superior-to-inferior axis, either sagittal or coronal slice images are commonly acquired with an in-plane resolution of approximately 2×2 mm^2 and a slice thickness of 5 mm (Tryggestad et al., 2013). Each image at a slice location is acquired multiple times by using either a consecutive or interleaved scheme, during which the respiratory surrogate is measured. Then, the images are sorted according to a given set of respiratory bins. To find a slice image corresponding to each of the bins, the imaging time allotted to a slice location needs to be at least one respiratory cycle. A longer scan will provide a better chance of improving image quality by averaging multiple images in the same bins (Tryggestad et al., 2013). Binning strategies have been evaluated in the context of 4D CT. Phase binning has been conventionally implemented in commercial scanners, but amplitude binning is potentially better than phase binning in terms of consistency and accuracy (Abdelnour et al., 2007; Li et al., 2012).

An alternative to retrospective sorting is prospective acquisition at predetermined respiratory states (Tokuda et al., 2008; Hu et al., 2013). Because all components (2D images or k-space segments) that compose a 3D volume at a respiratory state are supposedly in temporal coherency, the resultant 4D images are, in principle, free from mismatch of slices or aliasing artifacts. However, maintaining the triggering efficiency with a steady-state condition in magnetization is far from easy, particularly with irregular respiration. Thus, efforts have been directed at improving the efficiency of prospective 4D MRI (Du et al., 2015).

Besides sorting and binning strategies, a critical issue in 4D MRI is to derive respiratory surrogates that define respiratory states. A common method is to use external sensors such as respiratory bellows, video monitoring, or spirometers. External surrogates are measured at a high sampling rate, and the associated postprocessing is relatively straightforward. However, the method has some drawbacks, such as cost for the extra equipment, additional effort to prepare patients with the equipment, uncertainty of synchronization with image acquisition, and differences in the motions between external body surface and internal organs.

An alternative surrogate can be derived from internal MR signals. A feasible method is to acquire a navigator that is interleaved with the signals composing the 4D MRI. Then, the respiratory surrogate is derived from the navigator. Navigator echo (Tokuda et al., 2008) or pencil-beam navigator (Stam et al., 2012) are typical 1D navigators, and a full 2D image is also feasible for a navigator (von Siebenthal et al., 2007). A key limitation of using a navigator is that scanning efficiency is compromised due to the extra acquisition of a navigator. Another associated problem is the time gap between interleaved navigators and image acquisitions, during which non-negligible motion can occur. A potential solution to this limitation is simultaneous acquisition of navigator and image slices by using multiband RF pulses (Celicanin et al., 2015). Otherwise, an internal surrogate can be derived directly from the images without using an additional navigator, as demonstrated in recent studies (Cai et al., 2011; Yang et al., 2014; Uh et al., 2016).

The common challenge that remains in 4D MRI methods is to further increase the spatiotemporal resolutions by various undersampling techniques (Hansen et al., 2006; Lustig et al., 2007) while preserving image qualities. One way of addressing this is through the dynamic keyhole method (Lee et al., 2003), which improves image quality by using a previously acquired library of peripheral k-space datasets while central k-space datasets are acquired in near real time. Advances in such fast imaging techniques can contribute clinically feasible 4D MRI by reducing image artifacts, enabling delineation of small target volumes, and accommodating patients with fast or irregular respiratory patterns.

2.5 PERFUSION-WEIGHTED AND DIFFUSION-WEIGHTED IMAGING

Perfusion-weighted imaging provides information associated with blood supply to a tissue of interest and permeability of blood in the capillaries. The brain has been the most interesting organ analyzed in many perfusion studies because of its high demand for blood. Target parameters include cerebral blood volume (CBV), cerebral blood flow, and mean transit time. Dynamic susceptibility contrast (DSC) MRI has historically been the MRI protocol of choice for cerebral perfusion imaging because of its advantages with regard to contrast-to-noise and temporal resolution. DSC is typically acquired by a fast imaging protocol with a sampling frequency of 1 or 2 s during the first bolus passage of an intravenously administered gadolinium-chelated contrast agent. The voxelwise T2- or T2*-weighted temporal profiles are subsequently analyzed. Despite a long history of development and clinical use, DSC faces several challenges in the scenario of quantification for longitudinal studies or comparative studies across institutions, such as partial volume effect, determination of arterial input function, and compensation of extravasation (Knutsson et al., 2010; Willats and Calamante, 2013).

Dynamic contrast-enhanced (DCE) MRI is similar to DSC in terms of the use of contrast, but it usually refers to T1-weighted imaging with a lower temporal resolution. DCE has been used for capillary permeability imaging, for which the interstitial uptake of the contrast agent at a longer time scale is of primary interest. Targeted parameters include the volume-transfer coefficient (K^{trans}) and the interstitial volume fraction (v_e). Although some of the challenges associated with DCE are similar to those for DSC, the former has several advantages in quantification and geometric accuracy. DCE has been conventionally used for organs other than the brain, but it is currently replacing DSC for cerebral perfusion imaging because of recent advances in fast imaging techniques that incorporate high-performance gradient coils and multichannel receiver coils.

Arterial spin labeling (ASL) is another perfusion-imaging technique that is based on endogenous contrast (Alsop et al., 2014). In ASL, the blood flowing into a region of interest is electromagnetically labeled so that the perfused blood signal is differentiated from the extravascular signals. The advantages of ASL are that the scan time is not limited by bolus passage of the contrast agent, and the mathematical postprocessing is simpler. This technique is suitable especially when the exogenous contrast agent cannot be use in patients with compromised renal function or who require frequent repetitive follow-up. ASL has been studied for tumor grading (Wolf et al., 2005; Noguchi et al., 2008) and identification for radiation necrosis (Ozsunar et al., 2010). Because ASL has become available only recently in commercial scanners, the potential of this new technique for radiation oncology needs further evaluation.

DSC and DCE acquired before and during the course of radiation therapy have been studied for predicting response to radiation treatment (Cao et al., 2006b; Almeida-Freitas et al., 2014). These studies characterized the aggressiveness of tumors by using indices derived from the perfusion-weighted images, in which higher blood supply and enhanced capillary permeability reflect abnormal angiogenesis and the apoptosis of

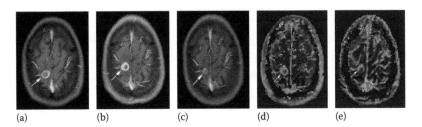

(a)   (b)   (c)   (d)   (e)

FIGURE 2.4 MR images obtained from a 55-year-old male patient with a right frontal lung metastasis (arrows). (a–c) T1-weighted images at baseline (a), 4-week (b), and 20-week (c) follow-up, respectively; (d and e) Parametric K^{trans} map at baseline (d) and 4-week (e) follow-up, respectively. At the 4-week follow-up, the tumor showed a 12% increase in volume, whereas the K^{trans} value reduced by 56% compared with the baseline. At the 21-week follow-up, the tumor volume decreased by 78% relative to the baseline.

endothelial cells. Figure 2.4 exemplifies MR images from a patient with brain metastasis who received stereotactic radiosurgery (Almeida-Freitas et al., 2014). The K^{trans} map in an early post-treatment period (4 weeks) (Figure 2.4e) showed lower permeability than that at baseline (Figure 2.4d). This reduced permeability predicted tumor control despite a 12% volume increase, which was confirmed because the tumor volume decreased by 78% at the midtreatment period (21 weeks) (Figure 2.4c). In contrast, the enhanced region in conventional T1-weighted images at the baseline (Figure 2.4a) and at the 4-week post-treatment (Figure 2.4b) did not show signs of tumor remission. A notable approach in recent perfusion imaging studies on radiotherapy is to identify subvolumes that respond to or are resistant to therapy within a tumor. The identification of such heterogeneous responses points to the need for patient-specific focal boost therapy (Groenendaal et al., 2012) or dose painting (van der Heide et al., 2012). Studies on patients with high-grade gliomas report that subvolumes with higher CBVs (Cao et al., 2006b) or severe vascular leakage (Cao et al., 2006a) before radiotherapy are a potential predictor of tumor progression. Tsien et al. (2010) showed that responders could be predicted compared to nonresponders by voxelwise analysis of CBV at 3 weeks from the initiation of radiotherapy but not by whole-tumor CBV. Recent studies introduced probabilistically defined subvolumes in order to increase the sensitivity (Farjam et al., 2013, 2014b). This probabilistic approach also avoids uncertainties arising from a fixed threshold and the need for registration between images. Compared with normal tissue, highly pronounced perfusion reflects aggressiveness of the tumor, whereas lower perfusion indicates hypoxia, which is known to be resistant to radiation therapy. A subsequent seemingly contradictory observation is that subvolumes with perfusion deficit are associated with local failure in patients with cervical or head and neck tumors who received radiation therapy (Yamashita et al., 2000; Wang et al., 2012). However, it is also possible that hypoxia and high perfusion coexist, leading to these conflicting findings (Lehtio et al., 2004).

Besides being used to assess tumor control, perfusion imaging has also been explored for other aspects of radiation oncology. For example, the side effects on nontargeted tissues have been studied by the relation between neurotoxicity and integrity of the cerebral vasculature (Cao et al., 2009). The post-treatment effect has been identified against the recurrent tumor using CBV (Hu et al., 2009).

Diffusion-weighted imaging (DWI) uses attenuation of the MR signal caused by molecular motion when it is encoded by magnetic field gradients. Brownian motion of water is the primary source of signal attenuation, but blood flow in a randomlike motion in capillaries also contributes to the attenuation (Le Bihan et al., 1988). Physiological motion induced by respiration and the cardiac cycle is another confounding factor in DWI. Because molecular motion in a tissue is affected by cell membranes and macromolecules in intracellular spaces, the measured "apparent" diffusion is lower than bulk diffusion and reflects the cellularity or structural integrity of the tissue of interest. DWI requires the selection of several imaging parameters according to the contrast and the type of tissue of interest. The time between dephasing and rephasing gradients and the gradient strength represented by the b-value are two primary parameters that determine the state of the observed diffusing water molecules.

The use of DWI for oncology studies is based on the fact that increased cellularity of neoplasmic tissues manifests as hyperintensity in the diffusion-weighted signal (hypointensity in the "PET-like" inverted gray scale) or a reduced apparent diffusion coefficient (ADC). Response to treatment is thus predicted by a decrease in the diffusion-weighted signal or an increase in the ADC, which reflects tumor necrosis or apoptosis. Unlike perfusion imaging, whole-body imaging is feasible with DWI, thereby making this technique attractive for detection of lesions (Padhani et al., 2011). ADC has been studied for grading brain tumors (Sugahara et al., 1999) before treatment and for evaluating responses (Huang et al., 2010) after radiotherapy. Recent studies also show its potential as a prognostic biomarker (Lambrecht et al., 2012; Zulfiqar et al., 2013). As in perfusion imaging, the heterogeneity of tumor response to therapy has also been studied using DWI via a voxel-based analysis (Moffat et al., 2005; Hamstra et al., 2008) or a probabilistic approach using diffusion abnormality indices (Farjam et al., 2014a).

A new approach in DWI studies is to use b-values higher than the conventional value (> 1,000 s/mm²). A higher b-value indicates a higher gradient strength that extensively attenuates the diffusion-weighted signal. This results in a clear differentiation of hypercellular tissues from edematous and normal tissues. Pramanik et al. (2015) revealed that the hypercellularity subvolume in glioblastoma patients that was not covered by the 95% prescribed dose volume was a significant negative predictor of progression-free survival, suggesting the potential of high b-value DWI for identifying boost volumes. Diffusion kurtosis imaging (DKI) is another unconventional DWI modality that uses a wide range of b-values, including high values, to take into account non-Gaussian diffusion (Jensen and Helpern, 2010). DKI shows high sensitivity and specificity for grading gliomas (Van Cauter et al., 2012) and for assessing response to microwave ablation and chemobilization treatment (Goshima et al., 2015). DKI has not yet been explored for assessing radiation treatment.

Diffusion tensor imaging (DTI) is an extended modality of DWI. DTI is estimated from a series of DWIs conducted with more than six directions of gradient encodings to find the diffusion tensor in each imaging voxel. Multiple indices can be derived from this tensor, including mean diffusivity and fractional anisotropy (FA), which represent the strength and directionality, respectively, of local diffusion. DTI has been studied extensively for the effect of radiation on nontargeted tissues during the treatment of brain cancers. DTI shows high sensitivity to microstructural integrity in the cerebral white matter, which is generally more prone than cortical gray matter to radiation-induced injury (Steen et al., 2001). Radiation-induced injury in childhood cancer survivors alters DTI-derived indices compared with those of healthy controls (Khong et al., 2003; Dellani et al., 2008). Longitudinal studies in adult patients with brain tumors have shown dose-dependent acute and subacute changes in DTI indices after therapy (Nagesh et al., 2008). Several studies have reported a correlation between changes in DTI indices and decreased neurocognitive functionality (Khong et al., 2006; Aukema et al., 2009; Chapman et al., 2012), suggesting that DTI is a potential imaging biomarker for treatment-related neurotoxicity.

An intriguing finding in these studies is that the effect of radiation dose depends on fiber tract or region (Chapman et al., 2013; Uh et al., 2013). If an association with observed neurotoxicity can be established, then information about differential radiation sensitivity will be valuable in treatment planning in order to protect more vulnerable brain regions. Figure 2.5 shows longitudinal FA maps of the pons for a medulloblastoma patient acquired at 18 and 45 months after radiotherapy. The decrease of FA in the transverse pontine fiber persists throughout follow-ups (white arrow), whereas the other white matter tracts, including the corticospinal tract and the medial lemniscus, show relatively smaller changes or recovery despite similar radiation doses being given to all these tracts (Uh et al., 2013). It has been speculated that the regional sensitivity to radiation is due to regional differences in vascularity (Qiu et al., 2007) or migration of oligodendrocyte progenitor cells (Bijl et al., 2005). Confounding factors such as hemorrhage (Qiu et al., 2006), hypercellularity (Ravn et al., 2013), and surgical defects (Uh et al., 2015) might contribute to the regional differences in observed responses.

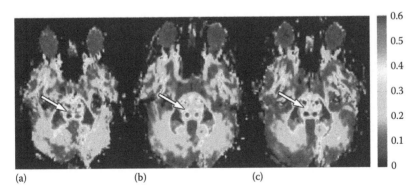

(a) (b) (c)

FIGURE 2.5 Fractional anisotropy (FA) maps of a medulloblastoma patient (male, baseline age 11 years) acquired at baseline (a) and the two follow-up times of 18 months (b) and 45 months (c) from baseline. The white arrows indicate a dorsal transverse pontine fiber showing more pronounced FA reduction than do other regions.

2.6 MAGNETIZATION TRANSFER AND CHEMICAL EXCHANGE SATURATION TRANSFER

A typical clinical MRI acquires signals from protons in water molecules (mobile or unbound protons) as opposed to those in macromolecules (bound or restricted protons). In general, the bound protons are not detectable because the transverse relaxation rate is too fast due to broadly distributed resonance frequencies (Henkelman et al., 2001; Tyler et al., 2007). However, they can be detected indirectly via the transfer of magnetization between bound and mobile protons. In magnetization transfer (MT) MRI (Wolff and Balaban, 1989; Grossman, 1994; Henkelman et al., 2001), protons bound to macromolecules are saturated by an RF pulse at an off-resonance frequency. Then, their influence on mobile protons in the vicinity (i.e., reduced signal from mobile protons) is measured. The effect of MT is quantified by the magnetization transfer ratio (MTR), which is computed as $(S_0 - S_{MT})/S_0$, where S_0 and S_{MT} represent the measured signals without and with the off-resonance RF pulse, respectively. MTR depends on the strength and length of the RF pulse. It should be noted that the mobile protons are also partially saturated by the off-resonance RF pulse, which results in extra signal reduction.

MT MRI has been useful for MR angiography (MRA) (Dagirmanjian et al., 1995) and for detecting white matter lesions associated with multiple sclerosis (Grossman, 1994; Pike et al., 2000; Chen et al., 2008). In the field of oncology, MT MRI has been used to distinguish low-grade (I and II) and high-grade (III and IV) astrocytomas and to detect collagen content in meningiomas, which suggests the potential of this technique to differentiate various types of brain tumors (Lundbom, 1992). A recent study on glioblastoma, meningiomas, and metastases found that quantitative MTR is not only heterogeneous within the tumor volume but also potentially different among the subdivisions of the tumor type (Garcia et al., 2015). In another study, MT MRI was used to assess postradiation fibrosis associated with rectal cancer treatment (Martens et al., 2014). This study found that areas of fibrosis had a significantly higher MTR than that for the residual tumor, the normal rectal wall, or the edematous rectal wall. The ability of MT MRI to estimate indirectly the macromolecular environment is potentially attractive in oncologic applications because it provides a unique noninvasive tool to investigate the biochemically active processes associated with cancer onset, progression, and regression. However, the changes in MTR are not specific to a single physiological process or a macromolecular structure, as indicated by a study on multiple sclerosis (Vavasour et al., 2011), which suggested a possible association of MTR with demyelination, inflammation, and edema. Therefore, MT MRI needs to be interpreted together with findings from other image contrasts and modalities.

Chemical exchange saturation transfer (CEST) MRI (Ward et al., 2000; Sherry and Woods, 2008) is similar to MT MRI, but it requires a sufficiently slow transfer between the mobile and bound proton pools and neglects relatively fast transfer such as dipolar cross-relaxation (van Zijl and Yadav, 2011). The CEST frequency specific to a given molecule can be found in the so-called Z-spectrum or CEST spectrum where the normalized water saturation is plotted against the RF, as demonstrated in Figure 2.6 (Chen et al. 2014). The CEST spectrum reflects attenuations in both mobile proton (i.e., water) and the molecule

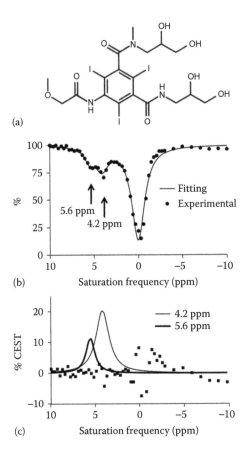

FIGURE 2.6 CEST of iopromide. (a) The chemical structure of iopromide. (b) A CEST spectrum of 200 mM iopromide at pH 6.69°C and 37.0°C with saturation applied at 2μT for 5 s. (c) Lorentzian line shapes fitted to the CEST spectrum show CEST effects at 4.2 ppm and 5.6 ppm. Squares represent the residuals of the fitting process.

of interest, but they can be separated using the symmetry of the saturation of mobile proton. The CEST frequency of a molecule is determined by that maximizing the asymmetry between positive and negative RF ranges, with which the CEST MR image is acquired.

CEST MRI provides flexible opportunities in terms of contrast agents (Ward et al., 2000). The contrast agent may be designed specifically to the desired biochemical features with less adverse effects than conventional gadolinium-based agents. In addition, preexisting molecules may be used as endogenous contrast agents without administration of exogenous agents. CEST MRI using endogenous agents has been used to detect urea in the kidney (Dagher et al., 2000) and amide protons of mobile proteins and peptides (Zhou et al., 2003).

CEST MRI is a promising in vivo molecular imaging tool to estimate pH, metabolites, proteins, temperature, and other distributes (van Zijl and Yadav, 2011; Liu et al., 2013). As such, CEST MRI has potential in applications for radiation oncology. The pH contrast based on CEST MRI has been used to estimate the tumor environment (Ward et al., 2000;

Chen et al., 2014) and for early detection of ischemia (Zhou et al., 2003). The ability of CEST MRI to separate edema from a brain tumor (Wen et al., 2010) and the effects of radiation treatment from tumor progression (Zhou et al., 2011) are also notable and underscore the possibility of tumor grading based on CEST MRI. In addition, a recent study used CEST MRI to detect the uptake of nonradioactive glucose analogs (2-deoxy-D-glucose and 2-fluoro-2-deoxy-D-glucose) in mice with orthotopic mammary tumors (Rivlin et al., 2013) and demonstrated its potential replacement for PET.

2.7 MAGNETIC RESONANCE SPECTROSCOPY

Magnetic resonance spectroscopy (MRS) measures the chemical shift of the NMR-active nuclei (typically ^{1}H; others include ^{13}C and ^{31}P) induced by the molecular environment, through which the relative amount of the corresponding metabolite can be determined. MRS can be conducted by single voxel spectroscopy (SVS) using either point-resolved spectroscopy or stimulated echo acquisition mode with a scan time of 2–5 min and a typical voxel size of 4–8 cm^3. Multivoxel spectroscopy, also known as MRS imaging (MRSI), provides spatially resolved spectra with a voxel size of 1–2 cm^3 with additional localization techniques. The scan time of MRSI needs to be longer than that for SVS. MRS has been studied to characterize tumors (Li et al., 2002) and assess tumor infiltration (Croteau et al., 2001). Typical targeted metabolites include N-acetylaspartate (NAA), choline (Cho), creatine (Cr), and lactate (Lac). NAA is a marker of neuronal integrity and function, Cho indicates cell membrane proliferation, Cr reflects cell energy metabolism, and Lac is associated with anaerobic glycolysis. In typical brain tumors, NAA and Cr are decreased whereas Cho and Lac are elevated (McKnight, 2004; Soares and Law, 2009).

The effect of radiation on the normal brain has been studied with MRS. The metabolite ratios of NAA:Cr and Cho:Cr decrease in the irradiated brain during early radiotherapy. Cr is used as a normalization factor under the assumption that it is stable (Sundgren et al., 2009). These alterations suggest radiation-induced neuronal dysfunction and membrane damage in the myelin, and these observations have been supported by earlier studies (Usenius et al., 1995; Chan et al., 2001). A related potential use of MRS for radiotherapy is to discriminate recurrent tumors from radiation injury in contrast-enhancing lesions (Zhang et al., 2014; Sundgren, 2009). Conventional MR imaging does not always differentiate reliably a progressing tumor from radiation-induced necrotic changes due to the presence of a nonenhancing tumor or enhancing necrosis (Byrne, 1994). A significantly higher Cho level, measured by elevated Cho:NAA and Cho:Cr ratios, suggests tumor recurrence, whereas a reduced Cho level indicates radiation necrosis (Chernov et al., 2006; Zeng et al., 2007). A sensitivity of 85% and a specificity of 69.2% can be achieved by MRS in the differential diagnosis of tumor recurrence from radiation necrosis (Smith et al., 2009). However, such a diagnosis might not be reliable in a tissue that contains mixed tumor and radiation necrosis (Rock et al., 2002). A recent meta-analysis of 18 studies showed that MRS alone has a moderate diagnostic performance and needs to be combined with other imaging techniques to improve diagnostic accuracy (Zhang et al., 2014).

MRS provides unique opportunities in advanced target definition. Early studies proposed the possibility of using MRS in guiding Gamma Knife® radiosurgery (Graves et al., 2000).

These studies revealed that patients who had no tumor pattern outside the targeted area, as assessed by MRS, showed a significantly better outcome than those with such abnormalities. A recent prospective phase II trial demonstrated that an MRS-guided radiosurgery boost was feasible with acceptable toxicity, and the survival of patients was higher than that for historical controls (Einstein et al., 2012). Pirzkall et al. (2001) reported that metabolically active high-grade glioma defined by MRI extended outside the T2-defined region, indicating the potential benefit of MRS-guided target delineation. A feasibility study showed that incorporating dose constraints based on the Cho:Cr ratio can guide dose painting in IMRT planning of gliomas (Narayana et al., 2007). Volumetric MRS revealed that metabolically active and infiltrative tumor areas are not covered by conventional treatment volumes during radiation therapy planning for glioblastoma multiforme, implying the potential utility of MRS in this process (Parra et al., 2014).

2.8 PET/MRI

PET/MRI is similar to PET/CT in that physiological, metabolic, and functional imaging by PET is combined by anatomical imaging at a higher resolution. However, PET/MRI has unique advantages over PET/CT. The lack of ionizing radiation in MRI is beneficial especially to pediatric or pregnant patients when multiple exams are needed. The average effective dose is approximately 25 mSv for PET/CT, compared with 3–4 mSv for the PET/MRI (Brix et al., 2005; Hirsch et al., 2013). Another advantage of PET/MRI is the higher soft-tissue contrast than for PET/CT, particularly in regions where PET suffers from lack of clear contrast due to the intrinsic high uptakes of widely used PET tracers. For instance, perineural spread of tumors and the infiltration of important anatomical landmarks such as the prevertebral fascia and great vessel walls can be visualized by PET/MRI (Queiroz and Huellner, 2015). The improved soft-tissue contrast in the pelvic region can improve the identification and evaluation of local spread in cervical cancer, metastatic lymph nodes, nodal involvement in endometrial cancer, and recurrent ovarian cancer (Bagade et al., 2015). Furthermore, perfusion-weighted and diffusion-weighted MRI, MT/CEST, and MRS reveal various pathophysiological features in the tumor in addition to anatomical images, as discussed in previous sections. A high-level, multiparametric imaging is feasible by integrating these features with the information gained from PET without uncertainties in spatial registration, which can facilitate an accurate target volume definition and adaptive dose painting for radiotherapy (Thorwarth et al., 2013).

A technical difficulty in merging PET and MRI is the interference between the high magnetic field for MRI and the PET detectors, which consist of the scintillator, photomultiplier tube, and associated electronics. This hurdle can be overcome by either sharing a patient table between the PET and MRI scanners arranged in tandem or using an MRI-compatible detector technology (e.g., the avalanche photodiode technology), which allows simultaneous acquisition of data from PET and MRI (Quick et al., 2013; Torigian et al., 2013; Shah and Huang, 2015). Another challenge is to estimate the PET attenuation map, which is not readily available without an additional CT scan. This limitation has been tackled by generating a pseudo-CT from MR images by various approaches, as described in Section 2.3 (Catana et al., 2010; Keereman et al., 2010; Hofmann et al., 2011; Schulz et al., 2011; Roy et al.,

2014). Other general disadvantages of MRI compared to CT, such as vulnerability to metal artifacts, the need of motion correction in chest or abdominal imaging due to slow imaging speed (Catana, 2015), and difficulties in imaging the lung because of very low spin density and high susceptibility effect, also need to be overcome for a successful PET/MRI (Boss et al., 2015).

2.9 SUMMARY

This chapter introduced several MRI techniques relevant to radiation oncology. The potential opportunities for MRI have been appreciated in recent years, and efforts have been made to overcome current challenges unique to MRI. Besides MR-compatible patient positioning systems, advanced imaging protocols have been proposed to suppress image artifacts and improve spatial accuracy. The improvements in image quality by using these protocols need to be considered with the associated trade-offs, such as modified receiver coil configurations, reduced image resolution, increased scan time, and any potential compromise to the safety and comfort of patients. Recent studies show the feasibility of MR-based simulation and 4D MRI. These techniques are expected to give advantageous alternatives in treatment planning after the detailed procedure is established by thorough evaluations. Various MRI contrasts can be used to quantitatively assess physiological and metabolic features in tumor and organs at risk via perfusion/diffusion and spectroscopic methods. Because these functional MRI methods have been developed for diagnostic purposes, further adjustment and optimization, such as suppressing distortions in perfusion/diffusion images, have been considered for the application to radiation oncology. In addition, a deliberate selection of image contrast and the design of MR protocols are essential to meet the specific needs of image-guided radiotherapy. PET/MRI is another example of the ongoing evolution of MRI scanners in conjunction with other imaging or treatment systems toward a higher level of image-guided radiation therapy. The evolution of advanced MRI techniques as well as efforts to have significant impacts on radiation oncology by using this emerging imaging modality will increase in the coming years.

REFERENCES

Abdelnour, A. F., S. A. Nehmeh, T. Pan, et al. 2007. Phase and amplitude binning for 4D-CT imaging. *Phys Med Biol* 52:3515–3529.

Almeida-Freitas, D. B., M. C. Pinho, M. C. Otaduy, H. F. Braga, D. Meira-Freitas, and C. da Costa Leite. 2014. Assessment of irradiated brain metastases using dynamic contrast-enhanced magnetic resonance imaging. *Neuroradiology* 56:437–443.

Alsop, D. C., J. A. Detre, X. Golay, et al. 2014. Recommended implementation of arterial spin-labeled perfusion MRI for clinical applications: A consensus of the ISMRM perfusion study group and the European consortium for ASL in dementia. *Magn Reson Med* 73:102–116.

Andreasen, D., K. Van Leemput, R. H. Hansen, J. A. Andersen, and J. M. Edmund. 2015. Patch-based generation of a pseudo CT from conventional MRI sequences for MRI-only radiotherapy of the brain. *Med Phys* 42:1596–1605.

Aukema, E. J., M. W. Caan, N. Oudhuis, et al. 2009. White matter fractional anisotropy correlates with speed of processing and motor speed in young childhood cancer survivors. *Int J Radiat Oncol Biol Phys* 74:837–843.

Bagade, S., K. J. Fowler, J. K. Schwarz, P. W. Grigsby, and F. Dehdashti. 2015. PET/MRI evaluation of gynecologic malignancies and prostate cancer. *Semin Nucl Med* 45:293–303.

Baldwin, L. N., K. Wachowicz, and B. G. Fallone. 2009. A two-step scheme for distortion rectification of magnetic resonance images. *Med Phys* 36:3917–3926.

Bernstein, M. A., K. F. King, and Z. J. Zhou. 2004. *Handbook of MRI Pulse Sequences*. Amsterdam, The Netherlands: Academic Press.

Bijl, H. P., P. van Luijk, R. P. Coppes, J. M. Schippers, A. W. Konings, and A. J. van Der Kogel. 2005. Regional differences in radiosensitivity across the rat cervical spinal cord. *Int J Radiat Oncol Biol Phys* 61:543–551.

Blackall, J. M., S. Ahmad, M. E. Miquel, J. R. McClelland, D. B. Landau, and D. J. Hawkes. 2006. MRI-based measurements of respiratory motion variability and assessment of imaging strategies for radiotherapy planning. *Phys Med Biol* 51:4147–4169.

Boss, A., M. Weiger, and F. Wiesinger. 2015. Future image acquisition trends for PET/MRI. *Semin Nucl Med* 45:201–211.

Brix, G., U. Lechel, G. Glatting, et al. 2005. Radiation exposure of patients undergoing whole-body dual-modality 18F-FDG PET/CT examinations. *J Nucl Med* 46:608–613.

Bydder, G. M., J. M. Pennock, R. E. Steiner, S. Khenia, J. A. Payne, and I. R. Young. 1985. The short TI inversion recovery sequence: An approach to MR imaging of the abdomen. *Magn Reson Imaging* 3:251–254.

Byrne, T. N. 1994. Imaging of gliomas. *Semin Oncol* 21:162–171.

Cai, J., Z. Chang, Z. Wang, W. Paul Segars, and F. F. Yin. 2011. Four-dimensional magnetic resonance imaging (4D-MRI) using image-based respiratory surrogate: A feasibility study. *Med Phys* 38:6384–6394.

Cao, Y., V. Nagesh, D. Hamstra, et al. 2006a. The extent and severity of vascular leakage as evidence of tumor aggressiveness in high-grade gliomas. *Cancer Res* 66:8912–8917.

Cao, Y., C. I. Tsien, V. Nagesh, et al. 2006b. Survival prediction in high-grade gliomas by MRI perfusion before and during early stage of RT. *Int J Radiat Oncol Biol Phys* 64:876–885.

Cao, Y., C. I. Tsien, P. C. Sundgren, et al. 2009. Dynamic contrast-enhanced magnetic resonance imaging as a biomarker for prediction of radiation-induced neurocognitive dysfunction. *Clin Cancer Res* 15:1747–1754.

Catana, C. 2015. Motion correction options in PET/MRI. *Semin Nucl Med* 45:212–223.

Catana, C., A. van der Kouwe, T. Benner, et al. 2010. Toward implementing an MRI-based PET attenuation-correction method for neurologic studies on the MR-PET brain prototype. *J Nucl Med* 51:1431–1438.

Celicanin, Z., O. Bieri, F. Preiswerk, P. Cattin, K. Scheffler, and F. Santini. 2015. Simultaneous acquisition of image and navigator slices using CAIPIRINHA for 4D MRI. *Magn Reson Med* 73:669–676.

Chan, Y. L., D. J. Roebuck, M. P. Yuen, et al. 2001. Long-term cerebral metabolite changes on proton magnetic resonance spectroscopy in patients cured of acute lymphoblastic leukemia with previous intrathecal methotrexate and cranial irradiation prophylaxis. *Int J Radiat Oncol Biol Phys* 50:759–763.

Chapman, C. H., V. Nagesh, P. C. Sundgren, et al. 2012. Diffusion tensor imaging of normal-appearing white matter as biomarker for radiation-induced late delayed cognitive decline. *Int J Radiat Oncol Biol Phys* 82:2033–2040.

Chapman, C. H., M. Nazem-Zadeh, O. E. Lee, et al. 2013. Regional variation in brain white matter diffusion index changes following chemoradiotherapy: A prospective study using tract-based spatial statistics. *PLoS One* 8:e57768.

Chen, J. T., D. L. Collins, H. L. Atkins, M. S. Freedman, D. L. Arnold, and M. S. B. M. T. Study Group Canadian. 2008. Magnetization transfer ratio evolution with demyelination and remyelination in multiple sclerosis lesions. *Ann Neurol* 63:254–262.

Chen, L. Q., C. M. Howison, J. J. Jeffery, I. F. Robey, P. H. Kuo, and M. D. Pagel. 2014. Evaluations of extracellular pH within in vivo tumors using acidoCEST MRI. *Magn Reson Med* 72:1408–1417.

Chernov, M. F., M. Hayashi, M. Izawa, et al. 2006. Multivoxel proton MRS for differentiation of radiation-induced necrosis and tumor recurrence after gamma knife radiosurgery for brain metastases. *Brain Tumor Pathol* 23:19–27.

Chung, N. N., L. L. Ting, W. C. Hsu, L. T. Lui, and P. M. Wang. 2004. Impact of magnetic resonance imaging versus CT on nasopharyngeal carcinoma: Primary tumor target delineation for radiotherapy. *Head Neck* 26:241–246.

Croteau, D., L. Scarpace, D. Hearshen, et al. 2001. Correlation between magnetic resonance spectroscopy imaging and image-guided biopsies: semiquantitative and qualitative histopathological analyses of patients with untreated glioma. *Neurosurgery* 49:823–829.

Dagher, A. P., A. Aletras, P. Choyke, and R. S. Balaban. 2000. Imaging of urea using chemical exchange-dependent saturation transfer at 1.5T. *J Magn Reson Imaging* 12:745–748.

Dagirmanjian, A., J. S. Ross, N. Obuchowski, et al. 1995. High resolution, magnetization transfer saturation, variable flip angle, time-of-flight MRA in the detection of intracranial vascular stenoses. *J Comput Assist Tomogr* 19:700–706.

Del Grande, F., F. Santini, D. A. Herzka, et al. 2014. Fat-suppression techniques for 3-T MR imaging of the musculoskeletal system. *Radiographics* 34:217–233.

Dellani, P. R., S. Eder, J. Gawehn, et al. 2008. Late structural alterations of cerebral white matter in long-term survivors of childhood leukemia. *J Magn Reson Imaging* 27:1250–1255.

Devic, S. 2012. MRI simulation for radiotherapy treatment planning. *Med Phys* 39:6701–6711.

Dietrich, O., M. F. Reiser, and S. O. Schoenberg. 2008. Artifacts in 3-T MRI: physical background and reduction strategies. *Eur J Radiol* 65:29–35.

Dinkel, J., C. Hintze, R. Tetzlaff, et al. 2009. 4D-MRI analysis of lung tumor motion in patients with hemidiaphragmatic paralysis. *Radiother Oncol* 91:449–454.

Doran, S. J., L. Charles-Edwards, S. A. Reinsberg, and M. O. Leach. 2005. A complete distortion correction for MR images: I. Gradient warp correction. *Phys Med Biol* 50:1343–1361.

Dowling, J. A., J. Lambert, J. Parker, et al. 2012. An atlas-based electron density mapping method for magnetic resonance imaging (MRI)-alone treatment planning and adaptive MRI-based prostate radiation therapy. *Int J Radiat Oncol Biol Phys* 83:e5–e11.

Du, D., S. D. Caruthers, C. Glide-Hurst, et al. 2015. High-quality t2-weighted 4-dimensional magnetic resonance imaging for radiation therapy applications. *Int J Radiat Oncol Biol Phys* 92:430–437.

Edmund, J. M., H. M. Kjer, K. Van Leemput, R. H. Hansen, J. A. Andersen, and D. Andreasen. 2014. A voxel-based investigation for MRI-only radiotherapy of the brain using ultra short echo times. *Phys Med Biol* 59:7501–7519.

Einstein, D. B., B. Wessels, B. Bangert, et al. 2012. Phase II trial of radiosurgery to magnetic resonance spectroscopy-defined high-risk tumor volumes in patients with glioblastoma multiforme. *Int J Radiat Oncol Biol Phys* 84:668–674.

Fallone, B. G., B. Murray, S. Rathee, et al. 2009. First MR images obtained during megavoltage photon irradiation from a prototype integrated linac-MR system. *Med Phys* 36:2084–2088.

Farjam, R., C. I. Tsien, F. Y. Feng, et al. 2013. Physiological imaging-defined, response-driven subvolumes of a tumor. *Int J Radiat Oncol Biol Phys* 85:1383–1390.

Farjam, R., C. I. Tsien, F. Y. Feng, et al. 2014a. Investigation of the diffusion abnormality index as a new imaging biomarker for early assessment of brain tumor response to radiation therapy. *Neuro Oncol* 16:131–139.

Farjam, R., C. I. Tsien, T. S. Lawrence, and Y. Cao. 2014b. DCE-MRI defined subvolumes of a brain metastatic lesion by principle component analysis and fuzzy-c-means clustering for response assessment of radiation therapy. *Med Phys* 41:011708.

Frahm, J., A. Haase, W. Hanicke, D. Matthaei, H. Bomsdorf, and T. Helzel. 1985. Chemical shift selective MR imaging using a whole-body magnet. *Radiology* 156:441–444.

Garcia, M., M. Gloor, O. Bieri, et al. 2015. Imaging of primary brain tumors and metastases with fast quantitative 3-dimensional magnetization transfer. *J Neuroimaging* 25:1007–1014.

Goshima, S., M. Kanematsu, Y. Noda, H. Kondo, H. Watanabe, and K. T. Bae. 2015. Diffusion kurtosis imaging to assess response to treatment in hypervascular hepatocellular carcinoma. *AJR Am J Roentgenol* 204:W543–W549.

Graves, E. E., S. J. Nelson, D. B. Vigneron, et al. 2000. A preliminary study of the prognostic value of proton magnetic resonance spectroscopic imaging in gamma knife radiosurgery of recurrent malignant gliomas. *Neurosurgery* 46:319–326; discussion 326–328.

Groenendaal, G., A. Borren, M. R. Moman, et al. 2012. Pathologic validation of a model based on diffusion-weighted imaging and dynamic contrast-enhanced magnetic resonance imaging for tumor delineation in the prostate peripheral zone. *Int J Radiat Oncol Biol Phys* 82:e537–e544.

Grossman, R. I. 1994. Magnetization transfer in multiple sclerosis. *Ann Neurol* 36 Suppl:S97–S99.

Gudur, M. S., W. Hara, Q. T. Le, L. Wang, L. Xing, and R. Li. 2014. A unifying probabilistic Bayesian approach to derive electron density from MRI for radiation therapy treatment planning. *Phys Med Biol* 59:6595–6606.

Halligan, S., J. C. Healy, and C. I. Bartram. 1998. Magnetic resonance imaging of fistula-in-ano: STIR or SPIR? *Br J Radiol* 71:141–145.

Hamstra, D. A., C. J. Galban, C. R. Meyer, et al. 2008. Functional diffusion map as an early imaging biomarker for high-grade glioma: Correlation with conventional radiologic response and overall survival. *J Clin Oncol* 26:3387–3394.

Hansen, M. S., C. Baltes, J. Tsao, S. Kozerke, K. P. Pruessmann, and H. Eggers. 2006. k-t BLAST reconstruction from non-Cartesian k-t space sampling. *Magn Reson Med* 55:85–91.

Henkelman, R. M., G. J. Stanisz, and S. J. Graham. 2001. Magnetization transfer in MRI: A review. *NMR Biomed* 14:57–64.

Hernando, D., D. C. Karampinos, K. F. King, et al. 2011. Removal of olefinic fat chemical shift artifact in diffusion MRI. *Magn Reson Med* 65:692–701.

Hirsch, F. W., B. Sattler, I. Sorge, et al. 2013. PET/MR in children. Initial clinical experience in paediatric oncology using an integrated PET/MR scanner. *Pediatr Radiol* 43:860–875.

Hofmann, M., I. Bezrukov, F. Mantlik, et al. 2011. MRI-based attenuation correction for whole-body PET/MRI: Quantitative evaluation of segmentation- and atlas-based methods. *J Nucl Med* 52:1392–1399.

Hofmann, M., F. Steinke, V. Scheel, et al. 2008. MRI-based attenuation correction for PET/MRI: A novel approach combining pattern recognition and atlas registration. *J Nucl Med* 49:1875–1883.

Hoogcarspel, S. J., J. M. Van der Velden, J. J. Lagendijk, M. van Vulpen, and B. W. Raaymakers. 2014. The feasibility of utilizing pseudo CT-data for online MRI based treatment plan adaptation for a stereotactic radiotherapy treatment of spinal bone metastases. *Phys Med Biol* 59:7383–7391.

Hsu, S. H., Y. Cao, K. Huang, M. Feng, and J. M. Balter. 2013. Investigation of a method for generating synthetic CT models from MRI scans of the head and neck for radiation therapy. *Phys Med Biol* 58:8419–8435.

Hsu, S. H., Y. Cao, T. S. Lawrence, et al. 2015. Quantitative characterizations of ultrashort echo (UTE) images for supporting air-bone separation in the head. *Phys Med Biol* 60:2869–2880.

Hu, L. S., L. C. Baxter, K. A. Smith, et al. 2009. Relative cerebral blood volume values to differentiate high-grade glioma recurrence from posttreatment radiation effect: Direct correlation between image-guided tissue histopathology and localized dynamic susceptibility-weighted contrast-enhanced perfusion MR imaging measurements. *AJNR Am J Neuroradiol* 30:552–558.

Hu, Y., S. D. Caruthers, D. A. Low, P. J. Parikh, and S. Mutic. 2013. Respiratory amplitude guided 4-dimensional magnetic resonance imaging. *Int J Radiat Oncol Biol Phys* 86:198–204.

Huang, C. F., S. Y. Chiou, M. F. Wu, H. T. Tu, W. S. Liu, and J. C. Chuang. 2010. Apparent diffusion coefficients for evaluation of the response of brain tumors treated by Gamma Knife surgery. *J Neurosurg* 113 Suppl:97–104.

Jensen, J. H., and J. A. Helpern. 2010. MRI quantification of non-Gaussian water diffusion by kurtosis analysis. *NMR Biomed* 23:698–710.

Johansson, A., M. Karlsson, and T. Nyholm. 2011. CT substitute derived from MRI sequences with ultrashort echo time. *Med Phys* 38:2708–2714.

Johansson, A., M. Karlsson, J. Yu, T. Asklund, and T. Nyholm. 2012. Voxel-wise uncertainty in CT substitute derived from MRI. *Med Phys* 39:3283–3290.

Jonsson, J. H., M. G. Karlsson, M. Karlsson, and T. Nyholm. 2010. Treatment planning using MRI data: An analysis of the dose calculation accuracy for different treatment regions. *Radiat Oncol* 5:62.

Kapanen, M., J. Collan, A. Beule, T. Seppala, K. Saarilahti, and M. Tenhunen. 2013. Commissioning of MRI-only based treatment planning procedure for external beam radiotherapy of prostate. *Magn Reson Med* 70:127–135.

Karlsson, M., M. G. Karlsson, T. Nyholm, C. Amies, and B. Zackrisson. 2009. Dedicated magnetic resonance imaging in the radiotherapy clinic. *Int J Radiat Oncol Biol Phys* 74:644–651.

Keereman, V., Y. Fierens, T. Broux, Y. De Deene, M. Lonneux, and S. Vandenberghe. 2010. MRI-based attenuation correction for PET/MRI using ultrashort echo time sequences. *J Nucl Med* 51:812–818.

Khong, P. L., D. L. Kwong, G. C. Chan, J. S. Sham, F. L. Chan, and G. C. Ooi. 2003. Diffusion-tensor imaging for the detection and quantification of treatment-induced white matter injury in children with medulloblastoma: A pilot study. *AJNR Am J Neuroradiol* 24:734–740.

Khong, P. L., L. H. Leung, A. S. Fung, et al. 2006. White matter anisotropy in post-treatment childhood cancer survivors: Preliminary evidence of association with neurocognitive function. *J Clin Oncol* 24:884–890.

Kim, T., S. Kim, Y. K. Park, K. K. Youn, P. Keall, and R. Lee. 2014a. Motion management within two respiratory-gating windows: Feasibility study of dual quasi-breath-hold technique in gated medical procedures. *Phys Med Biol* 59:6583–6594.

Kim, T., R. Pooley, D. Lee, P. Keall, R. Lee, and S. Kim. 2014b. Quasi-breath-hold (QBH) biofeedback in hated 3D thoracic MRI: Feasibility study. *Prog Med Phys* 25:72–77.

Knutsson, L., F. Stahlberg, and R. Wirestam. 2010. Absolute quantification of perfusion using dynamic susceptibility contrast MRI: Pitfalls and possibilities. *MAGMA* 23:1–21.

Korhonen, J., M. Kapanen, J. Keyrilainen, T. Seppala, and M. Tenhunen. 2014. A dual model HU conversion from MRI intensity values within and outside of bone segment for MRI-based radiotherapy treatment planning of prostate cancer. *Med Phys* 41:011704.

Lambert, J., P. B. Greer, F. Menk, et al. 2011. MRI-guided prostate radiation therapy planning: Investigation of dosimetric accuracy of MRI-based dose planning. *Radiother Oncol* 98:330–334.

Lambrecht, M., V. Vandecaveye, F. De Keyzer, et al. 2012. Value of diffusion-weighted magnetic resonance imaging for prediction and early assessment of response to neoadjuvant radiochemotherapy in rectal cancer: Preliminary results. *Int J Radiat Oncol Biol Phys* 82:863–870.

Lauterbur, P. C. 1973. Image formation by induced local interactions: Examples employing nuclear magnetic-resonance. *Nature* 242:190–191.

Le Bihan, D., E. Breton, D. Lallemand, M. L. Aubin, J. Vignaud, and M. Laval-Jeantet. 1988. Separation of diffusion and perfusion in intravoxel incoherent motion MR imaging. *Radiology* 168:497–505.

Lee, Y. K., M. Bollet, G. Charles-Edwards, et al. 2003. Radiotherapy treatment planning of prostate cancer using magnetic resonance imaging alone. *Radiother Oncol* 66:203–216.

Lehtio, K., O. Eskola, T. Viljanen, et al. 2004. Imaging perfusion and hypoxia with PET to predict radiotherapy response in head-and-neck cancer. *Int J Radiat Oncol Biol Phys* 59:971–982.

Li, H., C. Noel, J. Garcia-Ramirez, et al. 2012. Clinical evaluations of an amplitude-based binning algorithm for 4DCT reconstruction in radiation therapy. *Med Phys* 39:922–932.

Li, X., Y. Lu, A. Pirzkall, T. McKnight, and S. J. Nelson. 2002. Analysis of the spatial characteristics of metabolic abnormalities in newly diagnosed glioma patients. *J Magn Reson Imaging* 16:229–237.

Liney, G. P., and M. A. Moerland. 2014. Magnetic resonance imaging acquisition techniques for radiotherapy planning. *Semin Radiat Oncol* 24:160–168.

Liu, G., X. Song, K. W. Chan, and M. T. McMahon. 2013. Nuts and bolts of chemical exchange saturation transfer MRI. *NMR Biomed* 26:810–828.

Liu, Y., F. F. Yin, N. K. Chen, M. L. Chu, and J. Cai. 2015. Four-dimensional magnetic resonance imaging with retrospective k-space reordering: A feasibility study. *Med Phys* 42:534–541.

Lundbom, N. 1992. Determination of magnetization transfer contrast in tissue: An MR imaging study of brain tumors. *AJR Am J Roentgenol* 159:1279–1285.

Lustig, M., D. Donoho, and J. M. Pauly. 2007. Sparse MRI: The application of compressed sensing for rapid MR imaging. *Magn Reson Med* 58:1182–1195.

Martens, M. H., D. M. Lambregts, N. Papanikolaou, et al. 2014. Magnetization transfer ratio: A potential biomarker for the assessment of postradiation fibrosis in patients with rectal cancer. *Invest Radiol* 49:29–34.

McKnight, T. R. 2004. Proton magnetic resonance spectroscopic evaluation of brain tumor metabolism. *Semin Oncol* 31:605–617.

Merchant, T. E., G. R. Thelissen, H. C. Kievit, L. J. Oosterwaal, C. J. Bakker, and P. W. de Graaf. 1992. Breast disease evaluation with fat-suppressed magnetic resonance imaging. *Magn Reson Imaging* 10:335–340.

Moffat, B. A., T. L. Chenevert, T. S. Lawrence, et al. 2005. Functional diffusion map: A noninvasive MRI biomarker for early stratification of clinical brain tumor response. *Proc Natl Acad Sci U S A* 102:5524–5529.

Morgan, P. S., R. W. Bowtell, D. J. McIntyre, and B. S. Worthington. 2004. Correction of spatial distortion in EPI due to inhomogeneous static magnetic fields using the reversed gradient method. *J Magn Reson Imaging* 19:499–507.

Mutic, S. 2012. First Commercial Hybrid MRI-IMRT System. *Med Phys* 39:3934–3934.

Nagesh, V., C. I. Tsien, T. L. Chenevert, et al. 2008. Radiation-induced changes in normal-appearing white matter in patients with cerebral tumors: A diffusion tensor imaging study. *Int J Radiat Oncol Biol Phys* 70:1002–1010.

Narayana, A., J. Chang, S. Thakur, et al. 2007. Use of MR spectroscopy and functional imaging in the treatment planning of gliomas. *Br J Radiol* 80:347–354.

Noguchi, T., T. Yoshiura, A. Hiwatashi, et al. 2008. Perfusion imaging of brain tumors using arterial spin-labeling: Correlation with histopathologic vascular density. *AJNR Am J Neuroradiol* 29:688–693.

Ozsunar, Y., M. E. Mullins, K. Kwong, et al. 2010. Glioma recurrence versus radiation necrosis? A pilot comparison of arterial spin-labeled, dynamic susceptibility contrast enhanced MRI, and FDG-PET imaging. *Acad Radiol* 17:282–290.

Padhani, A. R., D. M. Koh, and D. J. Collins. 2011. Whole-body diffusion-weighted MR imaging in cancer: Current status and research directions. *Radiology* 261:700–718.

Park, Y. K., S. Kim, H. Kim, I. H. Kim, K. Lee, and S. J. Ye. 2011. Quasi-breath-hold technique using personalized audio-visual biofeedback for respiratory motion management in radiotherapy. *Med Phys* 38:3114–3124.

Parra, N. A., A. A. Maudsley, R. K. Gupta, et al. 2014. Volumetric spectroscopic imaging of glioblastoma multiforme radiation treatment volumes. *Int J Radiat Oncol Biol Phys* 90:376–384.

Pike, G. B., N. De Stefano, S. Narayanan, et al. 2000. Multiple sclerosis: magnetization transfer MR imaging of white matter before lesion appearance on T2-weighted images. *Radiology* 215:824–830.

Pipe, J. G., W. N. Gibbs, Z. Li, J. P. Karis, M. Schar, and N. R. Zwart. 2014. Revised motion estimation algorithm for PROPELLER MRI. *Magn Reson Med* 72:430–437.

Pirzkall, A., T. R. McKnight, E. E. Graves, et al. 2001. MR-spectroscopy guided target delineation for high-grade gliomas. *Int J Radiat Oncol Biol Phys* 50:915–928.

Pramanik, P. P., H. A. Parmar, A. G. Mammoser, et al. 2015. Hypercellularity components of glioblastoma identified by high b-value diffusion-weighted imaging. *Int J Radiat Oncol Biol Phys* 92:811–819.

Qiu, D., D. L. Kwong, G. C. Chan, L. H. Leung, and P. L. Khong. 2007. Diffusion tensor magnetic resonance imaging finding of discrepant fractional anisotropy between the frontal and parietal lobes after whole-brain irradiation in childhood medulloblastoma survivors: Reflection of regional white matter radiosensitivity? *Int J Radiat Oncol Biol Phys* 69:846–851.

Qiu, D., L. H. Leung, D. L. Kwong, G. C. Chan, and P. L. Khong. 2006. Mapping radiation dose distribution on the fractional anisotropy map: Applications in the assessment of treatment-induced white matter injury. *Neuroimage* 31:109–115.

Queiroz, M. A., and M. W. Huellner. 2015. PET/MR in cancers of the head and neck. *Semin Nucl Med* 45:248–265.

Quick, H. H., C. von Gall, M. Zeilinger, et al. 2013. Integrated whole-body PET/MR hybrid imaging: Clinical experience. *Invest Radiol* 48:280–289.

Raaymakers, B. W., J. J. Lagendijk, J. Overweg, et al. 2009. Integrating a 1.5 T MRI scanner with a 6 MV accelerator: Proof of concept. *Phys Med Biol* 54:N229–N237.

Rank, C. M., C. Tremmel, N. Hunemohr, A. M. Nagel, O. Jakel, and S. Greilich. 2013. MRI-based treatment plan simulation and adaptation for ion radiotherapy using a classification-based approach. *Radiat Oncol* 8:51.

Ravn, S., M. Holmberg, P. Sorensen, J. B. Frokjaer, and J. Carl. 2013. Differences in supratentorial white matter diffusion after radiotherapy: New biomarker of normal brain tissue damage? *Acta Oncol* 52:1314–1319.

Redpath, T. W., and S. C. Wayte. 1993. Fat suppressed magnetic resonance imaging at 0.5 T using binomial radiofrequency pulses. *Br J Radiol* 66:886–891.

Reinsberg, S. A., S. J. Doran, E. M. Charles-Edwards, and M. O. Leach. 2005. A complete distortion correction for MR images: II. Rectification of static-field inhomogeneities by similarity-based profile mapping. *Phys Med Biol* 50:2651–2661.

Remmert, G., J. Biederer, F. Lohberger, M. Fabel, and G. H. Hartmann. 2007. Four-dimensional magnetic resonance imaging for the determination of tumour movement and its evaluation using a dynamic porcine lung phantom. *Phys Med Biol* 52:N401–N415.

Rivlin, M., J. Horev, I. Tsarfaty, and G. Navon. 2013. Molecular imaging of tumors and metastases using chemical exchange saturation transfer (CEST) MRI. *Sci Rep* 3:3045.

Rock, J. P., D. Hearshen, L. Scarpace, et al. 2002. Correlations between magnetic resonance spectroscopy and image-guided histopathology, with special attention to radiation necrosis. *Neurosurgery* 51:912–919; discussion 919–920.

Rosen, B. R., V. J. Wedeen, and T. J. Brady. 1984. Selective saturation NMR imaging. *J Comput Assist Tomogr* 8:813–818.

Roy, S., W. T. Wang, A. Carass, J. L. Prince, J. A. Butman, and D. L. Pham. 2014. PET attenuation correction using synthetic CT from ultrashort echo-time MR imaging. *J Nucl Med* 55:2071–2077.

Schick, F., J. Forster, J. Machann, R. Kuntz, and C. D. Claussen. 1998. Improved clinical echo-planar MRI using spatial-spectral excitation. *J Magn Reson Imaging* 8:960–967.

Schreibmann, E., J. A. Nye, D. M. Schuster, D. R. Martin, J. Votaw, and T. Fox. 2010. MR-based attenuation correction for hybrid PET-MR brain imaging systems using deformable image registration. *Med Phys* 37:2101–2109.

Schulz, V., I. Torres-Espallardo, S. Renisch, et al. 2011. Automatic, three-segment, MR-based attenuation correction for whole-body PET/MR data. *Eur J Nucl Med Mol Imaging* 38:138–152.

Shah, S. N., and S. S. Huang. 2015. Hybrid PET/MR imaging: Physics and technical considerations. *Abdom Imaging* 40:1358–1365.

Sherry, A. D., and M. Woods. 2008. Chemical exchange saturation transfer contrast agents for magnetic resonance imaging. *Annu Rev Biomed Eng* 10:391–411.

Sjolund, J., D. Forsberg, M. Andersson, and H. Knutsson. 2015. Generating patient specific pseudo-CT of the head from MR using atlas-based regression. *Phys Med Biol* 60:825–839.

Smith, E. A., R. C. Carlos, L. R. Junck, C. I. Tsien, A. Elias, and P. C. Sundgren. 2009. Developing a clinical decision model: MR spectroscopy to differentiate between recurrent tumor and radiation change in patients with new contrast-enhancing lesions. *AJR Am J Roentgenol* 192:W45–W52.

Soares, D. P., and M. Law. 2009. Magnetic resonance spectroscopy of the brain: review of metabolites and clinical applications. *Clin Radiol* 64:12–21.

Stam, M. K., S. P. Crijns, B. A. Zonnenberg, et al. 2012. Navigators for motion detection during real-time MRI-guided radiotherapy. *Phys Med Biol* 57:6797–6805.

Stanescu, T., H. S. Jans, N. Pervez, P. Stavrev, and B. G. Fallone. 2008. A study on the magnetic resonance imaging (MRI)-based radiation treatment planning of intracranial lesions. *Phys Med Biol* 53:3579–3593.

Stanescu, T., K. Wachowicz, and D. A. Jaffray. 2012. Characterization of tissue magnetic susceptibility-induced distortions for MRIgRT. *Med Phys* 39:7185–7193.

Steen, R. G., B. S. M. Koury, C. I. Granja, et al. 2001. Effect of ionizing radiation on the human brain: White matter and gray matter T1 in pediatric brain tumor patients treated with conformal radiation therapy. *Int J Radiat Oncol Biol Phys* 49:79–91.

Stemkens, B., R. H. Tijssen, B. D. de Senneville, et al. 2015. Optimizing 4-dimensional magnetic resonance imaging data sampling for respiratory motion analysis of pancreatic tumors. *Int J Radiat Oncol Biol Phys* 91:571–578.

Sugahara, T., Y. Korogi, M. Kochi, et al. 1999. Usefulness of diffusion-weighted MRI with echo-planar technique in the evaluation of cellularity in gliomas. *J Magn Reson Imaging* 9:53–60.

Sundgren, P. C. 2009. MR spectroscopy in radiation injury. *AJNR Am J Neuroradiol* 30:1469–1476.

Sundgren, P. C., V. Nagesh, A. Elias, et al. 2009. Metabolic alterations: A biomarker for radiation-induced normal brain injury-an MR spectroscopy study. *J Magn Reson Imaging* 29:291–297.

Thomasson, D., D. Purdy, and J. P. Finn. 1996. Phase-modulated binomial RF pulses for fast spectrally-selective musculoskeletal imaging. *Magn Reson Med* 35:563–568.

Thorwarth, Daniela, Sara Leibfarth, and David Mönnich. 2013. Potential role of PET/MRI in radiotherapy treatment planning. *Clin Transl Imaging* 1:45–51.

Tokuda, J., S. Morikawa, H. A. Haque, et al. 2008. Adaptive 4D MR imaging using navigator-based respiratory signal for MRI-guided therapy. *Magn Reson Med* 59:1051–1061.

Torigian, D. A., H. Zaidi, T. C. Kwee, et al. 2013. PET/MR imaging: Technical aspects and potential clinical applications. *Radiology* 267:26–44.

Tryggestad, E., A. Flammang, S. Han-Oh, et al. 2013. Respiration-based sorting of dynamic MRI to derive representative 4D-MRI for radiotherapy planning. *Med Phys* 40:051909.

Tsien, C., C. J. Galban, T. L. Chenevert, et al. 2010. Parametric response map as an imaging biomarker to distinguish progression from pseudoprogression in high-grade glioma. *J Clin Oncol* 28:2293–2299.

Tyler, D. J., M. D. Robson, R. M. Henkelman, I. R. Young, and G. M. Bydder. 2007. Magnetic resonance imaging with ultrashort TE (UTE) PULSE sequences: Technical considerations. *J Magn Reson Imaging* 25:279–289.

Uh, J., M. A. Khan, and C. Hua. 2016. Four-dimensional MRI using an internal respiratory surrogate derived by dimensionality reduction. *Phys Med Biol* 61:7812–7832.

Uh, J., T. E. Merchant, Y. Li, et al. 2013. Differences in brainstem fiber tract response to radiation: A longitudinal diffusion tensor imaging study. *Int J Radiat Oncol Biol Phys* 86:292–297.

Uh, J., T. E. Merchant, Y. Li, X. Li, and C. Hua. 2014. MRI-based treatment planning with pseudo CT generated through atlas registration. *Med Phys* 41:051711.

Uh, J., T. E. Merchant, Y. Li, et al. 2015. Effects of surgery and proton therapy on cerebral white matter of craniopharyngioma patients. *Int J Radiat Oncol Biol Phys* 93:64–71.

Usenius, T., J. P. Usenius, M. Tenhunen, et al. 1995. Radiation-induced changes in human brain metabolites as studied by 1H nuclear magnetic resonance spectroscopy in vivo. *Int J Radiat Oncol Biol Phys* 33:719–724.

Van Cauter, S., J. Veraart, J. Sijbers, et al. 2012. Gliomas: diffusion kurtosis MR imaging in grading. *Radiology* 263:492–501.

van der Heide, U. A., A. C. Houweling, G. Groenendaal, R. G. Beets-Tan, and P. Lambin. 2012. Functional MRI for radiotherapy dose painting. *Magn Reson Imaging* 30:1216–1223.

van Zijl, P. C., and N. N. Yadav. 2011. Chemical exchange saturation transfer (CEST): What is in a name and what isn't? *Magn Reson Med* 65:927–948.

Vavasour, I. M., C. Laule, D. K. Li, A. L. Traboulsee, and A. L. MacKay. 2011. Is the magnetization transfer ratio a marker for myelin in multiple sclerosis? *J Magn Reson Imaging* 33:713–718.

von Siebenthal, M., G. Szekely, U. Gamper, P. Boesiger, A. Lomax, and P. Cattin. 2007. 4D MR imaging of respiratory organ motion and its variability. *Phys Med Biol* 52:1547–1564.

Voroney, J. P., K. K. Brock, C. Eccles, M. Haider, and L. A. Dawson. 2006. Prospective comparison of computed tomography and magnetic resonance imaging for liver cancer delineation using deformable image registration. *Int J Radiat Oncol Biol Phys* 66:780–791.

Wang, H., J. Balter, and Y. Cao. 2013. Patient-induced susceptibility effect on geometric distortion of clinical brain MRI for radiation treatment planning on a 3T scanner. *Phys Med Biol* 58:465–477.

Wang, P., A. Popovtzer, A. Eisbruch, and Y. Cao. 2012. An approach to identify, from DCE MRI, significant subvolumes of tumors related to outcomes in advanced head-and-neck cancer. *Med Phys* 39:5277–5285.

Ward, K. M., A. H. Aletras, and R. S. Balaban. 2000. A new class of contrast agents for MRI based on proton chemical exchange dependent saturation transfer (CEST). *J Magn Reson* 143:79–87.

Wen, Z., S. Hu, F. Huang, et al. 2010. MR imaging of high-grade brain tumors using endogenous protein and peptide-based contrast. *Neuroimage* 51:616–622.

Willats, L., and F. Calamante. 2013. The 39 steps: evading error and deciphering the secrets for accurate dynamic susceptibility contrast MRI. *NMR Biomed* 26:913–931.

Wolf, R. L., J. Wang, S. Wang, et al. 2005. Grading of CNS neoplasms using continuous arterial spin labeled perfusion MR imaging at 3 Tesla. *J Magn Reson Imaging* 22:475–482.

Wolff, S. D., and R. S. Balaban. 1989. Magnetization transfer contrast (MTC) and tissue water proton relaxation in vivo. *Magn Reson Med* 10:135–144.

Yamashita, Y., T. Baba, Y. Baba, et al. 2000. Dynamic contrast-enhanced MR imaging of uterine cervical cancer: Pharmacokinetic analysis with histopathologic correlation and its importance in predicting the outcome of radiation therapy. *Radiology* 216:803–809.

Yang, J., J. Cai, H. Wang, et al. 2014. Four-dimensional magnetic resonance imaging using axial body area as respiratory surrogate: initial patient results. *Int J Radiat Oncol Biol Phys* 88:907–912.

Yee, S. 2016. Segmentation of fat in MRI using a preparatory pair of rectangular RF pulses of opposite direction. *Magn Reson Imaging* 34:483–4491.

Zeng, Q. S., C. F. Li, K. Zhang, H. Liu, X. S. Kang, and J. H. Zhen. 2007. Multivoxel 3D proton MR spectroscopy in the distinction of recurrent glioma from radiation injury. *J Neurooncol* 84:63–69.

Zhang, H., L. Ma, Q. Wang, X. Zheng, C. Wu, and B. N. Xu. 2014. Role of magnetic resonance spectroscopy for the differentiation of recurrent glioma from radiation necrosis: A systematic review and meta-analysis. *Eur J Radiol* 83:2181–2189.

Zhou, J., J. F. Payen, D. A. Wilson, R. J. Traystman, and P. C. van Zijl. 2003. Using the amide proton signals of intracellular proteins and peptides to detect pH effects in MRI. *Nat Med* 9:1085–1090.
Zhou, J., E. Tryggestad, Z. Wen, et al. 2011. Differentiation between glioma and radiation necrosis using molecular magnetic resonance imaging of endogenous proteins and peptides. *Nat Med* 17:130–134.
Zulfiqar, M., D. M. Yousem, and H. Lai. 2013. ADC values and prognosis of malignant astrocytomas: Does lower ADC predict a worse prognosis independent of grade of tumor?—A meta-analysis. *AJR Am J Roentgenol* 200:624–629.

CHAPTER 3

Biological Imaging and Radiobiological Modeling for Treatment Planning and Response Assessment in Radiation Therapy

Vitali Moiseenko, Stephen R. Bowen, John P. Kirkpatrick, Robert Jeraj, and Lawrence B. Marks

CONTENTS

3.1 INTRODUCTION TO BIOLOGICAL IMAGING

Radiobiological descriptions of normal tissue and tumor response to radiation are often limited to fitting the mean or median of population-based data with little to no consideration of patient-specific variability of tumor or normal tissue properties that modulate this response. Biological imaging, including both molecular and functional imaging, noninvasively investigates properties that are spatially localized either to cancerous or functional tissue and that may dynamically vary with time. To use biological imaging for therapeutic applications, such as treatment planning and treatment response evaluation of spatial and temporal variations in biological properties that are unique to each individual patient, quantitative imaging biomarkers must be established. Quantitative imaging biomarkers may be associated with clinical endpoints, and they may complement tissues biomarkers, both potentially driving precision medicine (Mirnezami et al., 2012). Conceptually, predictive imaging biomarkers might guide initial therapy and subsequent adaptive therapies, thus facilitating administration of the right therapy to an individual patient at the optimal time. Within precision radiation therapy, quantitative biological imaging can define treatment-planning targets, define functional avoidance regions, and assess the response of targets and functional tissues during and after therapy (Jeraj et al., 2015).

Biological imaging provides a set of quantitative tools to probe radiobiological properties that may influence the response to radiotherapy. Several clinical reports have successfully related tumor control to biological properties probed by functional imaging prior to or during radiotherapy, for example, in patients with non-small-cell lung cancer (Ohri et al., 2015), rectal cancer (Joye et al., 2014), rhabdomyosarcoma (Casey et al., 2014), and cervical cancer (Mayr et al., 2012). These are only a few examples; overall, the literature strongly supports the premise that assessing tumor and normal tissue biological properties can predict treatment outcome, at least in some disease sites.

This evidence, however, does not answer a key question: "Can therapy be modified or adapted to offset the radiobiological effect of properties associated with an unfavorable outcome?" In the definitive setting, achieving higher rates of tumor control and lower rates of disease recurrence has spurred investigations of tumor heterogeneity metrics that drive differential response. Directly manipulating characteristics of radiation response, for example, by increasing oxygen concentration in hypoxic regions, appears appealing but logistically challenging. Thanks to modern radiotherapy planning and delivery technology, strategies for boosting the dose to regions identified as exhibiting characteristics associated with higher risk of recurrence compared to the remainder of the tumor volume are feasible (e.g., with intensity-modulated radiation therapy [IMRT]). This, of course, comes with the often conflicting requirement that such a boost be delivered without compromising normal tissue. Within the normal tissue, biological imaging may be helpful to assess distribution of functional or regenerative capacity and to enable the establishment of functional tissue dose constraints that may complement conventional dose-volume histogram objectives (Jeraj et al., 2010). Thus, there might be a potential benefit to using biological imaging to define subregions of tumors to be selectively "boosted" and/or to define subregions of normal tissues to be selectively "spared." A number of studies, including randomized trials, are currently in progress. An overview of challenges and opportunities for integrating biological imaging and radiobiological modeling into treatment planning and response assessment of radiation therapy are presented.

3.2 BIOLOGICAL IMAGING TO DEFINE TREATMENT PLANNING TARGETS AND AVOIDANCE REGIONS

Radiation therapy planning relies on region-of-interest definition for both tumor targets and normal tissue that are conventionally delineated on anatomic imaging modalities using computed tomography (CT) or magnetic resonance imaging (MRI). These provide structure and morphological information from which planning objectives are set by population-based prescriptions to achieve a certain degree of local tumor control at an acceptable risk of normal tissue complication. Biological imaging modalities such as positron emission tomography (PET), single photon emission tomography (SPECT), and functional MRI (fMRI) hold the promise to refine target and normal tissue definitions to incorporate additional patient-specific intrastructure variations. Several imaging biomarkers of cancer phenotype, as well as imaging biomarkers of normal tissue function, have dramatically altered the radiation therapy planning landscape for region definition (Thorwarth, 2015).

3.2.1 Imaging Biomarkers of Cancer Phenotype

3.2.1.1 Glucose Utilization

The workhorse imaging biomarker in oncology is fluorodeoxyglucose (FDG) radiolabeled with the positron-emitting ^{18}F nuclide for tomographic imaging with PET. It has rapidly and increasingly become integrated into radiotherapy planning (De Ruysscher and Kirsch, 2010, Nestle et al., 2006). Originally designed to map glucose metabolism and hexokinase activity, FDG uptake is tied to glucose transporter GLUT-1 and HIF-1α expression in certain disease sites (Kaira et al., 2014) as part of the glycolytic pathway that drives cancer cell metabolism (Semenza, 2009) through the Warburg effect and stands as one of many hallmarks of cancer (Hanahan and Weinberg, 2000, 2011). Baseline quantitative FDG uptake has been linked to increased risk of cancer-specific mortality and disease progression. Specifically, measures of magnitude of tumor FDG avidity (SUV_{max}, SUV_{mean});

physical extent of FDG avidity across a metabolic tumor volume (MTV); and, more recently, their product as the macroparameter for total lesion glycolysis (TLG) have demonstrated high prognostic value for patient risk stratification and outcome prediction, even in multicenter studies (Jhaveri and Linden, 2015). In addition, spatial concordance between baseline FDG PET avid regions and location of disease recurrence in non-small-cell lung cancer suggests that FDG may be used to spatially map high-risk areas for biological target definition (Aerts et al., 2009, 2012; Ohri et al., 2015). Collectively, these findings spurred the design of numerous radiation therapy planning studies and provided motivation to launch several clinical trials with integrated biological image-guided radiation therapy, which are discussed in greater detail below in sections describing modeling and clinical trials.

3.2.1.2 Hypoxia

Imaging biomarkers of tumor hypoxia, another hallmark of cancer progression and metastatic potential, provide more specific definitions of radiation-resistant subregions. Hypoxia-induced radioresistance follows from historical data demonstrating the oxygen effect (Hall and Giaccia, 2006) and the significant impact of baseline tumor oxygenation status on clinical outcomes (Nordsmark et al., 2005). Hypoxia imaging biomarkers are generally grouped into two distinct families of tracers: (1) nitroimidizoles radiolabeled with ^{18}F that include fluoromisonidizole (FMISO) and FAZA PET, as well as more recent optimization of HX4 imaging (van Loon et al., 2010, Zegers et al., 2013), and (2) metal complexes, including Cu-ATSM. Pretreatment FAZA PET could successfully stratify head and neck cancer patients by their disease-free survival (DFS) rates, with two-year DFS of 93% in non-hypoxic tumors compared to 60% in hypoxic ones (Mortensen, 2012). Combination FDG PET and FMISO PET in a multivariate analysis correlated best with long-term survival of head and neck cancer patients receiving radiation therapy (Thorwarth et al., 2006). High overlap of FMISO PET regions with local recurrences has further motivated feasibility studies for hypoxia image-guided dose escalation in head and neck cancer (Hendrickson et al., 2011), including advanced dose-painting-by-numbers approaches (Thorwarth et al., 2007a). In non-small-cell lung cancer, HX4 PET regions provide complementary targets that are generally smaller than conventional FDG PET regions (Zegers et al., 2014) for further individualized therapy. HX4 PET has also been investigated in head and neck, esophageal, and pancreatic cancers within the construct of prospective clinical trials designed in part to test its degree of repeatability (Klaassen et al., 2015, Zegers et al., 2015).

3.2.1.3 Perfusion

Imaging biomarkers of tumor perfusion have long been considered essential in characterizing the efficiency of chemotherapeutic and molecularly targeted drug delivery. Advances in quantitative dynamic image analysis have helped to identify prognostic and predictive parameters of tumor perfusion, both from dynamic contrast enhanced (DCE)-MRI and DCE-CT, for outcome prediction following radiotherapy (Zahra et al., 2007). Specifically, the volume threshold at low-intensity histogram quantiles that denote low tumor perfusion predicted poor local control and disease-specific survival in cervical cancer patients (Mayr et al., 2012, Wang et al., 2008). Parametric images of perfusion kinetics have also revealed negative correlation to apparent diffusion coefficient (ADC) maps in breast cancer patients and that their combination could best predict pathologic response of breast cancer patients (Li et al., 2014) as well as radiologic response of cervical cancer patients following radiotherapy (Zahra et al., 2009). In head and neck cancer, hypoperfused tumor subvolumes defined by a stochastic segmentation algorithm (global-initiated regularized local fuzzy clustering) on DCE-MRI were found to be significantly larger prior to treatment in patients with local failures (Wang et al., 2012), suggesting a potential target for treatment intensification.

3.2.1.4 Cell Proliferation
Imaging biomarkers of tumor cell proliferation, most commonly via thymidine kinase activity with fluorothymidine (FLT) PET imaging (Shields et al., 1998), also play a role in characterizing tumor radiation biology. While the uptake mechanism of FLT differs from purely radiolabeled thymidine and is not incorporated into DNA replication (Krohn et al., 2005), the macroparameter of FLT kinetics (K_{FLT}) was correlated with cellular markers of proliferation (Ki-67) in lung cancer (Muzi et al., 2005) and glioma patients (Muzi et al., 2006). However, FLT specificity is still limited in discriminating between reactive and metastatic lymph nodes of the oral cavity (Troost et al., 2009), making it a challenging imaging biomarker for clinical target definition outside the primary tumor.

3.2.1.5 Other
Additional imaging biomarkers with relevance to cancer therapy and the potential to refine the radiation therapy target definition include measures of osteoblastic and osteoclastic activity in bone dominant lesions ($[^{18}F]NaF$), membrane synthesis in prostate cancer ($[^{11}C]$ choline, $[^{11}C]$acetate), amino acid metabolism in brain cancer ($[^{18}F]FET$, $[^{18}F]FDOPA$), and estrogen receptor in breast cancer ($[^{18}F]FES$) (Linden et al., 2006, 2011). These approaches might help to define tumor subregions that are most important in determining treatment failure and thus might help to guide local and systemic therapies. For example, FET PET has altered the target definition of glioblastoma patients by identifying significantly larger gross tumor volumes and clinical target volumes (Munck Af Rosenschold et al., 2015). Large target volumes are especially critical in stage IV glioblastoma lesions whose pathological extent exhibits highly invasive and diffusive disease progression along white matter tracts and are severely underestimated by conventional images of T1- and T2-weighted MR sequences. A similar trend in patients with high-grade glioma was observed, with larger target volumes delineated on FDOPA PET compared to those defined on conventional MRI (Kosztyla et al., 2013). While the vast majority of recurrences were within the treatment field defined by the 95% isodose surface, many of these local recurrence regions extended outside the pretreatment functional target volume (Kosztyla et al., 2013). The authors concluded that a target definition based on FDOPA PET imaging may not improve local control of high-grade gliomas.

In the setting of locoregionally recurrent prostate cancer, choline PET (Picchio et al., 2014) has been used to define pelvic lymph node targets. In patients with breast cancer, FES PET has been shown to detect differences in pharmacodynamics of cancer therapies

TABLE 3.1   Example Imaging Biomarkers of Cancer Phenotype

Imaging Agent	Tracer/Contrast	Biologic Property	Modality
[^{18}F]FDG	Fluorodeoxyglucose	Glucose metabolism (hexokinase activity, glucose transport, glycolysis)	PET
[^{18}F]FMISO	Fluoromisonidazole	Hypoxia	PET
[^{18}F]FAZA	Fluoroazomycin arabinoside	Hypoxia	PET
[^{18}F]HX4	Flortanidazole	Hypoxia	PET
[^{61}Cu]Cu-ATSM	Diacetyl-methylsemithiocarbazone	Hypoxia	PET
DCE-CT	Iodinated contrast	Perfusion/Permeability	CT
DCE-MRI	Gadolinium contrast	Perfusion/Permeability	MRI
DW-MRI	—	Diffusion	MRI
[^{18}F]FLT	Fluorothymidine	Cell proliferation (thymidine kinase activity)	PET
[^{18}F]NaF	Sodium fluoride	Osteoblastic/osteoclastic enzyme activity	PET
[^{11}C]choline	Choline	Membrane synthesis	PET
[^{11}C]acetate	Acetate	Membrane synthesis	PET
[^{18}F]FET	Fluoroethyltyrosine	Amino acid metabolism	PET
[^{18}F]DOPA	Fluoro-dihydroxyphenylalanine	Amino acid metabolism	PET
[^{18}F]FES	Fluoroestradiol	Estrogen receptor activity	PET

and thereby better predict response (Linden et al., 2011). Breast cancer genotyping can precisely inform on indications for target definition using FES PET, which would further personalize dosing and sequencing of combination therapies. Table 3.1 provides an example set of imaging biomarkers and tumor biological properties probed by these biomarkers with potential application to radiation therapy planning. A set of comprehensive review articles and book chapters detailing multimodality cancer imaging biomarkers in radiation oncology can be found in the literature (Jeraj and Meyerand, 2008; Munley et al., 2013; Jeraj et al., 2011, 2013).

3.2.2 IMAGING BIOMARKERS OF NORMAL TISSUE FUNCTION

Biological imaging of normal tissue function complements cancer phenotype characterization for therapy selection and target definition. Regional mapping of "functional" normal tissue synergizes with regional mapping of the most "important" tumor subregions. This is especially true in settings where both tumor and dose-limiting normal tissue exhibit functional heterogeneity in spatial proximity to one another; for example, lung and liver parenchyma, and hematopoietic bone marrow.

3.2.2.1 Lung

Imaging biomarkers of pulmonary function center around blood perfusion and ventilation. Lung perfusion deficits assessed prior to radiation therapy with ^{99m}Tc-labeled macroaggregated albumin (MAA) SPECT have been linked with increased risk of pneumonitis and declines in pulmonary function test scores. In particular, mean perfused lung dose outperformed mean anatomic lung dose and other SPECT-based dose volume parameters for predicting CTCAE v4 grade 2 pneumonitis or higher in a retrospective NSCLC patient series (ROC AUC = 0.81) (Hoover et al., 2014b). Mean perfused lung dose can be parametrically defined as a treatment-planning objective to avoid functional lung irradiation. Higher-resolution lung perfusion images are achievable with phase matched and respiratory-correlated [^{68}Ga]MAA PET/CT (Siva et al., 2015), although translation of these imaging gains to pulmonary toxicity risk reduction has yet to be realized.

Pulmonary ventilation imaging spans multiple modalities, each with clear trade-offs in clinical utility: [^{99m}Tc]DTPA and [^{99m}Tc]Technegas SPECT, respiratory-correlated 4DCT (Nyeng et al., 2011; Yamamoto et al., 2014), hyperpolarized ^{3}He MRI (Lipson et al., 2002; Bauman et al., 2013; Ireland et al., 2010), and [^{68}Ga]Galligas PET (Siva et al., 2015). The spatial resolution with 4DCT, MRI, and PET is thought to be better than with SPECT. However, none of these pulmonary ventilation image parameters have been as strongly associated with pulmonary toxicity incidence as perfusion image parameters, and it is not clear if there is an incremental benefit of adding ventilation (e.g., ventilation plus perfusion) to perfusion assessments alone. One physiologic rationale derives from normal pulmonary vessel constriction in the presence of hypoxia (i.e., poor ventilation) that therefore reduces perfusion in unventilated regions. Conversely, the airways are less able to respond to poor perfusion; thus, observations of ventilation in unperfused regions are much more common, as is the case with a pulmonary embolus. Thus, imaging of perfusion is more sensitive than imaging of ventilation in areas of poor function defined by ventilation/perfusion mismatch.

3.2.2.2 Liver

Hepatic functional mapping has long been studied with planar imaging techniques but is rapidly evolving with the advent of quantitative volumetric imaging techniques. The colloid family of tracers, most prominently phytate and sulfur colloid SPECT, are highly correlated with clinical parameters of liver function defined either by composite scoring systems or serum marker clearance (Zuckerman et al., 2003). Imaging modalities have been investigated for diagnostic imaging of liver function, including PET with [^{18}F]fluoro-deoxygalactose (Sorensen et al., 2011a,b), and dynamic contrast-enhanced MRI (Cao et al., 2013) with gadoxetic acid (Sirlin et al., 2014) or gadoxetate disodium (Cruite et al., 2010; Ringe et al., 2010). SPECT radiotracers of liver function include [^{99m}Tc]hepatobiliary iminodiacetic acid (Wang and Cao, 2013), [^{99m}Tc]galactosyl-human serum albumin (Beppu et al., 2011), as well as [^{99m}Tc]sulfur and [^{99m}Tc]phytate colloid (Zuckerman et al., 2003). Sulfur colloid is taken up by the Kupffer cells of the reticuloendothelial system, which are intimately related to hepatocyte function. Sulfur colloid (SC) uptake has been shown to correlate to the blood serum marker indocyanine green, a well-established quantitative measure of liver function (Zuckerman et al., 2003). In addition, quantitative imaging parameters from colloid uptake such as the perfused hepatic mass were associated with explanted liver functional mass (Hoefs et al., 1997) and predicted clinical outcome (Everson et al., 2012).

Imaging of lung and liver, among many normal tissues, presents opportunities to define functional tissue avoidance region. Studies have demonstrated the technical feasibility of

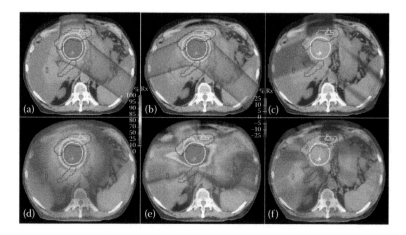

FIGURE 3.1   Differential hepatic avoidance radiation therapy (DHART) guided by [^{99m}Tc] sulfur colloid (SC) SPECT/CT. Conventional proton therapy (a) and volumetrically modulated arc therapy (d) plans do not account for spatial heterogeneity in liver function, whereas DHART plans with protons (b) and photons (e) conformally avoid functional liver regions through dose redistribution, with clear dose differences (difference images comparing the conventional versus the DHART plans), (c), (f). (Reproduced from Bowen, S.R. et al., *Radiother. Oncol.*, 115, 203–210, 2015. With permission.)

avoiding functional lung during radiation treatment planning (McGuire et al., 2006, 2010; Das et al., 2004; Das and Ten Haken, 2011; Siva et al., 2015). More recently, a similar concept was applied to avoid functional liver by redistributing liver dose at the voxel scale (Bowen et al., 2015; see Figure 3.1). These studies found that the degree to which functional tissue could be avoided was related to the geometric shape of the target and normal tissue. Specifically, lesions with high laterality that were partially surrounded by functional tissue yielded superior avoidance plans than did central lesions fully surrounded by functional tissue. Allowance of higher target dose heterogeneity and higher maximum dose permitted significant functional tissue dose reduction compared to strict target dose uniformity.

3.2.2.3 Bone Marrow

Recent attention has turned to noninvasive assessment of proliferating bone marrow, whereby FLT PET imaging can map the whole-body biodistribution in order to build a functional atlas for groups of patients according to age and gender (Campbell et al., 2015). Such a functional atlas can guide beam angle selection for radiation therapy as a first pass with individual FLT PET maps generated as necessary to account for further patient-specific variation. Treatment plans that incorporated FLT PET for bone marrow sparing with IMRT demonstrated a 40% reduction in the proliferating bone marrow volume receiving 20 Gy or greater compared to standard IMRT plans (McGuire et al., 2011), and, more recently, an even greater dose reduction using intensity-modulated proton therapy (IMPT) (Dinges et al., 2015). Table 3.2 summarizes imaging biomarkers and biological properties of normal tissues probed by these biomarkers.

TABLE 3.2 Example Imaging Biomarkers of Normal Tissue Function

Imaging Agent	Tracer/Contrast	Biologic Property	Modality
[99mTc]MAA	Macroaggregated albumin	Pulmonary perfusion (i.v.), hepatic perfusion (embolized)	SPECT
[68Ga]MAA	Macroaggregated albumin	Pulmonary perfusion (i.v.), hepatic perfusion (embolized)	PET
[99mTc]DTPA	Diethylene triamine pentaacetic acid	Pulmonary ventilation	SPECT
[99mTc]Technegas	Aerosol nanoparticles	Pulmonary ventilation	SPECT
[68Ga]Galligas	Aerosol nanoparticles	Pulmonary ventilation	PET
4DCT	—	Pulmonary ventilation (tissue deformation)	CT
Hyperpolarized Helium	^{3}He	Pulmonary ventilation	MRI
[18F]FDGal	Fluorodeoxygalactose	Hepatic galactose metabolism	PET
Gd-EOB-DTPA	Gadoxetic acid disodium	Hepatobiliary uptake/clearance	MRI
[99mTc]HIDA	Hepatobiliary iminodiacetic acid	Hepatobiliary clearance rate	SPECT
[99mTc]GSA	Galactosyl-human serum albumin	Hepatocyte activity	SPECT
[99mTc]SC	Sulfur colloid	Reticuloendothelial cell support of hepatocyte function	SPECT
[99mTc]PC	Phytate colloid	Reticuloendothelial cell support of hepatocyte function	SPECT
[18F]FLT	Fluorothymidine	Hematopoeitic cell proliferation	PET

3.3 BIOLOGICAL IMAGING TO ASSESS RESPONSE TO THERAPY

Biological imaging prior to therapy, as discussed above, may establish baseline risk for both normal tissue complication and tumor control. However, this may not account for other sources of variability in individual tumor or normal tissue *responses*. The consideration of patient-specific changes from baseline, longitudinally over time during therapy, might improve outcome prediction and guide adaptive therapies.

3.3.1 Imaging Biomarkers of Tumor Response

Tumor response assessment with longitudinal FDG PET during and after radiation therapy has demonstrated superior predictive power for clinical outcome compared to pretreatment FDG PET in several disease sites, including lung, head and neck (Farrag et al., 2010), and cervical cancer (Pallardy et al., 2010; Kidd et al., 2013; Schwarz et al., 2012). In the setting of induction chemotherapy and consolidation radiotherapy, qualitative FDG PET response assessment between modalities showed higher predictive power for survival than baseline FDG PET alone (Decoster et al., 2008). Quantitative tumor response assessment defined by SUV or TLG on midradiation treatment FDG PET at 2–4 weeks relative to baseline FDG PET was associated with poorer clinical outcomes (Yossi et al., 2015). In non-small-cell lung cancer, it was estimated that 60% of patients could be classified as early nonresponders on FDG PET and that this subpopulation had 2-year overall

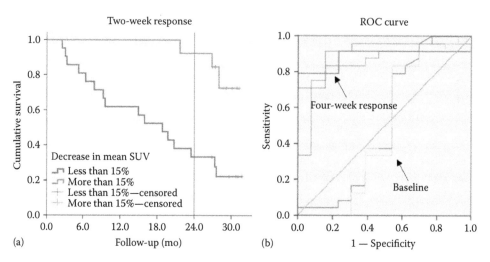

FIGURE 3.2 Mid-radiotherapy FDG PET/CT predicts outcomes of non-small-cell lung cancer patients. Change in mean standardized uptake value (SUV) at 2 weeks may stratify patients by prognosis, with responders having 92% 2-year overall survival and nonresponders having 33% 2-year overall survival (a). Change in FDG PET SUV and metabolic tumor volume (MTV) at 4 weeks was superior for outcome prediction compared to baseline imaging metrics (b). (Reproduced from van Elmpt, W. et al., *J. Nucl. Med.*, 53, 1514–1520, 2012b. With permission; Huang, W. et al., *Eur. J. Nucl. Med. Mol. Imaging.*, 38, 1628–1635, 2011. With permission.)

survival rates of 33% compared to 92% in the responder subpopulation (van Elmpt et al., 2012a). Four-week FDG PET response measures had significantly higher ROC AUC than baseline FDG PET measures (Huang et al., 2011; see Figure 3.2). This highlights both the prevalence of poor tumor FDG PET response early during treatment and its significant association with patient outcomes. Optimal FDG PET early imaging time points for response assessment appear to precede confounding signals from radiation-induced inflammation late in therapy. Later post-treatment recurrences detected by FDG PET are often seen following the resolution of this inflammation (e.g., beyond several months after therapy).

Similar PET tumor response assessment has been conducted with imaging of cell proliferation and hypoxia. In head and neck cancer, post-treatment response of FLT PET showed superior prediction of local control rates compared to post-treatment response of FDG PET (Kishino et al., 2012). FLT PET is especially sensitive to response assessment of myoablative therapies for patients with acute myelogenous leukemia, showing significant changes as early as 2 days into treatment (Vanderhoek et al., 2011). Early FLT PET response and high residual midtreatment FLT PET was predictive of local failure, showing that tumors with large residual proliferation may be most resistant to radiation therapy (Bradshaw et al., 2015). In patients with head and neck cancer, continued decrease in FLT SUV_{max} during the second and fourth weeks of radiotherapy was associated with better 3-year DFS (Hoeben et al., 2013). These changes in FLT PET uptake were shown to precede anatomic volume changes in radiotherapy targets and could inform adaptive therapy strategies to tumor subvolumes (Troost et al., 2010).

Other modalities for response assessment include perfusion and diffusion imaging with DCE-MRI (DCE-CT) and DWI-MRI, respectively, for improved outcome prediction. This is best demonstrated in patients with cervical and breast cancer undergoing definitive radiation therapy, whether external beam, brachytherapy, or both. The size of functional risk volumes defined at 2 or 4 weeks during therapy, and their lack of reduction, could segregate patients with poor local control and disease-specific survival (Mayr et al., 2010; Yuh et al., 2009; Huang et al., 2014). In head and neck cancer, tumor blood volume defined on DCE-MRI after 2 weeks of concurrent chemotherapy and radiation therapy increased significantly in patients with locally controlled disease compared to those with local failures (Cao et al., 2008). Similarly, 2-week increases during therapy in regional cerebral blood volume on DCE-MRI were associated with post-therapy tumor volume reduction in patients with brain metastases (Farjam et al., 2013). As with FDG PET, the timing of perfusion imaging for response assessment is critical to its clinical utility, and time-dependent morphological changes must be balanced with the complexity of the quantitative analysis.

A relatively modern class of tracers may provide insight into tumor response during ablative radiation therapy. They investigate various points along the apoptotic pathway, from upstream reversible apoptosis with ML-10 PET (Hoglund et al., 2011) to downstream irreversible apoptosis with ICMT-11 PET (Challapalli et al., 2013). However, these and other tracers that rely on active transport mechanisms pose challenges in assessment of response due to transient uptake levels in tissues of interest. Overcoming these limitations will require comprehensive pharmacokinetic modeling to decouple specific retention-dependent uptake from flow-dependent uptake. Despite these challenges, the recent introduction of apoptosis tracers into human imaging studies presents intriguing long-term potential for characterizing tumor response to high-dose hypofractionated radiation.

3.3.2 Imaging Biomarkers of Normal Tissue Response

Imaging biomarkers of treatment response have also been applied to normal tissues. In the lung, increased FDG PET signal during treatment has been linked to higher incidence of pneumonitis (Castillo et al., 2014) and pulmonary fibrosis (Umeda et al., 2015). In the bone marrow, decreased FLT PET uptake during therapy is associated with decline in neutrophil counts and tracks the longitudinal dynamics of bone marrow injury/recovery following specific treatment regimens (Leimgruber et al., 2014).

Comprehensive assessment of pulmonary perfusion response has been conducted with MAA SPECT, which is shown for an example patient in Figure 3.3. Response models of regional perfusion deficits have established a fairly linear relationship to radiation dose magnitude on the order of 1% reduction per Gy and have tracked the dynamics of perfusion reduction up to 10 years post-treatment (Marks et al., 2000; Zhang et al., 2010). The sum of these regional perfusion reductions has been linked, albeit weakly, to associated changes in global lung function (e.g., pulmonary function tests and pulmonary symptoms) (Kocak et al., 2007).

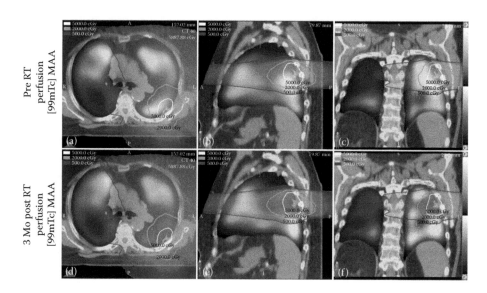

FIGURE 3.3   Lung perfusion imaging response to radiotherapy. Pre-radiotherapy [^{99m}Tc]MAA SPECT uptake distribution (hot metal colorwash) overlaid on the planning CT and radiation dose distribution (blue-green isodose lines) (a–c), and corresponding fused images at 3 months post-radiotherapy (d–f). Spatially colocalized changes in regional lung perfusion can be functionally linked to the radiation dose magnitude to derive dose–response relationships.

3.4 RADIOBIOLOGICAL MODELING TO GUIDE RADIATION THERAPY PLANNING

The evidence that biological imaging can predict tumor and normal tissue response, at least in some tumor types, is convincing. Incorporating this type of information into the planning process requires formal quantitative interpretation of the imaging/functional data to be included in predictive and optimization models. Tumor control probability (TCP) and normal tissue complication probability (NTCP) models in routine use convert dose-volume histograms (DVHs) into predicted probability of response and disregard spatial distribution of biological properties. Accounting for biological properties to improve therapeutic ratio has been explored in modeling and treatment planning studies. Broadly speaking, there are two approaches: one can apply the TCP and NTCP models to specific subregions of tumors and normal tissues, or one can attempt to "weight" the conventional TCP and NTCP models to consider functional heterogeneities. The former is typically easier. The latter is more challenging because the "weighting" functions needed are largely unknown and because planning systems do not readily enable the latter approach.

3.4.1 Radiobiological Models for Tumor Response

Incorporating tumor characteristics in treatment planning has been approached through segmenting regions for dose boosting and as dose-painting-by-numbers (Bentzen and Gregoire, 2011). The former is inviting because it can be utilized readily in commonly used treatment planning systems. The latter accounts more intricately for the spatial distribution of biological properties. In either approach, radiation dose magnitude must be connected to the tumor characteristic in question, and this relationship remains uncertain. The simplest way to address this problem is to assume a linear relationship between dose and voxel image intensity, I, ranging from the baseline dose, D_{min}, which has to be delivered to every voxel, and maximum dose that can be safely accommodated, D_{max}:

$$D(I) = D_{min} + \frac{I - I_{min}}{I_{max} - I_{min}} (D_{max} - D_{min})$$

However, one should be aware that such a linear relationship is an oversimplification and that more detailed modeling of the relationship between biological imaging and underlying biology (Bowen et al., 2011), as well as treatment outcomes, is needed. Because most treatment planning systems allow planning only on segmented structures, a hybrid approach to convert voxel-based prescription into a discrete number of dose levels has been proposed (Deveau et al., 2010; Korreman et al., 2010). Diminishing returns were noted as a larger number of levels was used, and the differences between DVH in optimized plans for 9 and 11 levels were small (Deveau et al., 2010; Korreman et al., 2010).

TCP models can be broadly divided into two categories: (1) empirical models, also known as phenomenological or statistical models, which neglect tumor biology and merely use known statistical approaches to describe observed data, and (2) mechanistic models, which account for underlying mechanisms of dose-response, albeit in a simplified manner. Empirical models typically describe the sigmoid curve for TCP as a function of dose using two parameters, D_{50}, dose at which 50% of tumors are controlled, and γ_{50}, normalized slope that is the percentage change in TCP per percentage change in dose at the 50% response level. Under the empirical formalism, tumor subvolumes exhibiting specific biological characteristics as visualized with functional imaging can be assigned model parameters reflecting their sensitivity or resistance to radiation (Kim and Tome, 2010). Dose-painting approaches can be further utilized and iso-TCP plans can be produced for clinical evaluation by a physician.

A popular mechanistic TCP model is based on Poisson statistics. To control the tumor, all clonogens, cells capable of regrowing the tumor, have to be eradicated. The expression for TCP can be readily written as:

$$\text{TCP} = \exp(-N_s),$$

$$N_s = \sum v_i \rho_i S_i,$$

where:
  N_s is the number of surviving clonogens
  v_i is partial volume
  ρ_i is clonogen density
  S_i is survival

The assumption is that, within the volume v_i dose, clonogen density and biological properties are homogenous. This formulation readily accounts for certain characteristics, for example, variation in clonogen density through the tumor volume. Characteristics affecting cellular response, such as environment and in particular hypoxia, or heterogeneity in sensitivity, can also be incorporated. Cell survival is commonly expressed using the linear-quadratic model:

$$S = \exp\left(-\alpha D - \beta G D^2\right)$$

where:

D is dose

α and β are model parameters associated with intra- and intertrack production of lethal lesions

G is the Lea-Catcheside factor accounting for protraction/fractionation of dose delivery.

Ruggieri et al. (2012) proposed a model to incorporate acute and chronic hypoxia into a TCP model and initially tested the model on preclinical observations in rat sarcoma subjects. The model assumes that the effect of hypoxia, quantified as the oxygen enhancement ratio (OER), can be introduced into the model as a dose-response modifying factor for model parameters established for oxygenated conditions: α_0 and β_0. Specifically, for chronic hypoxia (ch):

$$\mathrm{OER}_{ch} = \frac{\alpha_0}{\alpha_{ch}} = \left(\frac{\beta_0}{\beta_{ch}}\right)^{1/2}$$

A modifying factor was assumed for acute hypoxia in a similar fashion. This model has been generalized from preclinical experiments and translated to estimate TCP following stereotactic body radiotherapy in early stage non-small-cell lung cancer patients. Full details of modeling are beyond the scope of this chapter. In brief, the above narrative demonstrates how tumor characteristics as quantified by biological imaging can be incorporated into a commonly used Poisson-based TCP model. Other characteristics, for example, the presence of quickly proliferating pools of cells, can be also introduced as a separate term.

Jeong et al. (2013) proposed a more sophisticated model to account not only for hypoxia but also for the interplay between hypoxia and proliferation. This model assumes the presence of three compartments: (1) proliferative, (2) intermediate, and (3) hypoxic. This compartmentalization is based on access to oxygen and nutrients, and cell loss is assumed in the latter compartment without radiation. The model accounts for cell migration between compartments, response to radiation including correcting for OER, proliferation, and cell loss. The authors further applied the developed model to explore the relationship between FDG PET and radiobiological response by simulating various FDG profiles.

3.4.2 Radiobiological Models for Normal Tissue Response

Routinely used DVH to optimize and evaluate radiotherapy plans disregard spatial distribution of function through the volume of normal tissue, and this limitation has been appreciated. Volume is used as a surrogate of function. Location, functional burden, and regenerative capacity are ignored. Clinical data argue against this presumed uniform distribution of function. For example, risk of lung toxicity has been demonstrated to correlate with location even for the same mean lung dose (Seppenwoolde et al., 2002; Yorke et al., 2002). The concept of dose-function-histogram (DFH) proposed by Marks et al. (1995, 1999) provides a means to assess relative function rather than volume receiving at least a specific dose. In the example presented by Marks et al., each voxel in the normal lung can be ascribed a value from the SPECT scan that describes perfusion, and thereby function, in a quantitative manner.

Normal tissue function-driven planning objectives are commonly formulated either through segmenting functional volume (Lavrenkov et al., 2007) or calculating function-weighted dose metrics (Seppenwoolde et al., 2004). The former approach is similar to the use of dose-volume cutoffs, for example, the popular V_{20Gy} dosimetric parameter used for normal lung (Graham et al., 1999). In the dose-volume approach, volume receiving at least 20 Gy, V_{20Gy}, has to be limited to less than 30% to keep the incidence of radiation pneumonitis less than 20% (Marks et al., 2010; Bentzen et al., 2010). A similar objective is defined for functional volume rather than anatomic volume. For example, Lavrenkov et al. (2007) defined functional lung as a combination of subvolumes in the lung with at least 60% of maximum tracer uptake as seen in SPECT scan. The planning objective can be further formulated by limiting a proportion of functional lung, rather than whole normal lung, receiving 20 Gy or more, fV_{20Gy}. This concept can be readily incorporated into commonly used treatment planning systems that are designed to set objectives/constraints for defined volume.

The second approach based on function-weighted dose accounts for function distribution throughout the whole organ. This is in contrast to the segmentation-based method, which excludes a portion of the normal tissue volume from optimization objectives. This approach in many ways mirrors the dose-painting-by-numbers approach used for tumors. The optimal plan in terms of tumor control preferentially places a larger dose to tumor subvolumes exhibiting a characteristic associated with an unfavorable outcome, for example, hypoxia. The optimal plan for normal tissue sparing achieves the opposite—it drives high dose away from subvolumes exhibiting high functional burden (e.g., lung perfusion assessed with SPECT). Function-weighted metrics for normal tissue, such as SPECT-weighted mean dose, have also shown less sensitivity to variations in SPECT reconstruction. This is in contrast to segmenting volumes that are sensitive to scatter and attenuation corrections (Yin et al., 2010). The difficulty of this approach stems from incompatibility of treatment planning systems to include function-weighted dose objectives derived from an overlaid functional image. A recent investigation overcame this limitation by subdividing normal liver into several discrete levels of liver function and continuously redistributed the mean liver dose away from regions of relatively higher function on sulfur colloid SPECT/CT through variable dose optimization penalty weights (Bowen et al., 2015). A differential hepatic avoidance radiation therapy (DHART) plan is shown for an example patient in Figure 3.1. This is a natural mirror to the aforementioned planning technique that implemented PET-guided dose-painting-by-numbers at every tumor voxel in commercial treatment planning systems (Deveau et al., 2010).

3.4.3 Radiobiological Models for Adaptive Treatment Planning

Radiobiological modeling studies of outcomes for adaptive treatment planning following assessing response to radiation therapy using functional imaging are limited. Biological consequences of dose boosts can be readily assessed using simple models. For example, Gillham et al. (2008) compared two dose escalation strategies for NSCLC patients, one based on delivering a 12 Gy boost with shrinking fields after 66 Gy delivered in phase 1 for a total of 78 Gy, and the other based on repeat PET scan and replanning after 50 to 60 Gy were delivered. The boost dose in the latter approach was limited by normal tissue constraints. The authors used a log-logistic TCP model reported by Martel et al (1999). Gillham et al. concluded that adaptive boost planning would result in only modest outcomes improvement. This is not surprising because replanning was considered only after at least 50 Gy was delivered.

More complex TCP models, in particular the Poisson-based model, can be readily modified to account for biological properties including longitudinal data. For example, Thorwarth et al. (2007b) proposed a TCP model accounting for regional reoxygenation in head and neck tumors. The model was applied to interpret data for 10 head and neck cancer patients examined with FMISO–PET prior to radiotherapy and after 20 Gy of the prescribed dose of 70 Gy were delivered. The authors stated that the model was validated and advocated use of the model for hypoxia image-guided dose escalation. A common and unavoidable problem is that these models quickly become overparameterized. This limits their practical application, especially when model parameters are obtained by fitting to observed data and cannot be independently validated. Despite these limitations, model-based strategies can be developed and assessed. Sovik et al. (2007) proposed a model for dose redistribution to maximize TCP by selectively boosting hypoxic regions. However, all of the "bottom–up" models are yet to be applied in prospective clinical trials to assess their utility.

3.5 EMPIRICAL MODELS FOR INTEGRATING BIOLOGICAL IMAGING PROPERTIES INTO CLINICAL TRIALS

In contrast to bottom–up mechanistic radiobiological models, top–down empirical models extract parameters from statistical fits of clinical data. Empirical models utilize simple functions of few independent parameters that are best associated with the variation in observed clinical outcomes. Empirical models have integrated biological imaging into dose escalation trials to decrease local disease relapse risk or dose sparing trials to decrease normal tissue complication risk, and they can be grouped into three distinct categories: (1) discrete subvolume boosting or dose painting by contours (Ling et al., 2000), (2) continuous differential boosting or dose painting by numbers (Bentzen, 2005), and (3) functional tissue avoidance. Examples of such models within ongoing clinical trials are described.

3.5.1 Subvolume Boosting and Adaptive Dose Painting

No fewer than 6 early phase I/II trials in humans have been completed or are ongoing that escalate dose to discrete biological target volumes (Shi et al., 2014). All dose-painting trials were conducted in head and neck or non-small-cell lung cancer patients, and the vast majority utilized FDG PET/CT for target boost volume definition.

The FDG PET boost trial in the Netherlands utilizes discrete dose escalation to FDG PET avid regions based on pretreatment assessment and individualizes integral target dose boosts based on fixed integral lung dose (mean dose less than 20 Gy EQD$_2$). The FDG PET boost volumes are defined by a fixed threshold (50% SUV$_{max}$) and overlap with high-risk regions for local disease recurrence (van Elmpt et al. 2012b). On average, dose to FDG avid regions could be increased above 85 Gy in 24 fractions without significant increases in lung dose (van Elmpt et al. 2012b). The ARTFORCE trial boosts subvolumes in head and neck cancer with an accrual goal of 268 patients that tests superiority of locoregional control (80% power with 0.05 Type I error rate) and confirm whether baseline FDG PET is a clinically meaningful target in this patient population (Heukelom et al., 2013). A randomized phase II trial in head and neck cancer will test the superiority of a uniform dose boost directed at hypo-perfused tumor subvolumes defined on DCE MRI, presumed to be associated with a higher risk for local failure, relative to a standard-of-care dosing regimen (NCT02031250).

Other empirical models adaptively escalate dose based on midtreatment FDG PET in lung cancer (RTOG1106). The RTOG 1106 phase II trial is investigating whether tumor dose can be escalated to improve the local control rate based on FDG-PET/CT imaging at 40–46 Gy during treatment, which exploits higher achievable dose escalation at fixed normal tissue integral dose following tumor volume regression (Feng et al., 2009). Promising phase II trial results on dose escalation demonstrated improved 2-year local control rate of 84% after administration of 84 Gy median physical dose (BED$_{10}$ 108 Gy) (Kong et al., 2014). The RTOG 1106 trial can achieve up to 80.4 Gy to residual FDG avid regions in 30 fractions.

3.5.2 Dose Painting by Numbers

Dose painting by numbers delivers a continuously varying and nonuniform radiation dose prescription based on biological imaging. The transformation between image values and radiation dose, termed the prescription function (Bentzen, 2005), may have a unique shape that can determine the achievable planned dose (Bowen et al., 2009). As described conceptually above, investigations have tested the feasibility of dose painting by numbers under the assumption of a linear relationship between image intensity and dose (Vanderstraeten et al., 2006; Korreman et al., 2010; Deveau et al., 2010; Alber et al., 2003), providing a basis for clinical trial designs.

The adaptive dose-painting-by-numbers trial in head and neck cancer from Ghent University in Belgium delivers a boost within the GTV based on a linearly scaled relative FDG PET uptake at baseline, then adapts the nonuniform boost after 2 weeks based on a midtreatment FDG PET, and finally delivers the remaining standard dose of radiation based on the midtreatment CT. This upfront boost and subsequent adaptive boost leverage the radiobiological advantage of high-dose-per-fraction delivery prior to accelerated tumor cell repopulation. Under this trial schema, median doses in 2 Gy equivalent fraction sizes were 50 Gy to the elective nodal PTV, 90 Gy to the high-risk CTV, and 100 Gy to the GTV (Duprez et al., 2010), with 2-year locoregional control rates exceeding 90% and no grade 4 toxicity observed (Madani et al., 2011; Berwouts et al., 2013).

Beyond empirical models that modulate the dose-painting-by-numbers prescription based on normal tissue dose limits, other investigations have sought to characterize a prescription function from statistical fits of image voxel response to optimize an imaging surrogate endpoint. A first-order linear prescription function was derived from voxel regression of FDG PET response in canine sinonasal tumors to minimize 3-month posttreatment FDG PET uptake as an imaging surrogate of local failure risk (Bowen et al., 2012; Bradshaw et al., 2015). The function guaranteed minimum dose coverage to the PTV while painting dose to each voxel based on the prescription function of baseline relative FDG PET uptake, yielding maximum planned doses in excess of 90 Gy. Empirically or mechanistically derived dose-painting-by-numbers prescriptions guided by biological imaging represent an important evolutionary step in an increasingly precise radiation oncology paradigm.

3.5.3 Functional Tissue Avoidance

Trials of functional tissue avoidance radiation therapy are emerging. A phase I trial at the University of California at San Diego selectively spared dose to viable bone marrow defined by regions of high FDG PET avidity *and* low-fat fraction on IDEAL MRI (Liang et al., 2013). The study successfully demonstrated a significant reduction in grade 3 hemotologic toxicity from a 60% incidence following conventional radiation therapy to a 30% incidence after bone-marrow-sparing RT, and the group has since launched a phase II trial. Functional lung avoidance trials are also underway to improve outcomes of locally advanced non-small-cell lung cancer patients. The London Regional Cancer Center in London, Ontario, integrates hyperpolarized ^{3}He MRI for defining ventilated lung avoidance regions and will test differences in pulmonary toxicity and pulmonary quality-of-life metrics following functional lung avoidance RT against conventional RT in a randomized, double-blind trial (Hoover et al., 2014a). The University of Colorado, Denver, is conducting a similar early phase functional lung avoidance RT trial (NCT02528942) but defines avoidance regions from ventilation parameters extracted on 4DCT (Brennan et al., 2015).

3.6 CHALLENGES AND FUTURE DIRECTIONS

3.6.1 Current Challenges

Slow adoption of radiobiological target definition has been tied to a number of research questions that needed answering before safe clinical adoption of such technologies and methodologies. Several important questions to answer include the following:

- What biology to target? Are optimal treatment targets different for different tumor histologies?

- Are the biological targets stable? How often do we need to adapt the plans?

- How should the radiation dose be modulated in relation to the biology? How do we quantitatively translate from radiobiological heterogeneity to dose heterogeneity?

- Is it safe to deliver highly modulated radiation doses? How safely and effectively can we deliver extremely high doses to small areas?

- How accurately can the dose delivery follow biological heterogeneity?

- What other biomarkers beyond imaging do we need to consider?

- How can radiation therapy be integrated with other treatment modalities to increase the therapeutic ratio?

Significant research efforts have been invested into answering these fundamental questions, many combining the most advanced imaging and treatment technologies and requiring extraordinary efforts, particularly in the context of multicenter clinical trials. At the heart of these challenges with biological imaging for radiation therapy target definition is the spatial concordance or discordance between maps of different cancer phenotypes (Nyflot et al., 2012; Bradshaw et al., 2013). With dozens of segmentation algorithms and hundreds of biological imaging tracers, decoupling stochastic effects of image formation and image processing to establish definitive spatial and temporal stability of biological imaging targets remains a high priority (Buckler and Boellaard, 2011). This is best accomplished with standardized test-retest scans in early-phase clinical trials to establish baseline repeatability for quantitative analysis. Once stochastic variation from image formation is removed, underlying biological variations are revealed that affect target definition. For example, the spatial associations between tumor FDG, hypoxia, and perfusion uptake distributions have biological underpinnings but also lead to differences in target volumes and the achievable magnitude of radiation dose escalation. How does one effectively select and prioritize biological imaging targets for radiation therapy? Are correlations, or lack thereof, time dependent during and after courses of therapy? Studies have demonstrated that biological imaging correlation is greatly affected by tumor etiology (Bradshaw et al., 2013), but sources of biological variation for target definition can be difficult to identify and standardize. Work to date suggests the following:

- Radiobiology-guided targeting approaches will need to be tumor histology specific.

- Most potential radiobiology targets appear to be stable through the course of therapy; however, adaptation of the radiobiology treatment plans during treatment might be necessary in some cases.

- Radiobiology-guided targeting appears to require highly peaked doses to smaller regions that are analogous to brachytherapy and SBRT.

- Thus, advanced motion-management techniques are likely necessary, but they depend on the degree of spatial variation and margins used.

- Radiobiology-guided treatment plans can be delivered safely.

When utilizing biological imaging for response assessment, a critical challenge is the gap in spatial scale between the detectable response on images and the underlying tissue response

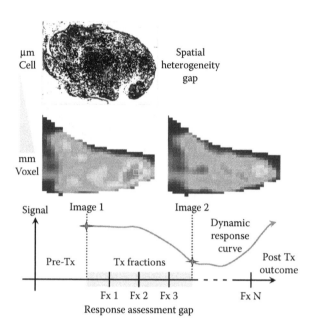

FIGURE 3.4 Spatial heterogeneity gap between cellular/image voxel scales and treatment response gap between imaging time points. These concepts are presented in a canine sinonasal cancer translational research model, showing immunohistochemical staining of an image-guided tissue biopsy slice, its corresponding spatial location within a pretreatment hypoxia PET image, and a midtreatment hypoxia PET image.

at the molecular scale, as well as the gap in temporal resolution between discrete imaging time points and the continuous dynamic response to therapy. Mechanistic models can bridge spatial gaps by including effects of intra-image voxel heterogeneity down at the cellular scale (Petit et al., 2009b) to modulate functional image response-defined tumor control (Petit et al., 2009a). Mechanistic and empirical models can interpolate between image time points to provide complete characterization of response dynamics, which would enable the design of virtual (*in silico*) trials prior to preclinical and clinical trials. The gap in resolving spatial heterogeneity and the gap in assessing dynamic response are illustrated in Figure 3.4.

Once spatial and temporal gaps are bridged, pairing the appropriate model with biological imaging modality to characterize a particular treatment response must be carefully considered due to complex trade-offs. For example, an empirical model may accurately predict outcomes from a cohort of patients but produce parameter fits with large statistical error and little biophysical interpretation. On the other hand, a mechanistic model may be composed of easily interpretable parameters with small errors but may not adequately account for additional variation when tested for prediction in future patients. The biological imaging modality must be chosen carefully to be sensitive to a treatment-specific response. For example, there is debate over whether FDG PET is an imaging biomarker for local disease recurrence or occult systemic disease in non-small-cell lung cancer patients. In locally advanced NSCLC, changes in FDG PET are linked to local radiation dose magnitude, and avid uptake regions co-localize with local failure regions. However, early-stage

NSCLC patient series demonstrate a correlation between FDG PET TLG and overall survival, despite the delivery of high doses to small anatomic tumor volumes, thus suggesting an elevated risk of occult disease as a function of increasing metabolic tumor burden (Vu et al., 2013). Imaging biomarkers of radiation resistance appear to depend on tumor histological subtype, as shown in canines with spontaneous sinonasal tumors (Bradshaw et al., 2013, 2015). In fact, the imaging biomarker selection likely depends on multifactorial genetics and degree of genomic heterogeneity.

Modeling the in vivo biological response of normal and malignant tissue to ionizing radiation delivered at high doses per fraction (greater than 8–10 Gy) during image-guided stereotactic radiosurgery (IGSRS) remains controversial. Whether one believes that these responses can be entirely explained by current model(s) (Brenner, 2008; Brown et al., 2014) versus the possibility that different response mechanisms evoked at high doses elicit "new" mechanisms may yield qualitatively different responses (Kirkpatrick et al., 2008; Sperduto et al., 2015), it is clear that IGSRS provides the spatial resolution required for irradiating a portion of a tumor while sparing adjacent tissues. For example, one might imagine utilizing SRS to preferentially irradiate the most metabolically active or rapidly proliferating volume of a tumor.

3.6.2 Future Directions

As noted above, considerable effort has been expended by researchers/clinicians to provide foundation for answers to the aforementioned challenges. Several approaches need to be considered to further develop this field.

Data standardization: Adoption of standardized data acquisition, analysis, and reporting will enable pooling of data and communication and thus facilitate convergence to consensus opinions on radiobiology-guided therapy. This begins with image acquisitions and reconstruction protocols that promote intrapatient repeatability and interpatient/interinstitution reproducibility. Data standardization is a common problem of medical data sharing.

Imaging and radiation dose co-localization: All imaging for radiotherapy planning and response assessment should (as much as possible) be conducted with the patient in the treatment position, which is facilitated by the use of customized immobilization devices to maximize patient comfort and reproducibility. Careful registration of biological imaging modalities with the planning CT and associated radiation therapy dose distribution should minimize distortion from nonphysical deformation as well as limit areas of misalignment to outside the target or normal tissue region of interest.

Target definition standardization: The literature suggests that target definition has the largest uncertainty. Standardized definition of biological target volumes through open-source segmentation tools (e.g., automatic stochastic algorithm) in the case of subvolume boosting or standardized dose prescription (e.g., first-order linear redistribution of fixed integral boost) in the case of dose painting by numbers, are essential to power meaningful meta-analyses and multicenter radiobiology-guided clinical trials.

Treatment planning flexibility: Treatment planning systems must accommodate the direct incorporation and processing of biological information. Taken to its logical conclusion, this will mean a complete overhaul of the DICOM RT standards to enable nonuniform prescriptions at the image voxel scale and objectives that incorporate spatially variant target and normal tissue radiosensitivities. This can be addressed formally through a task group or a working group.

Hybrid multiscale modeling: Future efforts to personalize prediction of radiobiological response may be built on the foundation of hybrid multiscale models that incorporate both mechanistic parameters and empirical imaging parameters. Such hybrid models predict tumor growth of glioblastomas and response to radiation therapy (Rockne et al., 2015), as well as tumor response to radiation therapy (Titz and Jeraj, 2008) and molecularly targeted antiangiogenic therapy (Titz et al., 2012) at the patient scale. These tumor response models can be coupled to normal tissue response models to explore synergy of image guidance for tumor dose escalation and functional tissue avoidance. An example of this concept is illustrated in Figure 3.5, which combines lung perfusion MAA SPECT/CT and tumor response on FDG PET/CT to generate functional lung avoidance and response-adaptive escalation (FLARE) RT plans in non-small-cell lung cancer patients. These approaches form the basis for increased

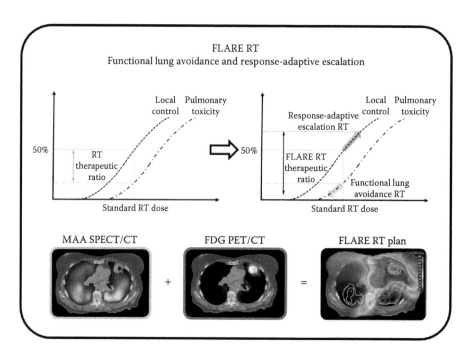

FIGURE 3.5 Functional lung avoidance and response-adaptive escalation (FLARE) RT. The concept is applied to an example stage III non-small-cell lung cancer patient, whose FLARE RT plan is personalized to avoid perfused lung regions defined on MAA SPECT/CT while targeting tumor subregions defined on FDG PET/CT. The clinical objective would be to reduce risk of pulmonary toxicity while increasing local tumor control.

precision in radiation therapy regimens that are tailored to each individual patient's variability in tumor and normal tissue response.

Dosing and sequencing of combination therapies: A better understanding of how best to integrate RT with systemic agents (e.g., immunological agents) might be fruitful. The recent advent of immunotherapy through checkpoint inhibition (anti-PD1/PD-L1, anti-CTLA4) has culminated in the Food and Drug Administration (FDA) approved pharmaceuticals for treatment of advanced stage melanoma (Hodi et al., 2010) and non-small-cell lung cancer (McCarthy et al., 2013; Brahmer et al., 2015). However, only subsets of patients benefit from favorable tumor response to these novel immunotherapies, opening investigative avenues of imaging biomarkers for therapy selection and synergy with biological image-guided RT. Our understating of immunogenic response to radiation now includes up- and downregulation of a host of cytokines (IL-1β, TNF-α, IL-6, IP-10, MIF, HMG-1, and neutrophils) at specific time intervals after irradiation (Siva et al., 2014) and levels of cytokine expression correlate with dose magnitude. Translational research to characterize this response with candidate imaging and tissue biomarkers can provide information about dosing and sequencing combinations of radiotherapy and immunotherapy. Promising molecular imaging of the immune system include tracers of upregulation and downregulation of antitumor immune response (HAC-PD1 PET, DOTA anti-CTLA-4 mAb PET) (Maute et al., 2015; Higashikawa et al., 2014). Figure 3.6 illustrates the specificity of PD1-PET in a preclinical setting whereby PD-L1-positive tumors, which are more likely to respond to anti-PD-L1 therapy, exhibit higher PET signal compared

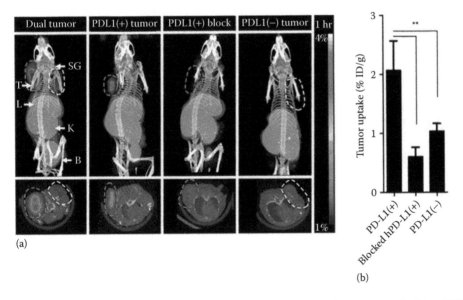

(a)

(b)

FIGURE 3.6 (a) Immune checkpoint receptor/antibody imaging of tumors with PD-1 PET has future applications to assess response and adapt therapy. (b) PD-1 PET uptake can be modulated by PD-L1 status, with an approximately twofold increase between PD-L1 negative and positive tumors. (Reproduced from Maute, R.L. et al., *Proc. Natl. Acad. Sci. USA.*, 112, E6506–E6514, 2015.)

TABLE 3.3   Candidate Imaging Biomarkers of Radiation and Immune-Mediated Response

Imaging Agent	Tracer/Contrast	Biologic Property	Modality
[^{18}F]ML-10	Fluoropentyl-methylmalonic acid	Reversible apoptosis	PET
[^{18}F]ICMT-11	Fluoropyrrolidinylsulfonyl-azaisatin	Irreversible apoptosis	PET
[^{18}F]HAC-PD1	Humanized programmed cell death receptor 1	Programmed cell death via immune checkpoint	PET
[^{64}Cu]anti-CTLA-4 mAb	DOTA-anti-CTLA-4 mAb	Amplification of tumor infiltrating T lymphocyte	PET

to PD-L1-negative tumors and blocked PD1 receptors. The main challenge with this approach is the relatively low dynamic range in PD1 expression (2:1 ratio between PD-L1-postive and PD-L1-negative tumors). Potential imaging biomarkers of the immune system have applications in patient selection/risk stratification, treatment planning, and response assessment. Table 3.3 lists example candidate imaging bio-markers of radiation response (ML-10, ICMT-11) and immune-mediated response. With optimization of delivery vehicles such as liposomes and other nanoparticles, future administration of theranostic agents for image-guided immunotherapy and radiotherapy is well within the realm of possibility. In this evolving landscape of biological imaging integration into therapy, demand for reliable multiscale models of response and outcome prediction will grow exponentially.

3.7 CONCLUSION

Biological imaging and radiobiological modeling seek to incorporate cancer and normal tissue-specific response to radiation therapy at the individual patient scale. Continued development and translation of quantitative imaging biomarkers into radiobiological mod-els may eventually overcome current challenges posed by limits in spatial heterogeneity resolution and complex temporal dynamics of therapeutic response. Unlike conventional randomized clinical trials designed to power superiority of one therapeutic intervention averaged over a patient population in relation to the standard of care, emerging precision medicine trials test efficacy of selective therapeutic strategies at the individual patient scale against case-matched controls. Measured and modeled quantitative imaging biomarkers will play a vital role in guiding selection, dosing, and sequencing of combination therapies for every patient as we treat in the age of precision radiation oncology.

REFERENCES

Aerts, H. J., J. Bussink, W. J. Oyen, W. van Elmpt, A. M. Folgering, D. Emans, M. Velders, P. Lambin, and D. De Ruysscher. 2012. Identification of residual metabolic-active areas within NSCLC tumours using a pre-radiotherapy FDG-PET-CT scan: A prospective validation. *Lung Cancer* 75 (1):73–76.

Aerts, H. J., A. A. van Baardwijk, S. F. Petit, C. Offermann, J. Loon, R. Houben, A. M. Dingemans et al. 2009. Identification of residual metabolic-active areas within individual NSCLC tumours using a pre-radiotherapy (18)Fluorodeoxyglucose-PET-CT scan. *Radiother Oncol* 91 (3):386–392.

Alber, M., F. Paulsen, S. M. Eschmann, and H. J. Machulla. 2003. On biologically conformal boost dose optimization. *Phys Med Biol* 48 (2):N31–N35.

Bauman, G., A. Scholz, J. Rivoire, M. Terekhov, J. Friedrich, A. de Oliveira, W. Semmler, L. M. Schreiber, and M. Puderbach. 2013. Lung ventilation- and perfusion-weighted Fourier decomposition magnetic resonance imaging: In vivo validation with hyperpolarized 3He and dynamic contrast-enhanced MRI. *Magn Reson Med* 69 (1):229–237.

Bentzen, S. M. 2005. Theragnostic imaging for radiation oncology: Dose-painting by numbers. *Lancet Oncol* 6 (2):112–117.

Bentzen, S. M., L. S. Constine, J. O. Deasy, A. Eisbruch, A. Jackson, L. B. Marks, R. K. Ten Haken, and E. D. Yorke. 2010. Quantitative analyses of normal tissue effects in the clinic (QUANTEC): An introduction to the scientific issues. *Int J Radiat Oncol Biol Phys* 76 (3 Suppl):S3–S9.

Bentzen, S. M., and V. Gregoire. 2011. Molecular imaging-based dose painting: A novel paradigm for radiation therapy prescription. *Semin Radiat Oncol* 21 (2):101–110.

Beppu, T., H. Hayashi, H. Okabe, T. Masuda, K. Mima, R. Otao, A. Chikamoto et al. 2011. Liver functional volumetry for portal vein embolization using a newly developed 99mTc-galactosyl human serum albumin scintigraphy SPECT-computed tomography fusion system. *J Gastroenterol* 46 (7):938–943.

Berwouts, D., L. A. Olteanu, F. Duprez, T. Vercauteren, W. De Gersem, W. De Neve, C. Van de Wiele, and I. Madani. 2013. Three-phase adaptive dose-painting-by-numbers for head-and-neck cancer: Initial results of the phase I clinical trial. *Radiother Oncol* 107 (3):310–316.

Bowen, S. R., R. J. Chappell, S. M. Bentzen, M. A. Deveau, L. J. Forrest, and R. Jeraj. 2012. Spatially resolved regression analysis of pre-treatment FDG, FLT and Cu-ATSM PET from post-treatment FDG PET: An exploratory study. *Radiother Oncol* 105:41–48.

Bowen, S. R., R. T. Flynn, S. M. Bentzen, and R. Jeraj. 2009. On the sensitivity of IMRT dose optimization to the mathematical form of a biological imaging-based prescription function. *Phys Med Biol* 54 (6):1483–1501.

Bowen, S. R., J. Saini, T. R. Chapman, R. S. Miyaoka, P. E. Kinahan, G. A. Sandison, T. Wong, H. J. Vesselle, M. J. Nyflot, and S. Apisarnthanarax. 2015. Differential hepatic avoidance radiation therapy: Proof of concept in hepatocellular carcinoma patients. *Radiother Oncol* 115:203–210.

Bowen, S. R., A. J. van der Kogel, M. Nordsmark, S. M. Bentzen, and R. Jeraj. 2011. Characterization of positron emission tomography hypoxia tracer uptake and tissue oxygenation via electro-chemical modeling. *Nucl Med Biol* 38 (6):771–780.

Bradshaw, T. J., S. R. Bowen, M. A. Deveau, L. Kubicek, P. White, S. M. Bentzen, R. J. Chappell, L. J. Forrest, and R. Jeraj. 2015. Molecular imaging biomarkers of resistance to radiation therapy for spontaneous nasal tumors in canines. *Int J Radiat Oncol Biol Phys* 91 (4):787–795.

Bradshaw, T. J., S. R. Bowen, N. Jallow, L. J. Forrest, and R. Jeraj. 2013. Heterogeneity in intratumor correlations of 18F-FDG, 18F-FLT, and 61Cu-ATSM PET in canine sinonasal tumors. *J Nucl Med* 54 (11):1931–1937.

Brahmer, J., K. L. Reckamp, P. Baas, L. Crino, W. E. Eberhardt, E. Poddubskaya, S. Antonia et al. 2015. Nivolumab versus docetaxel in advanced squamous-cell non-amall-cell lung cancer. *N Engl J Med* 373 (2):123–135.

Brennan, D., L. Schubert, Q. Diot, R. Castillo, E. Castillo, T. Guerrero, M. K. Martel et al. 2015. Clinical validation of 4-dimensional computed tomography ventilation with pulmonary function test data. *Int J Radiat Oncol Biol Phys* 92 (2):423–429.

Brenner, D. J. 2008. The linear-quadratic model is an appropriate methodology for determining isoeffective doses at large doses per fraction. *Semin Radiat Oncol* 18 (4):234–239.

Brown, J. M., D. J. Carlson, and D. J. Brenner. 2014. The tumor radiobiology of SRS and SBRT: Are more than the 5 Rs involved? *Int J Radiat Oncol Biol Phys* 88 (2):254–262.

Buckler, A. J., and R. Boellaard. 2011. Standardization of quantitative imaging: The time is right, and 18F-FDG PET/CT is a good place to start. *J Nucl Med* 52 (2):171–172.

Campbell, B. A., J. Callahan, M. Bressel, N. Simoens, S. Everitt, M. S. Hofman, R. J. Hicks, K. Burbury, and M. MacManus. 2015. Distribution atlas of proliferating bone marrow in non-small cell lung cancer patients measured by FLT-PET/CT imaging, with potential applicability in radiation therapy planning. *Int J Radiat Oncol Biol Phys* 92 (5):1035–1043.

Cao, Y., A. Popovtzer, D. Li, D. B. Chepeha, J. S. Moyer, M. E. Prince, F. Worden et al. 2008. Early prediction of outcome in advanced head-and-neck cancer based on tumor blood volume alterations during therapy: A prospective study. *Int J Radiat Oncol Biol Phys* 72 (5):1287–1290.

Cao, Y., H. Wang, T. D. Johnson, C. Pan, H. Hussain, J. M. Balter, D. Normolle et al. 2013. Prediction of liver function by using magnetic resonance-based portal venous perfusion imaging. *Int J Radiat Oncol Biol Phys* 85 (1):258–263.

Casey, D. L., L. H. Wexler, J. J. Fox, K. V. Dharmarajan, H. Schoder, A. N. Price, and S. L. Wolden. 2014. Predicting outcome in patients with rhabdomyosarcoma: Role of [(18)f]fluorodeoxyglucose positron emission tomography. *Int J Radiat Oncol Biol Phys* 90 (5):1136–1142.

Castillo, R., N. Pham, S. Ansari, D. Meshkov, S. Castillo, M. Li, A. Olanrewaju, B. Hobbs, E. Castillo, and T. Guerrero. 2014. Pre-radiotherapy FDG PET predicts radiation pneumonitis in lung cancer. *Radiat Oncol* 9:74.

Challapalli, A., L. M. Kenny, W. A. Hallett, K. Kozlowski, G. Tomasi, M. Gudi, A. Al-Nahhas, R. C. Coombes, and E. O. Aboagye. 2013. 18F-ICMT-11, a caspase-3-specific PET tracer for apoptosis: Biodistribution and radiation dosimetry. *J Nucl Med* 54 (9):1551–1556.

Cruite, I., M. Schroeder, E. M. Merkle, and C. B. Sirlin. 2010. Gadoxetate disodium-enhanced MRI of the liver: Part 2, protocol optimization and lesion appearance in the cirrhotic liver. *AJR Am J Roentgenol* 195 (1):29–41.

Das, S. K., M. M. Miften, S. Zhou, M. Bell, M. T. Munley, C. S. Whiddon, O. Craciunescu et al. 2004. Feasibility of optimizing the dose distribution in lung tumors using fluorine-18-fluorodeoxyglucose positron emission tomography and single photon emission computed tomography guided dose prescriptions. *Med Phys* 31 (6):1452–1461.

Das, S. K., and R. K. Ten Haken. 2011. Functional and molecular image guidance in radiotherapy treatment planning optimization. *Semin Radiat Oncol* 21 (2):111–118.

De Ruysscher, D., and C. M. Kirsch. 2010. PET scans in radiotherapy planning of lung cancer. *Radiother Oncol* 96 (3):335–338.

Decoster, L., D. Schallier, H. Everaert, K. Nieboer, M. Meysman, B. Neyns, J. De Mey, and J. De Greve. 2008. Complete metabolic tumour response, assessed by 18-fluorodeoxyglucose positron emission tomography (18FDG-PET), after induction chemotherapy predicts a favourable outcome in patients with locally advanced non-small cell lung cancer (NSCLC). *Lung Cancer* 62 (1):55–61.

Deveau, M. A., S. R. Bowen, D. C. Westerly, and R. Jeraj. 2010. Feasibility and sensitivity study of helical tomotherapy for dose painting plans. *Acta Oncol* 49 (7):991–996.

Dinges, E., N. Felderman, S. McGuire, B. Gross, S. Bhatia, S. Mott, J. Buatti, and D. Wang. 2015. Bone marrow sparing in intensity modulated proton therapy for cervical cancer: Efficacy and robustness under range and setup uncertainties. *Radiother Oncol* 115 (3):373–378.

Duprez, F., W. De Neve, W. De Gersem, M. Coghe, and I. Madani. 2010. Adaptive dose painting by numbers for head-and-neck cancer. *Int J Radiat Oncol Biol Phys.* 80 (4):1045–1055.

Everson, G. T., M. L. Shiffman, J. C. Hoefs, T. R. Morgan, R. K. Sterling, D. A. Wagner, S. Lauriski, T. M. Curto, A. Stoddard, and E. C. Wright. 2012. Quantitative liver function tests improve the prediction of clinical outcomes in chronic hepatitis C: Results from the Hepatitis C Antiviral Long-term Treatment Against Cirrhosis Trial. *Hepatology* 55 (4):1019–1029.

Farjam, R., C. I. Tsien, F. Y. Feng, D. Gomez-Hassan, J. A. Hayman, T. S. Lawrence, and Y. Cao. 2013. Physiological imaging-defined, response-driven subvolumes of a tumor. *Int J Radiat Oncol Biol Phys* 85 (5):1383–1390.

Farrag, A., G. Ceulemans, M. Voordeckers, H. Everaert, and G. Storme. 2010. Can 18F-FDG-PET response during radiotherapy be used as a predictive factor for the outcome of head and neck cancer patients? *Nucl Med Commun* 31 (6):495–501.

Feng, M., F. M. Kong, M. Gross, S. Fernando, J. A. Hayman, and R. K. Ten Haken. 2009. Using fluorodeoxyglucose positron emission tomography to assess tumor volume during radiotherapy for non-small-cell lung cancer and its potential impact on adaptive dose escalation and normal tissue sparing. *Int J Radiat Oncol Biol Phys* 73 (4):1228–1234.

Gillham, C., D. Zips, F. Ponisch, C. Evers, W. Enghardt, N. Abolmaali, K. Zophel et al. 2008. Additional PET/CT in week 5-6 of radiotherapy for patients with stage III non-small cell lung cancer as a means of dose escalation planning? *Radiother Oncol* 88 (3):335–341.

Graham, M. V., J. A. Purdy, B. Emami, W. Harms, W. Bosch, M. A. Lockett, and C. A. Perez. 1999. Clinical dose-volume histogram analysis for pneumonitis after 3D treatment for non-small cell lung cancer (NSCLC). *Int J Radiat Oncol Biol Phys* 45 (2):323–329.

Hall, E. J., and A. J. Giaccia. 2006. *Radiobiology for the Radiologist*. 6th ed. Philadelphia, PA: Lippincott Williams & Wilkins.

Hanahan, D., and R. A. Weinberg. 2000. The hallmarks of cancer. *Cell* 100 (1):57–70.

Hanahan, D., and R. A. Weinberg. 2011. Hallmarks of cancer: The next generation. *Cell* 144 (5):646–674.

Hendrickson, K., M. Phillips, W. Smith, L. Peterson, K. Krohn, and J. Rajendran. 2011. Hypoxia imaging with [F-18] FMISO-PET in head and neck cancer: Potential for guiding intensity modulated radiation therapy in overcoming hypoxia-induced treatment resistance. *Radiother Oncol* 101 (3):369–375.

Heukelom, J., O. Hamming, H. Bartelink, F. Hoebers, J. Giralt, T. Herlestam, M. Verheij et al. 2013. Adaptive and innovative Radiation Treatment FOR improving Cancer treatment outcomE (ARTFORCE): A randomized controlled phase II trial for individualized treatment of head and neck cancer. *BMC Cancer* 13:84.

Higashikawa, K., K. Yagi, K. Watanabe, S. Kamino, M. Ueda, M. Hiromura, and S. Enomoto. 2014. 64Cu-DOTA-anti-CTLA-4 mAb enabled PET visualization of CTLA-4 on the T-cell infiltrating tumor tissues. *PLoS One* 9 (11):e109866.

Hodi, F. S., S. J. O'Day, D. F. McDermott, R. W. Weber, J. A. Sosman, J. B. Haanen, R. Gonzalez et al. 2010. Improved survival with ipilimumab in patients with metastatic melanoma. *N Engl J Med* 363 (8):711–723.

Hoeben, B. A., E. G. Troost, P. N. Span, C. M. van Herpen, J. Bussink, W. J. Oyen, and J. H. Kaanders. 2013. 18F-FLT PET during radiotherapy or chemoradiotherapy in head and neck squamous cell carcinoma is an early predictor of outcome. *J Nucl Med* 54 (4):532–540.

Hoefs, J. C., F. Wang, and G. Kanel. 1997. Functional measurement of nonfibrotic hepatic mass in cirrhotic patients. *Am J Gastroenterol* 92 (11):2054–2058.

Hoglund, J., A. Shirvan, G. Antoni, S. A. Gustavsson, B. Langstrom, A. Ringheim, J. Sorensen, M. Ben-Ami, and I. Ziv. 2011. 18F-ML-10, a PET tracer for apoptosis: First human study. *J Nucl Med* 52 (5):720–725.

Hoover, D. A., D. P. Capaldi, K. Sheikh, D. A. Palma, G. B. Rodrigues, A. R. Dar, E. Yu et al. 2014a. Functional lung avoidance for individualized radiotherapy (FLAIR): Study protocol for a randomized, double-blind clinical trial. *BMC Cancer* 14:934.

Hoover, D. A., R. H. Reid, E. Wong, L. Stitt, E. Sabondjian, G. B. Rodrigues, J. K. Jaswal, and B. P. Yaremko. 2014b. SPECT-based functional lung imaging for the prediction of radiation pneumonitis: A clinical and dosimetric correlation. *J Med Imaging Radiat Oncol* 58 (2):214–222.

Huang, W., T. Zhou, L. Ma, H. Sun, H. Gong, J. Wang, J. Yu, and B. Li. 2011. Standard uptake value and metabolic tumor volume of (1)(8)F-FDG PET/CT predict short-term outcome early in the course of chemoradiotherapy in advanced non-small cell lung cancer. *Eur J Nucl Med Mol Imaging* 38 (9):1628–1635.

Huang, Z., K. A. Yuh, S. S. Lo, J. C. Grecula, S. Sammet, C. L. Sammet, G. Jia et al. 2014. Validation of optimal DCE-MRI perfusion threshold to classify at-risk tumor imaging voxels in heterogeneous cervical cancer for outcome prediction. *Magn Reson Imaging* 32 (10):1198–1205.

Ireland, R. H., O. S. Din, J. A. Swinscoe, N. Woodhouse, E. J. van Beek, J. M. Wild, and M. Q. Hatton. 2010. Detection of radiation-induced lung injury in non-small cell lung cancer patients using hyperpolarized helium-3 magnetic resonance imaging. *Radiother Oncol* 97 (2):244–248.

Jeong, J., K. I. Shoghi, and J. O. Deasy. 2013. Modelling the interplay between hypoxia and proliferation in radiotherapy tumour response. *Phys Med Biol* 58 (14):4897–4919.

Jeraj, R., S. R. Bowen, M. Deveau, P. Galavis, and N. Jallow. 2011. Biologically conformal radio-therapy: Targeting based on biological imaging. In *Uncertainties in External Beam Radiation Therapy*, J. Palta and T. R. Mackie (Eds.), pp. 15–43. Madison, WI: Medical Physics Publishing.

Jeraj, R., S. R. Bowen, N. Jallow, M. J. Nyflot, and M. Vanderhoek. 2013. Molecular imaging in radiation oncology. In *The Modern Technology of Radiation Oncology*, J. Van Dyk (ed.), pp. 25–58. Madison, WI: Medical Physics Publishing.

Jeraj, R., T. Bradshaw, and U. Simoncic. 2015. Molecular imaging to plan radiotherapy and evaluate its efficacy. *J Nucl Med* 56 (11):1752–1765.

Jeraj, R., Y. Cao, R. K. Ten Haken, C. Hahn, and L. Marks. 2010. Imaging for assessment of radiation-induced normal tissue effects. *Int J Radiat Oncol Biol Phys* 76 (3 Suppl):S140–S144.

Jeraj, R., and M. E. Meyerand. 2008. Molecular and functional imaging in radiation oncology. *Cancer Treat Res* 139:63–95.

Jhaveri, K., and H. Linden. 2015. Measuring tumor metabolism by 18F-FDG PET predicts outcome in a multicenter study: A step off in the right direction. *J Nucl Med* 56 (1):1–2.

Joye, I., C. M. Deroose, V. Vandecaveye, and K. Haustermans. 2014. The role of diffusion-weighted MRI and (18)F-FDG PET/CT in the prediction of pathologic complete response after radio-chemotherapy for rectal cancer: A systematic review. *Radiother Oncol* 113 (2):158–165.

Kaira, K., M. Serizawa, Y. Koh, T. Takahashi, A. Yamaguchi, H. Hanaoka, N. Oriuchi et al. 2014. Biological significance of 18F-FDG uptake on PET in patients with non-small-cell lung cancer. *Lung Cancer* 83 (2):197–204.

Kidd, E. A., M. Thomas, B. A. Siegel, F. Dehdashti, and P. W. Grigsby. 2013. Changes in cervical cancer FDG uptake during chemoradiation and association with response. *Int J Radiat Oncol Biol Phys* 85 (1):116–122.

Kim, Y., and W. A. Tome. 2010. Dose-painting IMRT optimization using biological parameters. *Acta Oncol* 49 (8):1374–1384.

Kirkpatrick, J. P., J. J. Meyer, and L. B. Marks. 2008. The linear-quadratic model is inappropriate to model high dose per fraction effects in radiosurgery. *Semin Radiat Oncol* 18 (4):240–243.

Kishino, T., H. Hoshikawa, Y. Nishiyama, Y. Yamamoto, and N. Mori. 2012. Usefulness of 3'-deoxy-3'-18F-fluorothymidine PET for predicting early response to chemoradiotherapy in head and neck cancer. *J Nucl Med* 53 (10):1521–1527.

Klaassen, R., R. J. Bennink, G. van Tienhoven, M. F. Bijlsma, M. G. Besselink, M. I. van Berge Henegouwen, J. W. Wilmink et al. 2015. Feasibility and repeatability of PET with the hypoxia tracer [(18)F]HX4 in oesophageal and pancreatic cancer. *Radiother Oncol* 116 (1):94–99.

Kocak, Z., G. R. Borst, J. Zeng, S. Zhou, D. R. Hollis, J. Zhang, E. S. Evans et al. 2007. Prospective assessment of dosimetric/physiologic-based models for predicting radiation pneumonitis. *Int J Radiat Oncol Biol Phys* 67 (1):178–186.

Kong, F. M., J. Zhao, J. Wang, and C. Faivre-Finn. 2014. Radiation dose effect in locally advanced non-small cell lung cancer. *J Thorac Dis* 6 (4):336–347.

Korreman, S. S., S. Ulrich, S. Bowen, M. Deveau, S. M. Bentzen, and R. Jeraj. 2010. Feasibility of dose painting using volumetric modulated arc optimization and delivery. *Acta Oncol* 49 (7):964–971.

Kosztyla, R., E. K. Chan, F. Hsu, D. Wilson, R. Ma, A. Cheung, S. Zhang, V. Moiseenko, F. Benard, and A. Nichol. 2013. High-grade glioma radiation therapy target volumes and patterns of fail-ure obtained from magnetic resonance imaging and 18F-FDOPA positron emission tomog-raphy delineations from multiple observers. *Int J Radiat Oncol Biol Phys* 87 (5):1100–1106.

Krohn, K. A., D. A. Mankoff, M. Muzi, J. M. Link, and A. M. Spence. 2005. True tracers: Comparing FDG with glucose and FLT with thymidine. *Nucl Med Biol* 32 (7):663–671.

Lavrenkov, K., J. A. Christian, M. Partridge, E. Niotsikou, G. Cook, M. Parker, J. L. Bedford, and M. Brada. 2007. A potential to reduce pulmonary toxicity: The use of perfusion SPECT with IMRT for functional lung avoidance in radiotherapy of non-small cell lung cancer. *Radiother Oncol* 83 (2):156–162.

Leimgruber, A., A. Moller, S. J. Everitt, M. Chabrot, D. L. Ball, B. Solomon, M. MacManus, and R. J. Hicks. 2014. Effect of platinum-based chemoradiotherapy on cellular proliferation in bone marrow and spleen, estimated by (18)F-FLT PET/CT in patients with locally advanced non-small cell lung cancer. *J Nucl Med* 55 (7):1075–1080.

Li, X., H. Kang, L. R. Arlinghaus, R. G. Abramson, A. B. Chakravarthy, V. G. Abramson, J. Farley, M. Sanders, and T. E. Yankeelov. 2014. Analyzing spatial heterogeneity in DCE- and DW-MRI parametric maps to optimize prediction of pathologic response to neoadjuvant chemotherapy in breast cancer. *Transl Oncol* 7 (1):14–22.

Liang, Y., M. Bydder, C. M. Yashar, B. S. Rose, M. Cornell, C. K. Hoh, J. D. Lawson et al. 2013. Prospective study of functional bone marrow-sparing intensity modulated radiation therapy with concurrent chemotherapy for pelvic malignancies. *Int J Radiat Oncol Biol Phys* 85 (2):406–414.

Linden, H. M., B. F. Kurland, L. M. Peterson, E. K. Schubert, J. R. Gralow, J. M. Specht, G. K. Ellis et al. 2011. Fluoroestradiol positron emission tomography reveals differences in pharmacodynamics of aromatase inhibitors, tamoxifen, and fulvestrant in patients with metastatic breast cancer. *Clin Cancer Res* 17 (14):4799–4805.

Linden, H. M., S. A. Stekhova, J. M. Link, J. R. Gralow, R. B. Livingston, G. K. Ellis, P. H. Petra et al. 2006. Quantitative fluoroestradiol positron emission tomography imaging predicts response to endocrine treatment in breast cancer. *J Clin Oncol* 24 (18):2793–2799.

Ling, C. C., J. Humm, S. Larson, H. Amols, Z. Fuks, S. Leibel, and J. A. Koutcher. 2000. Towards multidimensional radiotherapy (MD-CRT): Biological imaging and biological conformality. *Int J Radiat Oncol Biol Phys* 47 (3):551–560.

Lipson, D. A., D. A. Roberts, J. Hansen-Flaschen, T. R. Gentile, G. Jones, A. Thompson, I. E. Dimitrov et al. 2002. Pulmonary ventilation and perfusion scanning using hyperpolarized helium-3 MRI and arterial spin tagging in healthy normal subjects and in pulmonary embolism and orthotopic lung transplant patients. *Magn Reson Med* 47 (6):1073–1076.

Madani, I., F. Duprez, T. Boterberg, C. Van de Wiele, K. Bonte, P. Deron, W. De Gersem, M. Coghe, and W. De Neve. 2011. Maximum tolerated dose in a phase I trial on adaptive dose painting by numbers for head and neck cancer. *Radiother Oncol* 101 (3):351–355.

Marks, L. B., D. Hollis, M. Munley, G. Bentel, M. Garipagaoglu, M. Fan, J. Poulson et al. 2000. The role of lung perfusion imaging in predicting the direction of radiation-induced changes in pulmonary function tests. *Cancer* 88 (9):2135–2141.

Marks, L. B., G. W. Sherouse, M. T. Munley, G. C. Bentel, and D. P. Spencer. 1999. Incorporation of functional status into dose-volume analysis. *Med Phys* 26 (2):196–199.

Marks, L. B., D. P. Spencer, G. W. Sherouse, G. Bentel, R. Clough, K. Vann, R. Jaszczak, R. E. Coleman, and L. R. Prosnitz. 1995. The role of three dimensional functional lung imaging in radiation treatment planning: The functional dose-volume histogram. *Int J Radiat Oncol Biol Phys* 33 (1):65–75.

Marks, L. B., R. K. Ten Haken, and M. K. Martel. 2010. Guest editor's introduction to QUANTEC: A users guide. *Int J Radiat Oncol Biol Phys.* 76 (3 Suppl):S1–S2.

Martel, M. K., R. K. Ten Haken, M. B. Hazuka, M. L. Kessler, M. Strawderman, A. T. Turrisi, T. S. Lawrence, B. A. Fraass, and A. S. Lichter. 1999. Estimation of tumor control probability model parameters from 3-D dose distributions of non-small cell lung cancer patients. *Lung Cancer* 24 (1):31–37.

Maute, R. L., S. R. Gordon, A. T. Mayer, M. N. McCracken, A. Natarajan, N. G. Ring, R. Kimura et al. 2015. Engineering high-affinity PD-1 variants for optimized immunotherapy and immuno-PET imaging. *Proc Natl Acad Sci U S A* 112 (47):E6506–E6514.

Mayr, N. A., Z. Huang, J. Z. Wang, S. S. Lo, J. M. Fan, J. C. Grecula, S. Sammet et al. 2012. Characterizing tumor heterogeneity with functional imaging and quantifying high-risk tumor volume for early prediction of treatment outcome: Cervical cancer as a model. *Int J Radiat Oncol Biol Phys* 83 (3):972–979.

Mayr, N. A., J. Z. Wang, S. S. Lo, D. Zhang, J. C. Grecula, L. Lu, J. F. Montebello, J. M. Fowler, and W. T. Yuh. 2010. Translating response during therapy into ultimate treatment outcome: A personalized 4-dimensional MRI tumor volumetric regression approach in cervical cancer. *Int J Radiat Oncol Biol Phys* 76 (3):719–727.

McCarthy, F., R. Roshani, J. Steele, and T. Hagemann. 2013. Current clinical immunotherapy targets in advanced nonsmall cell lung cancer (NSCLC). *J Leukoc Biol* 94 (6):1201–1206.

McGuire, S. M., L. B. Marks, F. F. Yin, and S. K. Das. 2010. A methodology for selecting the beam arrangement to reduce the intensity-modulated radiation therapy (IMRT) dose to the SPECT-defined functioning lung. *Phys Med Biol* 55 (2):403–416.

McGuire, S. M., Y. Menda, L. L. Ponto, B. Gross, M. Juweid, and J. E. Bayouth. 2011. A methodology for incorporating functional bone marrow sparing in IMRT planning for pelvic radiation therapy. *Radiother Oncol* 99 (1):49–54.

McGuire, S. M., S. Zhou, L. B. Marks, M. Dewhirst, F. F. Yin, and S. K. Das. 2006. A methodology for using SPECT to reduce intensity-modulated radiation therapy (IMRT) dose to functioning lung. *Int J Radiat Oncol Biol Phys* 66 (5):1543–1552.

Mirnezami, R., J. Nicholson, and A. Darzi. 2012. Preparing for precision medicine. *N Engl J Med* 366 (6):489–491.

Mortensen, L. S., J. Johansen, J. Kallehauge, H. Primdahl, M. Busk, P. Lassen, J. Alsner et al. 2012. FAZA PET/CT hypoxia imaging in patients with squamous cell carcinoma of the head and neck treated with radiotherapy: Results from the DAHANCA 24 trial. *Radiother Oncol* 105(1):14–20.

Munck af Rosenschold, P., J. Costa, S. A. Engelholm, M. J. Lundemann, I. Law, L. Ohlhues, and S. Engelholm. 2015. Impact of [18F]-fluoro-ethyl-tyrosine PET imaging on target definition for radiation therapy of high-grade glioma. *Neuro Oncol* 17 (5):757–763.

Munley, M. T., G. C. Kagadis, K. P. McGee, A. S. Kirov, S. Jang, S. Mutic, R. Jeraj, L. Xing, and J. D. Bourland. 2013. An introduction to molecular imaging in radiation oncology: A report by the AAPM Working Group on Molecular Imaging in Radiation Oncology (WGMIR). *Med Phys* 40 (10):101501.

Muzi, M., A. M. Spence, F. O'Sullivan, D. A. Mankoff, J. M. Wells, J. R. Grierson, J. M. Link, and K. A. Krohn. 2006. Kinetic analysis of 3'-deoxy-3'-18F-fluorothymidine in patients with gliomas. *J Nucl Med* 47 (10):1612–1621.

Muzi, M., H. Vesselle, J. R. Grierson, D. A. Mankoff, R. A. Schmidt, L. Peterson, J. M. Wells, and K. A. Krohn. 2005. Kinetic analysis of 3'-deoxy-3'-fluorothymidine PET studies: Validation studies in patients with lung cancer. *J Nucl Med* 46 (2):274–282.

Nestle, U., S. Kremp, and A. L. Grosu. 2006. Practical integration of [(18)F]-FDG-PET and PET-CT in the planning of radiotherapy for non-small cell lung cancer (NSCLC): The technical basis, ICRU-target volumes, problems, perspectives. *Radiotherapy and Oncology* 81 (2):209–225.

Nordsmark, M., S. M. Bentzen, V. Rudat, D. Brizel, E. Lartigau, P. Stadler, A. Becker et al. 2005. Prognostic value of tumor oxygenation in 397 head and neck tumors after primary radiation therapy: An international multi-center study. *Radiother Oncol* 77 (1):18–24.

Nyeng, T. B., J. F. Kallehauge, M. Hoyer, J. B. Petersen, P. R. Poulsen, and L. P. Muren. 2011. Clinical validation of a 4D-CT based method for lung ventilation measurement in phantoms and patients. *Acta Oncol* 50 (6):897–907.

Nyflot, M. J., P. M. Harari, S. Yip, S. B. Perlman, and R. Jeraj. 2012. Correlation of PET images of metabolism, proliferation and hypoxia to characterize tumor phenotype in patients with cancer of the oropharynx. *Radiother Oncol* 105 (1):36–40.

Ohri, N., B. Piperdi, M. K. Garg, W. R. Bodner, R. Gucalp, R. Perez-Soler, S. M. Keller, and C. Guha. 2015. Pre-treatment FDG-PET predicts the site of in-field progression following concurrent chemoradiotherapy for stage III non-small cell lung cancer. *Lung Cancer* 87 (1):23–27.

Pallardy, A., C. Bodet-Milin, A. Oudoux, L. Campion, E. Bourbouloux, C. Sagan, C. Ansquer et al. 2010. Clinical and survival impact of FDG PET in patients with suspicion of recurrent cervical carcinoma. *Eur J Nucl Med Mol Imaging* 37 (7):1270–1278.

Petit, S. F., H. J. Aerts, J. G. van Loon, C. Offermann, R. Houben, B. Winkens, M. C. Ollers, P. Lambin, D. De Ruysscher, and A. L. Dekker. 2009a. Metabolic control probability in tumour subvolumes or how to guide tumour dose redistribution in non-small cell lung cancer (NSCLC): An exploratory clinical study. *Radiother Oncol* 91 (3):393–398.

Petit, S. F., A. L. Dekker, R. Seigneuric, L. Murrer, N. A. van Riel, M. Nordsmark, J. Overgaard, P. Lambin, and B. G. Wouters. 2009b. Intra-voxel heterogeneity influences the dose prescription for dose-painting with radiotherapy: A modelling study. *Phys Med Biol* 54 (7):2179–2196.

Picchio, M., G. Berardi, A. Fodor, E. Busnardo, C. Crivellaro, C. Giovacchini, C. Fiorino et al. 2014. (11)C-Choline PET/CT as a guide to radiation treatment planning of lymph-node relapses in prostate cancer patients. *Eur J Nucl Med Mol Imaging* 41 (7):1270–1279.

Ringe, K. I., D. B. Husarik, C. B. Sirlin, and E. M. Merkle. 2010. Gadoxetate disodium-enhanced MRI of the liver: Part 1, protocol optimization and lesion appearance in the noncirrhotic liver. *AJR Am J Roentgenol* 195 (1):13–28.

Rockne, R. C., A. D. Trister, J. Jacobs, A. J. Hawkins-Daarud, M. L. Neal, K. Hendrickson, M. M. Mrugala et al. 2015. A patient-specific computational model of hypoxia-modulated radiation resistance in glioblastoma using 18F-FMISO-PET. *J R Soc Interface* 12 (103):20141174.

Ruggieri, R., N. Stavreva, S. Naccarato, and P. Stavrev. 2012. Applying a hypoxia-incorporating TCP model to experimental data on rat sarcoma. *Int J Radiat Oncol Biol Phys* 83 (5): 1603–1608.

Schwarz, J. K., B. A. Siegel, F. Dehdashti, and P. W. Grigsby. 2012. Metabolic response on post-therapy FDG-PET predicts patterns of failure after radiotherapy for cervical cancer. *Int J Radiat Oncol Biol Phys* 83 (1):185–190.

Semenza, G. L. 2009. Regulation of cancer cell metabolism by hypoxia-inducible factor 1. *Semin Cancer Biol* 19 (1):12–16.

Seppenwoolde, Y., K. De Jaeger, L. J. Boersma, J. S. Belderbos, and J. V. Lebesque. 2004. Regional differences in lung radiosensitivity after radiotherapy for non-small-cell lung cancer. *Int J Radiat Oncol Biol Phys* 60 (3):748–758.

Seppenwoolde, Y., M. Engelsman, K. De Jaeger, S. H. Muller, P. Baas, D. L. McShan, B. A. Fraass et al. 2002. Optimizing radiation treatment plans for lung cancer using lung perfusion information. *Radiother Oncol* 63 (2):165–177.

Shi, X., X. Meng, X. Sun, L. Xing, and J. Yu. 2014. PET/CT imaging-guided dose painting in radiation therapy. *Cancer Lett* 355 (2):169–175.

Shields, A. F., J. R. Grierson, B. M. Dohmen, H. J. Machulla, J. C. Stayanoff, J. M. Lawhorn-Crews, J. E. Obradovich, O. Muzik, and T. J. Mangner. 1998. Imaging proliferation in vivo with [F-18] FLT and positron emission tomography. *Nat Med* 4 (11):1334–1336.

Sirlin, C. B., H. K. Hussain, E. Jonas, M. Kanematsu, J. Min Lee, E. M. Merkle, M. Peck-Radosavljevic, S. B. Reeder, J. Ricke, and M. Sakamoto. 2014. Consensus report from the 6th International forum for liver MRI using gadoxetic acid. *J Magn Reson Imaging* 40 (3):516–529.

Siva, S., M. MacManus, T. Kron, N. Best, J. Smith, P. Lobachevsky, D. Ball, and O. Martin. 2014. A pattern of early radiation-induced inflammatory cytokine expression is associated with lung toxicity in patients with non-small cell lung cancer. *PLoS One* 9 (10):e109560.

Siva, S., R. Thomas, J. Callahan, N. Hardcastle, D. Pham, T. Kron, R. J. Hicks, M. P. MacManus, D. L. Ball, and M. S. Hofman. 2015. High-resolution pulmonary ventilation and perfusion PET/CT allows for functionally adapted intensity modulated radiotherapy in lung cancer. *Radiother Oncol* 115(2):157–162.

Sorensen, M., K. Frisch, D. Bender, and S. Keiding. 2011a. The potential use of 2-[(1)(8)F]fluoro-2-deoxy-D-galactose as a PET/CT tracer for detection of hepatocellular carcinoma. *Eur J Nucl Med Mol Imaging* 38 (9):1723–1731.

Sorensen, M., K. S. Mikkelsen, K. Frisch, L. Bass, B. M. Bibby, and S. Keiding. 2011b. Hepatic galactose metabolism quantified in humans using 2-18F-fluoro-2-deoxy-D-galactose PET/CT. *J Nucl Med* 52 (10):1566–1572.

Sovik, A., E. Malinen, O. S. Bruland, S. M. Bentzen, and D. R. Olsen. 2007. Optimization of tumour control probability in hypoxic tumours by radiation dose redistribution: A modelling study. *Phys Med Biol* 52 (2):499–513.

Sperduto, P. W., C. W. Song, J. P. Kirkpatrick, and E. Glatstein. 2015. A hypothesis: Indirect cell death in the radiosurgery era. *Int J Radiat Oncol Biol Phys* 91 (1):11–13.

Thorwarth, D. 2015. Functional imaging for radiotherapy treatment planning: Current status and future directions-a review. *Br J Radiol* 88 (1051):20150056.

Thorwarth, D., S. M. Eschmann, F. Holzner, F. Paulsen, and M. Alber. 2006. Combined uptake of [18F]FDG and [18F]FMISO correlates with radiation therapy outcome in head-and-neck cancer patients. *Radiother Oncol* 80 (2):151–156.

Thorwarth, D., S. M. Eschmann, F. Paulsen, and M. Alber. 2007a. Hypoxia dose painting by numbers: A planning study. *Int J Radiat Oncol Biol Phys* 68 (1):291–300.

Thorwarth, D., S. M. Eschmann, F. Paulsen, and M. Alber. 2007b. A model of reoxygenation dynamics of head-and-neck tumors based on serial 18F-fluoromisonidazole positron emission tomography investigations. *Int J Radiat Oncol Biol Phys* 68 (2):515–521.

Titz, B., and R. Jeraj. 2008. An imaging-based tumour growth and treatment response model: Investigating the effect of tumour oxygenation on radiation therapy response. *Phys Med Biol* 53 (17):4471–4488.

Titz, B., K. R. Kozak, and R. Jeraj. 2012. Computational modelling of anti-angiogenic therapies based on multiparametric molecular imaging data. *Phys Med Biol* 57 (19):6079–6101.

Troost, E. G., J. Bussink, A. L. Hoffmann, O. C. Boerman, W. J. Oyen, and J. H. Kaanders. 2010. 18F-FLT PET/CT for early response monitoring and dose escalation in oropharyngeal tumors. *J Nucl Med* 51 (6):866–874.

Troost, E. G., J. Bussink, W. J. Oyen, and J. H. Kaanders. 2009. 18F-FDG and 18F-FLT do not discriminate between reactive and metastatic lymph nodes in oral cancer. *J Nucl Med* 50 (3):490–491.

Umeda, Y., Y. Demura, M. Morikawa, M. Anzai, M. Kadowaki, S. Ameshima, T. Tsuchida et al. 2015. Prognostic value of dual-time-point 18F-FDG PET for idiopathic pulmonary fibrosis. *J Nucl Med* 56 (12):1869–1875.

van Elmpt, W., D. De Ruysscher, A. van der Salm, A. Lakeman, J. van der Stoep, D. Emans, E. Damen, M. Ollers, J. J. Sonke, and J. Belderbos. 2012a. The PET-boost randomised phase II dose-escalation trial in non-small cell lung cancer. *Radiother Oncol* 104 (1):67–71.

van Elmpt, W., M. Ollers, A. M. Dingemans, P. Lambin, and D. De Ruysscher. 2012b. Response assessment using 18F-FDG PET early in the course of radiotherapy correlates with survival in advanced-stage non-small cell lung cancer. *J Nucl Med* 53 (10):1514–1520.

van Loon, J., M. H. Janssen, M. Ollers, H. J. Aerts, L. Dubois, M. Hochstenbag, A. M. Dingemans et al. 2010. PET imaging of hypoxia using [18F]HX4: A phase I trial. *Eur J Nucl Med Mol Imaging* 37 (9):1663–1668.

Vanderhoek, M., M. B. Juckett, S. B. Perlman, R. J. Nickles, and R. Jeraj. 2011. Early assessment of treatment response in patients with AML using [(18)F]FLT PET imaging. *Leuk Res* 35 (3):310–3106.

Vanderstraeten, B., W. De Gersem, W. Duthoy, W. De Neve, and H. Thierens. 2006. Implementation of biologically conformal radiation therapy (BCRT) in an algorithmic segmentation-based inverse planning approach. *Phys Med Biol* 51 (16):N277–N286.

Vu, C. C., R. Matthews, B. Kim, D. Franceschi, T. V. Bilfinger, and W. H. Moore. 2013. Prognostic value of metabolic tumor volume and total lesion glycolysis from (1)(8)F-FDG PET/CT in patients undergoing stereotactic body radiation therapy for stage I non-small-cell lung cancer. *Nucl Med Commun* 34 (10):959–963.

Wang, H., and Y. Cao. 2013. Spatially resolved assessment of hepatic function using 99mTc-IDA SPECT. *Med Phys* 40 (9):092501.

Wang, J. Z., N. A. Mayr, D. Zhang, and W. T. Yuh. 2008. MRI for cervical cancer not only correlates with tumor hypoxia, but also predicts ultimate outcome: In regard to Lim et al. (*Int J Radiat Oncol Biol Phys* 2008;70:126–133). *Int J Radiat Oncol Biol Phys* 71 (5):1602–1603; author reply p. 1603.

Wang, P., A. Popovtzer, A. Eisbruch, and Y. Cao. 2012. An approach to identify, from DCE MRI, significant subvolumes of tumors related to outcomes in advanced head-and-neck cancer. *Med Phys* 39 (8):5277–5285.

Yamamoto, T., S. Kabus, C. Lorenz, E. Mittra, J. C. Hong, M. Chung, N. Eclov et al. 2014. Pulmonary ventilation imaging based on 4-dimensional computed tomography: Comparison with pulmonary function tests and SPECT ventilation images. *Int J Radiat Oncol Biol Phys* 90 (2):414–422.

Yin, L. S., L. Tang, G. Hamarneh, B. Gill, A. Celler, S. Shcherbinin, T. F. Fua et al. 2010. Complexity and accuracy of image registration methods in SPECT-guided radiation therapy. *Phys Med Biol* 55 (1):237–246.

Yorke, E. D., A. Jackson, K. E. Rosenzweig, S. A. Merrick, D. Gabrys, E. S. Venkatraman, C. M. Burman, S. A. Leibel, and C. C. Ling. 2002. Dose-volume factors contributing to the incidence of radiation pneumonitis in non-small-cell lung cancer patients treated with three-dimensional conformal radiation therapy. *Int J Radiat Oncol Biol Phys* 54 (2):329–339.

Yossi, S., S. Krhili, J. P. Muratet, A. L. Septans, L. Campion, and F. Denis. 2015. Early assessment of metabolic response by 18F-FDG PET during concomitant radiochemotherapy of non-small cell lung carcinoma is associated with survival: A retrospective single-center study. *Clin Nucl Med* 40 (4):e215–e221.

Yuh, W. T., N. A. Mayr, D. Jarjoura, D. Wu, J. C. Grecula, S. S. Lo, S. M. Edwards et al. 2009. Predicting control of primary tumor and survival by DCE MRI during early therapy in cervical cancer. *Invest Radiol* 44 (6):343–350.

Zahra, M. A., K. G. Hollingsworth, E. Sala, D. J. Lomas, and L. T. Tan. 2007. Dynamic contrast-enhanced MRI as a predictor of tumour response to radiotherapy. *Lancet Oncol* 8 (1):63–74.

Zahra, M. A., L. T. Tan, A. N. Priest, M. J. Graves, M. Arends, R. A. Crawford, J. D. Brenton, D. J. Lomas, and E. Sala. 2009. Semiquantitative and quantitative dynamic contrast-enhanced magnetic resonance imaging measurements predict radiation response in cervix cancer. *Int J Radiat Oncol Biol Phys* 74 (3):766–773.

Zegers, C. M., W. van Elmpt, B. Reymen, A. J. Even, E. G. Troost, M. C. Ollers, F. J. Hoebers et al. 2014. In vivo quantification of hypoxic and metabolic status of NSCLC tumors using [18F] HX4 and [18F]FDG-PET/CT imaging. *Clin Cancer Res* 20 (24):6389–6397.

Zegers, C. M., W. van Elmpt, K. Szardenings, H. Kolb, A. Waxman, R. M. Subramaniam, D. H. Moon et al. 2015. Repeatability of hypoxia PET imaging using [F]HX4 in lung and head and neck cancer patients: A prospective multicenter trial. *Eur J Nucl Med Mol Imaging* 42:1840–1849.

Zegers, C. M., W. van Elmpt, R. Wierts, B. Reymen, H. Sharifi, M. C. Ollers, F. Hoebers et al. 2013. Hypoxia imaging with [(1)(8)F]HX4 PET in NSCLC patients: Defining optimal imaging parameters. *Radiother Oncol* 109 (1):58–64.

Zhang, J., J. Ma, S. Zhou, J. L. Hubbs, T. Z. Wong, R. J. Folz, E. S. Evans, R. J. Jaszczak, R. Clough, and L. B. Marks. 2010. Radiation-induced reductions in regional lung perfusion: 0.1-12 year data from a prospective clinical study. *Int J Radiat Oncol Biol Phys* 76 (2):425–432.

Zuckerman, E., G. Slobodin, E. Sabo, D. Yeshurun, J. E. Naschitz, and D. Groshar. 2003. Quantitative liver-spleen scan using single photon emission computerized tomography (SPECT) for assessment of hepatic function in cirrhotic patients. *J Hepatol* 39 (3):326–332.

CHAPTER 4

Multimodality Imaging for Planning and Assessment in Radiation Therapy

Matthias Guckenberger, Geoffrey Hugo, and Elisabeth Weiss

CONTENTS

4.1 MULTIMODALITY IMAGING IN RADIOTHERAPY: CURRENT STATUS

Modern radiotherapy planning and delivery provide highly conformal radiation treatment with focused targeting of the tumor and avoidance of sensitive normal tissue structures. High-quality three-dimensional (3D) and, more and more often, four-dimensional (4D) imaging data are prerequisites for spatially accurate radiation treatment. Compared to diagnostic radiology, imaging in radiotherapy has to meet particular requirements:

1. As images will be used for dose calculation, geometrical accuracy of the displayed patient anatomy is of high importance for spatially correct dose distributions.

2. For geometrical reproducibility during treatment, imaging the patient in the same position as the one used for treatment planning is necessary for precise dose delivery during each treatment fraction.

3. Short image acquisition times reduce patient discomfort associated with specific patient positioning devices and decrease associated patient shifts and displacements. In addition, short imaging times minimize physiological organ motion during the imaging session, such as bowel movements, bladder filling, and respiratory motion, which can result in imaging artifacts and compromise the interpretation of images.

Computed tomography (CT) fulfills these needs and has served for many years as the imaging workhorse in radiotherapy. Dedicated CT scanners in radiation oncology departments provide the specific information required for the radiation planning process. Radiotherapy-specific requirements are laser alignment systems for virtual simulation, big-bore systems for fitting patient immobilization devices and respiration-correlated imaging. In contrast, imaging dose-reduction technologies are of lesser relevance in radiotherapy planning compared to diagnostic imaging. CT images display the geometrically accurate 3D information of the full axial circumference of the patient's anatomy as needed for external beam radiotherapy planning. Radiation dose calculation is based on electron density information provided by CT images. CT images show the macroscopic structure of the tumor, the tumor volume, and its position relative to critical organs, and are therefore used for delineation of therapy-relevant structures by the physician during treatment planning, a process that is typically supported by diagnostic quality images. Iodinated contrast can be applied during CT acquisition to enhance soft-tissue contrast and better demarcate the tumor extent over background, for example, in brain and liver, or relative to vascular structures, such as in the neck and mediastinum. In addition to the planning situation, CT images or cone-beam CT images may also be obtained routinely before and as indicated during each treatment session for image guidance. These on-treatment images are acquired in the treatment position to assess patient setup and to check on the correct position of the tumor and normal tissue relative to the reference planning CT image. Depending on the therapeutic setting, variations in tumor position or tumor volume detected during treatment could trigger replanning and adaptation of the treatment plan to the new geometrical information. During follow-up visits after the end of therapy, diagnostic CT imaging is commonly used to assess treatment outcome based on established assessment criteria (Eisenhauer et al., 2009).

While CT imaging clearly provides important information for the radiotherapy processes, lack of soft-tissue contrast associated with limited diagnostic sensitivity and specificity results in uncertainties concerning the identification of the tumor and tumor extent. As expected, tumor delineation has been found to cause one of the largest systematic errors in radiotherapy (Arnesen et al., 2014; Kirisits et al., 2014). In addition, CT image information is essentially limited to macroscopic geometrical information of the patient's anatomy, and it lacks features that describe normal tissue function or tumor biology, predict tumor response during therapy, and allow a prognosis of the patient's outcome.

For this reason, additional information from magnetic resonance imaging (MRI) and 18-fluorodeoxyglucose positron emission tomography (FDG PET) have been integrated into the radiotherapy workflow in recent years. Morphological MRI has better soft-tissue contrast than CT and is associated with excellent diagnostic sensitivity and specificity for many tumors such as brain tumors, prostate cancer, cervix cancer, and others (Hricak, 2005; Jansen et al., 2000; Chang et al., 2013). The added benefit of superior tumor visibility is also used for tumor delineation in many tumor sites. While there are no significant reductions in contouring uncertainty between multiple observers using MRI, the integration of MRI in the tumor delineation process provides higher confidence in the target definition during planning and reassessment of the target during treatment (Weltens et al., 2001; Wu et al., 2005). So far, MRI has been integrated most successfully into the radiotherapy planning of brain tumors. Compared to CT, MRI provides detailed information about tumor volume, location, and functional characteristics, and is associated with only limited distortions and minimal positional variability, which allows straightforward image fusion with CT.

FDG PET imaging is the standard functional imaging method to investigate glucose metabolism in tumors using radioactively labeled glucose molecules and has been established for characterization and staging of many tumors, such as lung, head and neck, and pelvic tumors (Cuaron et al., 2013; Mirpour et al., 2013). Although PET has inferior image resolution compared to CT and MRI, the displayed metabolic information is now routinely used for tumor delineation and response assessment during and after radiotherapy in a variety of tumors.

The use of both morphological MR and FDG PET has become standard practice for treatment planning in many radiation oncology centers. Integration of MR and FDG PET in the radiotherapy process has been accompanied by multiple modifications of a CT-only workflow. This includes, among others, training of all members of the radiotherapy team with regard to the strengths and pitfalls of multimodality imaging, including image interpretation, delineation uncertainties and registration errors. Use of multimodality imaging necessitates the implementation of new procedural protocols and other safety and quality measures to account for specific challenges, such as correction for MRI distortions or determination of standard uptake value (SUV) thresholds for contouring. In addition, new software tools enable image registration of the planning CT with the diagnostic MR and PET images that are often acquired at different time points and in different patient positions. While MR and FDG PET complement the radiotherapy planning and delivery process in several tumor entities, CT is still used as the standard for dose calculation.

At present, a strong focus of research and development in radiotherapy aims to modulate radiation treatment further to provide individually tailored treatment plans depending on identified patient, tumor, and normal tissue features. Various imaging methods are currently investigated for their impact on individualized radiotherapy planning and response assessment before, during, and after therapy. Among these are F-18 39-deoxy-39-fluorothymidine (FLT) PET (proliferation), dynamic contrast-enhanced (DCE) MRI (perfusion, vascular density, vessel permeability, monitoring of antiangiogenic therapies), diffusion-weighted (DW) MRI (water diffusion, cellularity), blood oxygen level–dependent (BOLD) MRI, magnetic resonance spectroscopy (MRS; tumor metabolism, pH, hypoxia, and others), single-photon emission CT (SPECT; for various applications, e.g., perfusion), and others. Some biomarkers derived from these imaging tests have shown initial promise for characterizing the individual tumor prior to and/or early during therapy. For biomarkers to be useful for individualized therapy modification, they must be specific to the parameter measured. Variations in biomarkers prior to therapy or as a result of therapy must be quantifiable, and reproducibly measured and correlated with the patient outcome parameters such as tumor control and survival. In some instances, the combined information from multiple imaging biomarkers, for example, multiparametric imaging of prostate cancer, improves the assessment of the patient's individual risk profile and direct therapeutic options. Based on validated biomarkers, individual radiation dose levels could be optimized at the appropriate time points, for example, via radiation boost delivery or dose painting for radioresistant subvolumes, or dose reduction in tumors that respond well. Individualized dose prescriptions are currently under investigation in prospective trials, mostly based on FDG PET, and are subjects of ongoing research for other biomarkers.

In addition to characterizing the tumor itself, current investigations also examine the ability to assess the functionality of normal tissue before and during therapy. For example, to test the effect of radiotherapy-related lung changes such as pneumonitis and decrease in ventilated volume, 4D ventilation CT is currently evaluated relative to hyperpolarized MRI, 68-gallium PET and 99m-technetium SPECT (Kipritidis et al., 2014; Siva et al., 2014). If validated, 4D ventilation CT could be used to adjust treatment fields through adaptive sparing of healthy lung during radiotherapy of the chest. Another area of interest, one that is highly relevant for radiotherapy planning, is the measurement of physiological respiration-related motion of patient anatomy. Whereas in current routine, respiration-correlated (4D) CTs are used successfully to assess tumor motion in the chest due to the high contrast differences between tumor and lung, fast MR imaging sequences are presently being investigated to measure intraabdominal tumor motion such as tumors in the liver and pancreas (4D MRI), where poor soft-tissue contrast limits the application of CT (Yang et al., 2014).

With the wide spectrum of MRI applications in radiotherapy, ranging from imaging tumor morphology and assessing tumor motion, to analyzing tumor and normal tissue function and even measuring radiation dose, multiple efforts have been undertaken to bring dedicated MRI scanners to the radiation oncology departments themselves. Integrated MR simulators and MR-linacs or MRI-cobalt machines are being developed and installed that can be used during patient simulation for treatment planning and for soft-tissue-based image guidance and motion assessment during treatment delivery (Ménard and van der Heide 2014).

In this chapter, multimodality imaging in radiotherapy is described in more detail using two specific tumor sites as examples: FDG PET for lung cancer and MRI for cervix cancer. These examples demonstrate the current use of multimodality imaging for treatment planning, response assessment, and adaptive treatment opportunities. Technical aspects of integrating multimodality imaging into the radiotherapy workflow are addressed, and future directions of multimodality imaging are explored.

4.2 MULTIMODALITY IMAGE REGISTRATION, FUSION, AND SEGMENTATION

4.2.1 Basics of Image Registration and Fusion

To be of use for radiotherapy applications, images must first be brought into spatial alignment with each other. The process of determining such spatial alignment is called *image registration*. In general, registration proceeds by identifying corresponding points, features, or regions between images and then solving for a transformation that would map the locations of these regions from one image to another based on this information.

This process can be performed manually or it can be automated through a computer algorithm. A wide variety of image registration algorithms are available (Hill et al., 2001; Sotiras et al., 2013), but many automated algorithms have a common form in that they set up the registration problem as a numerical optimization problem. This general algorithm

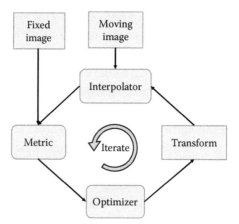

FIGURE 4.1 Typical form of a general image registration algorithm setup as an optimization problem. See the text for a detailed description of the components.

to register two images is shown in Figure 4.1 (Johnson et al., 2015). Typically, one image is used as the reference image, which is denoted in this figure as the fixed image. This image remains unchanged throughout the registration process. The other image is called the moving image; it is the image transformed to match the fixed image. A transformation is a spatial mapping that converts locations in one image space to the other. There are numerous types of transformations, although these can be generalized into rigid (translations and/or rotations only), affine, and nonrigid or deformable. Deformable transformations are often parameterized by a set of basic functions such as radial basic functions or b-splines. This parameterization helps to keep the transform smooth and to reduce the complexity of the optimization problem by representing the transform with fewer variables. When the transformation is applied to the moving image, often an interpolator is needed to apply the transformation because it is unlikely that the transformation maps voxel centers in the moving image directly to the voxel centers in the fixed image. Once the moving image has been transformed into the space of the fixed image, a metric is applied to compare the two images. The metric can be composed of a variety of image similarity (or dissimilarity) measures as well as penalty measures. The image similarity term determines how closely the transformed moving and fixed images are aligned, while the penalty term can be used to help prevent unrealistic transformations such as folding or tearing of space.

For multimodality image registration, the selection of an appropriate metric is critical. The metric provides a single fitness value. The goal of the optimizer is to adjust the transformation iteratively in order to either minimize or maximize the metric, depending on the formulation of the metric. Standard optimization algorithms are generally based on gradient descent approaches. To prevent the optimizer from descending to a local minimum (or ascending to a local maximum), registration algorithms often employ a hierarchical, or multiresolution, approach. First, either the images and/or the transformation is smoothly downsampled to a coarse resolution. The optimizer proceeds until a solution is found, then the images are upsampled and registration is performed at this higher resolution. The upsampling and registration process can be done for any number of resolution levels. The multiresolution approach tends to emphasize large, clear features and large spatial differences in the lower resolution levels, and more appropriately deals with fine, subtle features and spatial differences in the higher resolution levels.

Pairwise registration denotes the process when image registration is performed on a pair of images; groupwise registration denotes that a group of more than two images are simultaneously registered. Alternatively, a group of images could be registered separately to a single reference image through the pairwise approach.

Once aligned, two or more images can be overlaid and visualized simultaneously in a process known as *image fusion*. A variety of techniques for visualizing two or more images simultaneously are available. Some of the more common strategies for the fusion of multimodality images are discussed below.

4.2.2 Applications of Registration and Fusion and Task Requirements

Registration is used for a wide variety of tasks in medical image analysis, including orienting patient images into a common coordinate system for population comparison studies, building models of physiological processes such as respiration, and fusion of functional and morphological imaging to localize physiological signatures to anatomy. Some of these applications, for example, respiration motion models, are also used in radiotherapy. The use of multimodality imaging can be generally divided into three major tasks: segmentation to define the target and critical region, response or change detection, and plan adaptation.

4.2.2.1 Target and Critical Region Definition

The goal for structure definition is to align anatomical regions between two or more images of different modality so that disparate information (visible on one image but not another) can be use simultaneously by the physician to define the radiotherapy target and/or critical normal tissue regions. *Segmentation* is the process by which specific regions of an image are delineated or labeled. Computer-assisted segmentation in radiation therapy is still an emerging area, although approaches using an atlas of prelabeled images registered to the image to be segmented are available commercially, mainly for well-defined normal tissue organs. For target definition, however, manual segmentation remains the standard due to the often subtle and ill-defined appearance of the target challenging rule-based computer segmentation.

For non-small-cell lung cancer, 2-deoxy-2-(18F)fluoro-D-glucose positron emission tomography (FDG PET) is widely used to supplement CT-based target definition (Nestle et al., 1999). While lung tumors have high contrast on CT relative to normal lung parenchyma, their appearance closely resembles that of other normal tissue such as the soft tissue of the hilum and mediastinum, fluid in pleural effusion, and abnormal parenchyma such as collapsed lung (atelectasis and consolidation). Thus, when tumors abut such regions, defining the tumor boundary can be challenging. FDG PET provides high contrast of the active cancerous regions relative to all of these benign tissues so PET images can assist in boundary definition. FDG PET can also be used to identify pathologically involved lymph nodes on the CT image.

Another example of multimodality imaging for region definition in lung cancer is the use of PET (Vidal Melo et al., 2003; De Ruysscher et al., 2009), SPECT (Marks et al., 1993), or CT-based ventilation imaging (Guerrero et al., 2006; Reinhardt et al., 2008; Ding et al., 2010) to identify functional regions of the lung parenchyma for avoidance during treatment planning. Such ventilation images must be registered to a planning CT image so that radiation dose can be calculated in the identified functional regions. Dose response models (which are still in development) for functional and nonfunctional lung are being investigated in an effort to generate better predictors for lung toxicity after radiation therapy.

4.2.2.2 Response Detection

In assessing early (i.e., during the first few fractions or weeks of therapy) or late (i.e., post-treatment) response to radiation therapy, a common strategy is to register follow-up images to a baseline (pretreatment) image in order to measure the response globally and to identify any potential radioresistive subvolumes of the original target. Follow-up to baseline registration is generally monomodal, where the same imaging modality is used for both imaging sessions. However, in assessing the spatial location of responding subvolumes relative to the dose distribution, the baseline-follow-up pair must be registered to the planning image to enable fusion of the dose distribution to the response image.

In lung cancer, FDG PET has been investigated as a potential measure of early response (van Baardwijk et al., 2007). Most investigators have used a rigid and highly manual registration process to compensate for the challenges introduced by the shrinkage of the lung tumor (Aerts et al., 2008; Ding et al., 2010). In cervix cancer, a number of modalities, including morphological MRI (Mayr et al., 1996), DW MRI (Liu et al., 2009), DCE MRI (Zahra et al., 2007), and PET (Kidd et al., 2007), have been used to assess response to therapy. (See section 4.3.2.1 and 4.4.2.1 for more detail.)

4.2.2.3 Plan Adaptation

Adaptive radiotherapy is the process of feeding imaging information acquired during treatment back into the planning process to adjust the treatment plan in response to geometrical, anatomical, or functional changes in the patient. Adaptive radiotherapy is often implemented using a set of action levels, or tolerances, on some parameters that can be evaluated and then used to initiate plan adaptation if exceeded. A reasonable set of parameters includes the cumulative doses received by the targets and critical structures. To calculate cumulative dose, "delivered" dose must be calculated on images acquired during treatment. The during-treatment images are then registered to a reference image (often the planning CT) to accumulate the delivered doses. This registration process can involve multimodality images if localization images such as cone-beam CT or megavoltage CT are used to calculate the delivered doses, and therefore these images must be registered to the planning CT. Adaptive radiotherapy has been applied for non-small-cell lung cancer (NSCLC) using these localization CT images (Balik et al., 2013), simulation CT images

(Guckenberger et al., 2010; Weiss et al., 2013), and also using functional images such as FDG PET (Feng et al., 2009; De Ruysscher et al., 2012).

4.2.3 Challenges and Strategies for Multimodality Registration

4.2.3.1 Dealing with Large Deformations and Pose Changes

Applying image registration to multimodality imaging for the tasks described above has many challenges. Several of these challenges are also present for monomodal image registration. One of the most common challenges involves large deformations that can occur when trying to register an image acquired on a diagnostic scanner to a planning CT image acquired in the treatment position. Because immobilization is rarely used for diagnostic imaging, the patient can be in a dramatically different position in the diagnostic image compared to the planning image (Figure 4.2). A conceptually simple solution is to image the patient in the treatment position for all images that are to be registered and fused for radiotherapy applications. However, this is not always a practical solution. For example, with the use of MRI, many of the existing immobilization systems are not MR-compatible. In addition, the immobilization equipment may not physically fit into the scanner. Finally, this solution is not practical for registration of diagnostic images already acquired or acquired at an outside institution.

An alternative solution is to employ a deformable image registration algorithm. As mentioned above, deformable registration is based on a transformation that can model nonrigid changes between the images to be registered. When dealing with large deformations typical in registering diagnostic to radiotherapy images, one should take care to refine the registration region to only the region that is needed. This strategy can be implemented by selecting a region of interest for the registration or by cropping one or both images.

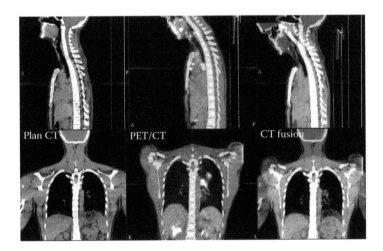

FIGURE 4.2 Issues with registration of diagnostic and radiotherapy planning multimodality imaging. Left column: Planning CT in the treatment position, with arms down. Middle column: Diagnostic PET/CT, with head on a rest and arms up. Right column: Rigid fusion of diagnostic CT and planning CT. Deformable registration may be challenging, so rigid registration near the region of interest (chest) was performed.

Otherwise, large deformations outside the useful area can sometimes influence the registration in the important regions.

4.2.3.2 Images Acquired at Different Times

Large deformations can also occur when attempting to register images acquired at different times. Most deformable registration algorithms are designed on the assumption that all locations in one image have a corresponding location in the other. Mathematically, this assumption means that the two images can be transformed to each other through a continuous deformation that preserves topology. In other words, the deformation may stretch, bend, or warp one image, but it does not require adding or removing tissue or other objects from one image nor does it involve tearing or merging of space in order to generate an accurate match. If this assumption is violated—by tumor growth/shrinkage, weight gain/loss, or surgery, for example—care must be taken to ensure such changes are handled by the algorithm appropriately. For the purposes of this chapter, we'll term these issues as the *correspondence problems*.

In cervix cancer radiotherapy, many of these challenges are commonly present where the corresponding problems occur for the image pairs to be registered. Cervical tumors are highly responsive to radiation: They often shrink dramatically in response to therapy. Pelvic organs such as the bladder and rectum can also have large variations due to filling and emptying. While the organs themselves have little tissue gain or loss, changes in their contents introduce the correspondence problems into the image registration task. Sliding, looping, and other complex changes in the pelvic organs—for example, loops of bowel present in one image and not in the other—can also introduce challenges to the registration process. Finally, the registration of images acquired for brachytherapy (BT) planning to each other or to an image acquired during external beam radiotherapy may suffer from a major correspondence problem if the BT applicator is present in one image and not in the other.

In lung cancer radiotherapy, correspondence problems are also common. Lung tumors also demonstrate shrinkage in response to radiation during therapy, which can similarly challenge image registration. Noncancerous pathologies such as pleural effusion, partially collapsed lung such as consolidation and atelectasis, and pneumonia can appear on one image but be resolved on another. Such pathologies can dramatically alter the appearance of the normal lung parenchyma, thus introducing correspondence problems (Sonke and Belderbos, 2010; Hugo et al., 2013).

At this time, limited solutions are available clinically for solving the correspondence problems described above. For very large changes, it may be appropriate to register the images manually by, for example, manually identifying corresponding landmarks in the image pair and then applying a fitting algorithm, such as a thin plate spline, to generate the full transformation. In this circumstance, the transformation between landmarks is simply an interpolation of the transformation from nearby landmarks and therefore may not sufficiently represent complicated deformations. Similarly, some registration packages allow for user interaction to adjust areas that lack registration manually.

Registering images where the tumor has shrunk or grown is a challenging task, with solutions depending heavily on the application. If one wishes simply to transfer the tumor delineation from one image to another, it may be sufficient to use a typical topologically preserving deformable registration algorithm, which will generally try to match the visible boundaries of the tumor. For change detection, a rigid registration allows one to visualize intensity changes, while a deformable registration allows for volume change to be measured (Bral et al., 2009; Brink et al., 2014). For dose accumulation, one must consider the potential loss of tumor mass because this may have an impact on the accumulation of dose (Siebers and Zhong, 2008).

In cervix cancer, registration of images containing BT applicators can be difficult, as described above. Some solutions to this problem include segmentation of the applicator to exclude it from the registration (Zhen et al., 2015) and registering both images to an intermediate image such as an atlas or template (Christensen et al., 2001). However, these works serve as proof of principle and represent an active area of research.

4.2.3.3 Challenges and Strategies for Appearance Changes

In addition to the above general challenges for all imaging modalities, unique challenges arise specifically for multimodality image registration. Certain image modalities may have different appearance for the same tissue under different imaging conditions. Two examples are MRI and cone-beam CT. In MRI, surface coil location can introduce intensity variations across the image due to the bias field. Depending on the accuracy in relocating the receiver coils relative to the patient anatomy, the intensity variation may not be repeatable from one scan to another in the same patient. Such nonuniformity in intensity can often be removed, however, with postprocessing (Tustison et al., 2010).

Cone-beam CT used for radiotherapy localization imaging is prone to artifact due to lag and ghosting in the detection, and cupping or capping due to beam hardening and scatter. Some of these artifacts may vary in location and intensity depending on the patient position and imaging technique, and may therefore change tissue appearance in the same patient from one scan to the next. Calibration of the intensity to a known reference using a phantom, postprocessing scatter and beam-hardening corrections or using modifications to the imaging systems to reduce scatter are viable strategies to remove nonuniformity and reduce artifacts that might otherwise influence registration accuracy (Rührnschopf and Klingenbeck, 2011).

4.2.3.4 Metrics for Multimodality Image Registration

When registering images of different modality to each other, some features or information are present in one image and not in another. Regions that lack correspondence cannot be used in the image registration process itself because registration requires correspondence of regions or features present in both images in the pair. Many traditional registration metrics—such as sum of square differences in image intensity and cross correlation—rely on a linear mapping of intensity between images. However, such a mapping is not always present when registering images of differing modality. For example, bone appears bright on CT and air dark, whereas both air and bone can appear dark on conventional T1- and T2-weighted MRI. These linear metrics should be avoided for multimodality image registration. Several metrics are available to handle such nonlinear intensity mappings between multimodality images, including mutual information (Wells et al. 1996), correlation ratio (Roche et al. 2001), and others.

4.2.3.5 Registration and Fusion of Functional and Morphological Images
Functional images, by definition, do not contain information on anatomy, so registration of functional images directly to morphological images can be challenging. Therefore, it is not recommended to register a functional image to an anatomical image directly because it can result in an artificial and incorrect transformation. One common and successful solution to this problem is the use of a "companion" image. The archetypical example here is PET/CT. The PET image is the functional image that is intrinsically registered to the CT image through integration in the scanner. To register the PET/CT to, for example, a planning CT, one would first register the CT part of the PET/CT to the planning CT using a monomodal registration algorithm, then simply apply the transform to the PET image to align it to the planning CT. Such a strategy can also be employed for functional MRI (fMRI) such as DW MRI or dynamic contrast enhanced MRI. In this case, the fMR image is acquired in the same study with an appropriate morphological MRI such as a T1- or T2-weighted scan. This T1- or T2-weighted companion image can then be registered to, for example, a planning CT using a multimodality image registration algorithm, and the resulting transformation can be applied to the fMR image.

4.3 MULTIMODALITY-BASED TREATMENT PLANNING: CLINICAL APPLICATION IN LUNG CANCER

4.3.1 FDG PET Imaging for Advanced Radiotherapy Treatment of Non-Small-Cell Lung Cancer

4.3.1.1 FDG PET in Non-Small-Cell Lung Cancer
The prognosis of patients diagnosed with NSCLC is highly stage dependent. Radiotherapy, alone or as part of a multimodal concept, plays an important role in the curative and palliative treatment at all stages.

In general, overall survival is favorable when the tumor is detected at an early stage without lymph node metastases. Standard treatment has been surgical lobectomy and mediastinal and hilar lymph node dissection. Radiotherapy has been used only in patients who are medically inoperable because of comorbidities, mostly due to lung and heart diseases. Despite radiotherapy being delivered with curative intent, overall survival has been poor as a result of (1) death due to comorbidities and/or (2) inability of conventional radiotherapy to control the tumor locally despite the administering of treatment at an early stage and small cancer volumes. This has changed dramatically in the last few years with the establishment of stereotactic-body radiotherapy (SBRT), where multiple advanced radiotherapy technologies are combined for accurate and high-dose irradiation. SBRT is now the standard of care in medically inoperable patients and achieves improved overall survival (OS) compared to conventional radiotherapy (Vansteenkiste et al., 2013). Today, SBRT even challenges the results of surgery and might become a treatment alternative for operable patients in the future (Chang et al., 2015). Advanced PET imaging plays an essential role in many key steps of SBRT, as described in more detail below.

In strong contrast, OS is poor when NSCLC is detected at a locally advanced stage with the presence of lymph node metastases. This poor prognosis remains true despite aggressive multimodality treatment using surgery, radiotherapy, and chemotherapy (Van Meerbeeck et al., 2007; Albain et al., 2009). Consequently, combined radiotherapy and chemotherapy without surgical resection is recommended as the primary treatment option in patients with clinical N2 and N3 NSCLC. The observation that local tumor progression frequently persists for patients treated with radiochemotherapy only compared to trimodality treatment indicates the need of a radiotherapy component with increased intensity (Albain et al., 2009). However, a recent randomized phase III trial testing the high-dose (74 Gy) compared to low-dose (60 Gy) radiation component (Bradley et al., 2015) showed that local tumor control and overall survival were worse for the high-dose arm. This "inverse" result evoked intense discussions, leading to the conclusion that dose-escalated radiotherapy cannot be recommended when conventional radiotherapy technologies are used. It also brings into question whether advanced PET imaging might improve the therapeutic ratio in locally advanced NSCLC.

4.3.1.2 FDG PET for Staging in Non-Small-Cell Lung Cancer

Selection of the appropriate patients for any radical local treatment is essential, whether or not patients are treated with surgery, radiotherapy, or multimodal approaches. FDG PET has become the standard of care for staging of nodal and distant metastases based on international guidelines. In a prospective study of mostly locally advanced stage NSCLC patients, the initial staging was performed using CT imaging followed by a second staging procedure using FDG PET imaging. The FDG PET findings changed the treatment intent from curative to palliative in one-quarter of the patients because of disease upstaging (Kalff et al., 2001). A similar study has been performed in a patient cohort referred for radical radiotherapy (Mac Manus et al., 2013). Seventy-six patients with mostly cN+ were included after CT-based staging, only. After additional FDG PET staging, a radical radiochemotherapy was performed in only 66% of the patients. Treatment intent changed to palliative in 34% of the patients with the detection of distant metastases and more extensive nodal disease. Overall survival was 35.6% and 4.1% at four years after treatment with curative and palliative intent, respectively, showing the importance of using FDG PET staging to select the high-risk population accurately.

It is essential to perform timely FDG PET staging before the start of radical treatment. Staging should be repeated if the interval between the initial staging and start of treatment becomes too long due to the risk of tumor progression. Everitt et al. (2010) evaluated staging based on two sequential FDG PET/CT images in 82 patients acquired with a median interval of 24 days. Interscan disease progression (TNM stage) was detected in 11 (39%) patients, and the treatment intent changed from curative to palliative in 8 cases (29%) because of detection of newly developed distant metastases, indicating the importance of timely and repeat FDG PET staging if the staging interval becomes longer than 4 weeks. In addition, the average SUV increased by 16% and maybe prognostic for worse overall survival.

Nodal staging is of fundamental importance for the decision of the appropriate treatment strategy: curative versus palliative, surgical versus nonsurgical, conventionally fractionated radiotherapy versus stereotactic body radiotherapy. Accurate nodal staging has substantial influence on the target volume definition in radiotherapy with curative intent. It is well documented that nodal staging using CT imaging has low sensitivity and specificity. This is especially true in patients with cN0 disease. D'Cunha et al. (2005) reported the results of the CALGB 9761 study of 502 patients with clinical stage I using CT staging. After surgical resection and mediastinal lymph node dissection, 14% of the patients had pathologic stage 2 disease and another 13.5% had stage 3 disease.

In patients with clinical stage IA disease, two studies evaluated the accuracy of FDG PET nodal staging. Park et al. (2010) reported a retrospective study of 147 patients with clinical stage IA disease based on FDG PET staging. After systematic lymph node dissection, 14.3% of patients had occult nodal (N1 or N2) metastasis. Total N1 and N2 involvement was 9.5% and 4.8%, respectively. Multivariate analysis demonstrated that a primary tumor SUV_{max} greater than 7.3 was an independent predictor of occult nodal metastasis. Another retrospective study by Stiles et al. (2009) evaluated 266 patients with stage IA based on CT and FDG PET staging. After mediastinal lymph node dissection, N1 and N2 disease was detected in 6.8% and 4.9% of the patients, respectively (Stiles et al., 2009). Tumor size greater than 2 cm and FDG PET positivity were risk factors for understaging using FDG PET.

Overall mediastinal staging accuracy has been evaluated in an extensive number of studies. Results of one meta-analysis support the superiority of FDG PET staging compared with CT staging. For CT imaging alone, the median sensitivity and specificity of mediastinal staging were 61% and 79%, respectively. For FDG PET staging, median sensitivity and specificity were 85% and 90%, respectively. However, specificity of FDG PET staging was decreased when CT showed enlarged lymph nodes (Gould et al., 2003). Consequently, all positive nodal findings in FDG PET need to be confirmed pathologically, using endobronchial ultrasound (EBUS) or mediastinoscopy.

4.3.1.3 Consequences of FDG PET Imaging for Target Definition and Characterization

In patients with stage I NSCLC based on FDG PET staging, SBRT only targets the primary tumor, and no elective irradiation of the hilar or mediastinal lymph node regions is performed. The omission of elective nodal irradiation combined with other high-precision technologies (respiration correlated imaging, image guidance, intrafractional motion management) allows for irradiation with escalated and hypofractionated doses. Biological equivalent doses (BEDs) greater than 100 Gy are delivered in 1–8 fractions. As a consequence, local tumor control is significantly improved compared to conventionally fractionated radiotherapy and reaches more than 90% in the majority of the studies, values similar to the best surgical results. Simultaneously, regional failures are rare after FDG PET staging despite omission of elective nodal irradiation: Senthi et al. (2012) reported a 7.8% regional failure rate at 2 years in a large cohort of 676 patients. The improved local control achieved with SBRT combined with accurate nodal staging using FDG PET translates into overall survival, which is a significant improvement compared to conventional radiotherapy (Shirvani et al., 2012). Today, OS is mainly limited by comorbidities of the patients as well as systemic progression of disease.

In locally advanced NSCLC, involved field irradiation without elective nodal irradiation is currently practiced in most studies and centers, if FDG PET had been performed for staging purposes. This involved field irradiation reduces the irradiated volume, making either substantial reduction of toxicity or substantial isotoxic dose escalation possible (Belderbos et al., 2008). However, a recent systematic review reported identical rates of elective nodal failure regardless of the use of FDG PET staging: Elective nodal failures were reported in 6.3% and 6.6% of the patients with and without FDG PET/CT staging, respectively (Kepka and Socha, 2015). Therefore, the results of the PET Plan study (NCT00697333) are eagerly awaited, where the value of involved field irradiation after FDG PET staging is evaluated in a randomized fashion.

Accurate analysis of breathing-induced tumor motion and its integration into the target volume definition process are essential steps in the treatment process of lung cancer. Respiration correlated 4D-CT is the current method of choice, despite its limitations due to residual motion artifacts and short image acquisition time. FDG PET imaging with its long image acquisition time has been suggested as a tool similar to slow CT scanning to evaluate and quantify patient individual tumor motion. Although this concept appears straightforward, studies showed only a poor correlation between FDG PET and the gold standard of 4D-CT (Hanna et al., 2012). FDG PET for assessment of tumor motion amplitude and motion patterns is therefore not recommended.

Several 4D technologies have been developed to acquire and/or reconstruct respiration-correlated 4D PET images (Kruis et al., 2013). Using such technologies should allow for more precise tumor characterization and delineation even in the presence of large tumor motion.

Definition of the target volume is associated with substantial uncertainties, especially in locally advanced stage NSCLC. Using CT imaging only, lung cancer appears similar to other soft-tissue structures, such as mediastinum or atelectasis. Outlining the tumor can therefore be affected by larger uncertainties. This variability in the target contours between multiple experts in lung cancer treatment has been quantified in several studies. Steenbackers et al. (2006) evaluated 22 cases with early and locally advanced stage NSCLC. Twenty experts delineated the gross tumor volume using CT images only. One year later, delineation was repeated using coregistered FDG PET/CT images. The delineation variability between the 22 experts was reduced from 1 cm (one standard deviation) using CT imaging only to 0.4 cm after adding FDG PET imaging.

Automatic segmentation of the gross tumor volume in FDG PET images has been analyzed for many years, and multiple segmentation algorithms have been evaluated (Sridhar et al., 2014). However, none proved to be robust and reliable enough to replace manual segmentation on CT images.

4.3.1.4 Opportunities for Plan Adaptation

Whether high FDG-uptake of NSCLC is a prognostic marker is a controversial issue in the literature. If true, and if pretreatment, intratreatment, or early post-treatment FDG PET characteristics are correlated with outcome, then such FDG PET information could be used for patient stratification and subsequent treatment adaptation.

The large uncertainty in the literature is at least partially explained by differences in the methodology of outcome modeling. First, different endpoints were used for correlation with FDG PET characteristics: local tumor control, progression-free survival, and overall survival. Second, different metrics of FDG PET images were used for outcome correlation: SUV max; metabolic tumor volume; and, more recently, texture characteristics like coarseness, contrast, and busyness. Tumor volume alone is a well-known and independent prognostic marker, which could confound analysis using FDG PET as an additional prognostic marker (Reymen et al. 2013). It has also been cautioned, with some controversy, whether or how inflammatory reactions during or shortly after radio(chemo)therapy influence FDG measurements.

A meta-analysis of 21 studies by Paesmans et al. (2010) evaluated the prognostic value of pretreatment FDG uptake of the NSCLC, with an endpoint of overall survival. Patient individual data were not available, but the median SUV value of each study was used as a threshold for correlation calculations. This study reported a poor prognostic value for high SUV compared with low SUV; the overall combined hazard ratio was 2.08. However, no optimal SUV threshold was observed for the differentiation of a good and bad prognostic patient group. This may be caused by the limitations of the meta-analysis or because there were not two groups of patients but rather a continuous increase in the hazard as SUV increases.

Machtay et al. (2013) reported a large prospective study of 250 patients with locally advanced NSCLC. All patients were treated with conventionally fractionated radiotherapy and concurrent platinum-based chemotherapy without surgery. Pretreatment as well as post-treatment FDG PET characteristics were analyzed. The 2-year survival rate for the entire population was 42.5%, and pretreatment SUV_{peak} and SUV_{max} were not correlated with overall survival, which is in contrast to the meta-analysis of Paesmans et al. (2010). In contrast, SUV_{peak} and SUV_{max} in FDG PET images acquired about 14 weeks after treatment were significantly correlated with overall survival. Overall survival was worse for patients with higher residual FDG PET uptake.

Aerts et al. (2009) analyzed the spatial correlation between pre- and post-treatment SUV in 55 patients treated with chemoradiation for locally advanced NSCLC. The pretreatment and post-treatment FDG PET was acquired about 2 weeks prior to and 12 weeks after radiotherapy, respectively. Patients with residual metabolic-active areas within the tumor had a significantly worse survival compared to patients with a complete metabolic response. Most important, the location of residual metabolic-active areas within the primary tumor after therapy correlated with the location of high FDG uptake before radiotherapy. Consequently, pretreatment FDG PET imaging could be used to identify subvolumes of the primary tumor that are at risk for incomplete metabolic response.

A similar study was performed by the University of Michigan group (Kong et al., 2007). In 15 stages I–III patients NSCLC treated with a definitive dose of fractionated radiotherapy, pretreatment (2 weeks), intratreatment (after approximately 45Gy), and post-treatment (3–4 months) FDG PET/CTs were recruited. A significant correlation between metabolic tumor response during radiotherapy and metabolic tumor response 3 months post-RT was observed.

If either pretreatment or intratreatment FDG PET characteristics are correlated with local tumor control and/or survival, it would be an attractive strategy to adapt radiotherapy to these individual functional and biological tumor characteristics to overcome the negative prognostic factor of poor metabolic response. This concept of biologically adapted radiotherapy was first proposed more than 10 years ago (Ling et al., 2000) and is currently under clinical investigation.

Until today, key questions in biologically adapted radiotherapy have remained unanswered and have prevented broad adoption of biologically adaptive radiotherapy. Detailed discussion of these issues is beyond the scope of this chapter, but here are some key questions:

- Which functional imaging modality, PET-tracer, and image characteristic are most appropriate for tumor characterization?

- What are the relevant time points for image acquisition and subsequent treatment adaptation?

- What are the dynamics of functional and biological tumor characteristics, and is multiple-step adaptation required?

- Are biological characteristics representative for one patient and one tumor as a whole, or can tumors be subdivided into areas of different biology and function?

- How does one establish a correlation between biology, treatment, and outcome?

- How does one translate biological information into a robust radiotherapy treatment plan?

- How does one evaluate the outcome of such biological adaptive radiotherapy?

- How does one integrate functional and biological markers other than imaging into treatment, for example, genomic RNA expression?

Two active studies about biological adapted radiotherapy for locally advanced NSCLC will be discussed next. The University of Michigan group reported a pilot study where radiotherapy treatment was adapted during treatment after delivery of 40–50 Gy based on FDG PET/CT (Feng et al., 2009). In 14 patients with stages I–III NSCLC, the midtreatment FDG PET/CT showed a metabolic complete response in 2 cases and increased FDG uptake in the adjacent lung tissue in another 2 patients. In the remaining 10 cases, CT-morphological and FDG PET metabolic volumes decreased by 26% (range, +15% to −75%) and 44% (range, +10% to −100%), respectively. Replanning of radiotherapy based on a reduced metabolic volume was then performed in 6 patients and allowed a substantial intensification of radiotherapy of 30–102 Gy (mean 58 Gy) without increased risk of toxicity. Even this small study by Feng et al. clearly demonstrates that (1) biologically adapted radiotherapy will not be feasible in all patients, (2) not all patients will benefit from biologically adapted radiotherapy, and (3) the potential benefit of normal tissue sparing or treatment intensification is huge in some patients. This concept of biologically adapted radiotherapy using midtreatment FDG PET/CT is currently investigated in a randomized phase II trial (NCT01507428).

Another study is based on the experience of Aerts et al. (2009). A randomized phase II study is conducted in patients with inoperable stage IB to stage III NSCLC (NCT01024829) (Van Elmpt et al., 2012). The patients were randomized to conventional radiotherapy with a total dose of 66 Gy given in 24 fractions of 2.75 Gy, with an integrated boost to the primary tumor as a whole (Arm A) or with an integrated boost to the 50% SUV_{max} area of the primary tumor (of the pretreatment FDG PET scan) (Arm B). The primary endpoint of this study is progression-free survival. No results have been published so far; this study is ongoing.

4.3.2 FDG PET Imaging for Post-Therapy Assessment of Radiotherapy Effects

4.3.2.1 Follow-up Response Assessment of SBRT Using FDG PET

The majority of the patients treated with SBRT for stage I NSCLC will develop radiation-induced Grade I pneumonitis with fibrotic changes in the high-dose region. These fibrotic changes are known to change in shape, size, and volume for many years after radiotherapy. Consequently, differentiation of normal tissue response and local recurrence is difficult using morphological CT imaging only. The use of invasive procedures to verify or exclude local recurrence should be practiced with caution in the usually fragile patient population with few salvage options. FDG PET has been recommended as a key tool for differentiation between fibrotic changes and local tumor recurrence (Huang et al., 2013). SUV values greater than 5 or above pretreatment values have been suggested as highly suspicious for local recurrence. However, FDG PET is not recommended shortly after high-dose radiotherapy like SBRT because active inflammatory changes are frequently seen in the first few months after SBRT.

4.3.2.2 Potential for Normal Tissue Avoidance and Identification of Treatment Sequelae

FDG uptake due to inflammatory processes in the tumor or normal tissue close to the target volume may negatively influence tumor imaging and characterization. However, it may help to visualize, quantify, or predict inflammatory radiation-induced toxicity. Pneumonitis is the most relevant toxicity in radiotherapy of lung cancer. Its development is correlated to radiotherapy dose, for example, mean lung dose or the volume of the lung exposed to 20 Gy or more. However, this correlation is rather weak, which might be explained by interpatient variability in the radiosensitivity of the lungs. Early detection of radiation-induced lung damage might therefore allow better individualization and adaptation of radiotherapy to the differences in pulmonary radiosensitivity or the adaptation of follow-up procedure.

The Melbourne group correlated post-treatment (median 70 days) FDG uptake in the lungs with development of radiation-induced pneumonitis (Mac Manus et al., 2011).

The authors reported a significant association between the worst Radiation Therapy Oncology Group (RTOG) pneumonitis grade occurring at any time after radiotherapy and severity of FDG uptake in the lung. No association between FDG uptake and the duration of pneumonitis was observed. The authors concluded that FDG PET may be useful in the prediction, diagnosis, and therapeutic monitoring of radiation pneumonitis.

However, post-treatment prediction of pneumonitis or lung toxicity has no value in terms of modifying radiotherapy to reduce the risk of pneumonitis, for example, by redistribution of radiotherapy dose or dose deescalation. Petit et al. (2011) reported a promising study where pretreatment FDG uptake in the lungs was correlated with post-radiotherapy development of radiation-induced lung toxicity. The hypothesis was that pretreatment inflammation in the lungs makes pulmonary tissue more susceptible to radiation damage. The authors reported that the 95th percentile of the FDG uptake in the lungs, excluding the clinical tumor volume, was significantly correlated with risk of developing radiation-induced pneumonitis. Pretreatment FDG PET could therefore be used for risk stratification and selection of patients for closer follow-up.

4.3.3 Novel PET Tracers and Their Application in Lung Cancer

As discussed in detail above, FDG PET imaging in NSCLC is associated with several limitations. Novel radiotracers may overcome some of these limitations. For example, hypoxia is a well-established biomarker for increased. Several tracers (e.g., FMISO, FAZA, HX4, and Cu–ATSM) are currently under evaluation to quantity hypoxia and to integrate this information about radioresistance into the treatment process, either by modifying the radiotherapy component or by sensitizing hypoxia with systemic treatments.

4.4 MULTIMODALITY-BASED TREATMENT PLANNING: CLINICAL APPLICATION IN CERVIX CANCER

4.4.1 MRI for Combined External Beam Radiotherapy and Brachytherapy of Cervix Cancer

4.4.1.1 MRI in Cervix Cancer

Cervix cancer is the fourth most common cancer in women and represents about 8% of all cancer diagnoses in women (http://www.wcrf.org/int/cancer-facts-figures/worldwide-data). Radiotherapy combined with chemotherapy (CRT) is the standard treatment approach for patients with locally advanced stage IB to stage IVA disease. These tumors are typically confined to the pelvis, including potential involvement of paraortic lymph nodes, but they are too advanced for surgery. The survival rate depends on the extent of the disease at the time of diagnosis. The 5-year overall survival rate in the United States is 68% for all tumor stages and 57% for locally advanced disease (http://www.cancer.net/cancer-types/cervical-cancer/statistics). Despite excellent pelvic control for early-stage disease, recurrence in the pelvis involving the primary tumor area and areas of direct tumor invasion in the parametria and vagina as well as pelvic lymph nodes have been frequent in locoregionally advanced tumors. In a large single institution report, 24% of all patients had locoregional recurrences following standard CRT (Beadle et al., 2010).

Definitive radiation therapy of cervix cancer is one of the most challenging treatments in radiation oncology: Application of high radiation doses is required to achieve tumor control; however, the geometrical distribution of target volumes in the pelvis with a central tumor positioned directly between radiosensitive organs (bladder, rectum, sigmoid, and small bowel) and involved or at-risk lymph nodes neighboring the small bowel are challenging for external beam therapy. High radiation doses cannot typically be delivered by external beam radiotherapy alone and necessitate an additional BT boost. External beam treatment (EBT) covers large parts of the pelvic content, including the uterus, parametria, lymph nodes, and the cervix tumor, as well as parts of the vagina, bladder, rectum, sigmoid, and small bowel. For many years BT has been performed through intracavitary insertion of standard systems of tandem applicators combined with ovoids or rings that deliver mostly pear-shaped dose distributions determined by the specific applicator configurations. Treatment planning for radiotherapy of cervix cancer has been based on gynecological exam, ultrasound, X-ray and CT images. While all these methods have their proven value in the treatment process, only through the increasing availability of MRI has it been possible in recent years to achieve significant improvements of patient outcome in cervix cancer. With its excellent soft-tissue imaging characteristics, MRI now plays an important role in many aspects of tumor staging, planning and delivery of radiotherapy, and post-treatment response assessment and surveillance.

4.4.1.2 MRI for Staging in Cervix Cancer

MRI has been used primarily for local tumor staging to assess the extent of primary tumor disease, particularly to determine tumor diameter and infiltration of the parametria. MRI was found to be superior to both clinical exam and CT imaging in correctly assessing the local tumor situation (Hricak, 2005; Mitchell et al., 2006; see Figure 4.3). MRI is often combined with FDG PET-CT because the latter is preferred for whole body staging and abdominal and pelvic lymph node assessment compared to CT imaging (Lin et al., 2003) and MRI (Rizzo et al., 2014). Information about both lymph node involvement and local tumor extent are required at the time of tumor diagnosis to decide on the primary tumor treatment: surgical tumor resection versus primary radiochemotherapy. MRI is therefore recommended for cervix cancer staging in various guidelines (Balleguier et al., 2010).

 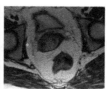 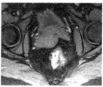

| CT | PET-CT | T2-weighted MRI | T1-weighted MRI |

FIGURE 4.3 Multimodality imaging for cervix carcinoma. Comparison of pretreatment assessment of the tumor extent on CT, PET–CT, and MRI. MRI showing right parametrial infiltration (*) that is not clearly shown on other image modalities.

4.4.1.3 Consequences of MRI for Target Definition and Characterization for External Beam Radiotherapy and Brachytherapy

4.4.1.3.1 Radiotherapy Planning   MRI has become an essential imaging method, particularly for target definition. Various pathology studies have documented excellent correspondence between the tumor extent shown on MRI images and the pathologic specimen with regard to internal cervical os invasion (De Boer et al., 2013), measurement of stromal invasion (Rizzo et al., 2014), and assessment of parametrial invasion (Lien et al., 1993). Overall, the tumor presentation on MRI is well validated for use in radiotherapy planning.

Both European (GEC ESTRO) and U.S. (ABS) BT societies issued guidelines on the use of MRI for radiotherapy planning, particularly for the use in BT (Dimopoulos et al., 2012; Haie Meder et al., 2005; Hellebust et al., 2010; Viswanathan et al., 2012a, 2012b, 2014). These guidelines address the multiple benefits but also the potential pitfalls of introducing MRI to radiotherapy and BT planning, in particular. Challenging areas include the need for image registration with CT images that are needed for dose calculation, the effect of MR image distortion on geometrical accuracy, problems with intracavitary applicator visibility, susceptibility artifacts and reconstruction errors, and the need for development of MRI-compatible titanium or plastic applicators (Kirisits et al., 2014). Some of these issues have been addressed in Section 4.2.3 of this chapter.

4.4.1.3.2 External Beam Treatment   The most clinically relevant benefit of MRI for cervix cancer planning lies in the clear visualization of the cervix tumor in multiple planes, which allows for a reliable volumetric definition of the target volume. Early studies investigated the benefit of MRI imaging to assess tumor coverage with standard fields. It was observed that pelvic box fields resulted in missing the tumor in up to 66% of the treatments (Weiss et al., 2002; Zunino et al., 1999). In particular, the standard rectum block in the lateral fields was responsible for a high rate of tumor miss (Kim et al., 2014). Based on these findings, MRI in the treatment position has become a standard imaging procedure for EBT planning (Thomas et al., 1997).

Compared to other factors influencing uncertainties of the radiotherapy process, variabilities in target definition are the largest uncertainties for both EBT and BT (Kirisits et al., 2014). In an early CT-based contouring study of primary tumor and lymph node clinical target volume (CTV) in cervix cancer, the ratio between smallest and largest contoured volumes ranged between a factor of 1.3 and 4.9 (Weiss et al., 2003). Similar studies investigating MRI-based contouring of only the primary cervix tumor and central organs, excluding lymph node areas, also found significant variations between observers. While contouring the actual tumor in the cervix appears rather reproducible, determining soft-tissue boundaries to neighboring organs and infiltration of the parametria can result in large disagreement (Wu et al., 2005). To reduce these wide range variations even with MRI, various groups have developed guidelines for contouring external beam CTV (Lim et al., 2011; Toita et al., 2011).

Another factor that strongly influences EBT precision is the large inter- and intrafraction mobility of pelvic organs that affect both target volume dose coverage and dose to sensitive normal tissue organs. Compared to CT, MRI is particularly useful to measure motion variability due to its excellent soft-tissue visualization, the absence of radiation, the availability of multiplanar imaging and fast 4D imaging. Performing repeated daily or weekly MRI scans, investigators found that the uterine fundus is rather flexible and moves more than the cervix and cervix tumor. Uterus motion is particularly dependent on the bladder filling, whereas the cervix position is influenced by rectum filling. Repeated MRI showed large mobility of the uterus in the superior-inferior direction with average position differences of 7.1 mm (SD 6.8) (Taylor and Powell, 2008). In a large meta-analysis, cervix motion was mostly in the anteroposterior direction and varied between 2.3 and 16 mm. In addition, pelvic lymph node translations of 7 to 30 mm were observed (Jadon et al., 2014). Even for intrafraction variability on repeat MRIs, large CTV position changes of 10 mm were found (Kerkhof et al., 2009). To compensate for the observed organ motion, margins of up to 4 cm in the uterine fundus area and 1.5 cm at the cervical os were recommended to encompass 90% of all motion variations (Van de Bunt et al., 2008). Such large clinical target volume-planning target volume (CTV-PTV) margins result in inclusion of large volumes of normal tissue in the target volume, causing potentially detrimental side effects that reduce the overall therapeutic ratio. Various solutions have been suggested, for example, anisotropic and tapered population-based margins with larger margins at the fundus compared to the cervix (Gordon et al., 2011), individualized margins based on the observation of large interpatient variations, repeated soft-tissue-based image guidance, and adaptation of the treatment plan to adjust for the observed variations during a treatment series (Chan et al., 2008).

The target motion variability observed on MRI has resulted in a slow implementation of intensity-modulated radiotherapy (IMRT) for cervix cancer compared to other tumors. While higher conformality and steeper dose fall-off with IMRT are desirable, uncertainties in the target position require at least daily soft-tissue-based image guidance, or even during treatment, to keep safety margins small, ensure sufficient target coverage, and keep normal tissue doses low (Kerkhof et al., 2009). Despite promising early results, the benefit of IMRT is still controversial (Gandhi et al., 2013; Hasselle et al., 2011; Wright et al., 2013) and might lie mostly in a reduction of side effects.

4.4.1.3.3 Brachytherapy   While MRI for EBT planning is still in development, the recent availability of MRI has transformed BT for cervix cancer. BT delivers a boost dose to the central cervix tumor, including its extensions to the vagina and uterus, and direct infiltration of the parametria and other organs. With the ability to contour the target in 3D, it has become clear that previous applicator-based methods delivered either too large dose volumes for small tumors or too small dose volumes for large tumors, resulting in the risk of overdosing of normal tissues and underdosing the tumor, respectively (Takenaka et al., 2012; Tanderup et al., 2010). A tumor volume of 31 cc was determined as a threshold for 2D planning still delivering sufficient target dose coverage. For larger tumors, 3D image guidance is required to deliver sufficient target dose coverage. Point A, which has been the classical reference dose point for tumor dose, was found to be an insufficient surrogate for the actual delivered tumor dose (Tanderup et al., 2010). Even in small tumors, MRI-guided BT was found to reduce the dose to the highest dose volumes (D2cc) to normal tissues by

12% to 32% (Zwahlen et al., 2009). Because of the introduction of MRI and the improved visualization of the residual tumor extent at the time of BT, new applicators with interstitial needles have been added to the existing repertoire of applicators to cover particularly lateral extensions of the tumor into the parametria, which would have been underdosed with classical applicator configurations (Haack et al., 2009). In addition, a new MRI-based nomenclature of target volumes has been introduced in parallel with new guidelines for dose reporting (GEC ESTRO). The new guidelines and MR imaging facilitate individualized treatment planning for BT, a process called image-guided adaptive BT (IGABT).

Despite the improved soft-tissue visualization with MRI, large delineation uncertainties exist that are even more relevant for BT planning than for EBT planning given the sharp dose fall-off. In some studies, CT-based contours were found to produce larger tumor volumes and better operator agreement compared to MRI (Hegazy et al., 2013). Contouring for high-risk CTVs was found to be more reliable than contouring other target structures with volumetric conformity indices between 0.58 and 0.77 (Petric et al., 2013). Resulting dose variations stemming from contouring uncertainties affect both the target and normal tissue (Arnesen et al., 2014), resulting in 10% variations of the D90 dose (Hellebust et al., 2013). Adding EBT and BT contouring uncertainties results in 5 Gy errors for the target and 2–3 Gy errors for normal tissues (Hellebust et al., 2013). To improve contouring and to reduce motion during imaging, imaging planes that are orthogonal to the applicator axis, and 3D image acquisition has been recommended (Petric et al., 2012).

One of the problems with MRI for radiotherapy planning comes from the usually limited access to MR scanners. While real-time MRI at the time of BT applicator insertion would be desirable for guidance, MRI is typically obtained once in a diagnostic radiology department after the applicator has already been inserted prior to the first BT session. Also, repeated MRI for each BT session may be of advantage in large tumors that change volume between BT sessions and for plans with high-dose exposure of normal tissue due to unfavorable location. For small residual tumors, one MRI at the time of first BT appears sufficient when CT based planning is otherwise performed (Nesvacil et al., 2013).

Tumor control rates with MRI-based image-guided adaptive BT show a clear dose effect response. A D90 of the high-risk CTV of 87 Gy or higher resulted in 4% local recurrences compared with 20% for less than 87 Gy (Dimopoulos et al. 2009a). Poor coverage of large target volumes during IGABT was observed as an important factor for impaired local control (Charra-Brunaud et al., 2012). Compared to conventional BT, IGABT with MRI combined with standard EBT reduced local recurrences from 32% to 7%, reduced treatment-related side effects, and improved 3-year overall survival from 51% to 86% (Rijkmans et al., 2014). Similar results of high local control rates greater than 90%, even for tumors larger than 5 cm, improved overall survival, and reduced toxicity were also reported by other groups (Lindegaard et al, 2013; Mazeron et al., 2013; Nomden et al., 2013; Pötter et al., 2011). Results with larger patient groups are expected from the international effects of MRI guided brachytherapy in locally advanced cervical cancer (EMBRACE) trial that investigates MRI for BT planning in a prospective observational study (https://www.embracestudy.dk/).

4.4.1.4 Opportunities for Plan Adaptation

Cervix carcinoma offers many opportunities for adaptive radiotherapy approaches with the goals of improving dose conformality, reducing normal tissue toxicity, and potentially increasing tumor dose to underdosed areas or areas of residual disease. These goals can be achieved by adjusting the dose plan based on individual patient characteristics of tumor volume regression, tumor motion, and setup reproducibility. MRI plays an important role in the development of adaptive radiotherapy because it provides 3D information about organ motion (see above) and tumor volume shrinkage. MRI-guided adaptive BT has essentially incorporated the adaptive concept where the MRI-based BT dose plan is adjusted to the individual patient's 3D imaging information obtained prior to each fraction, which might include repeated CT scans (Rey et al., 2013). For external beam radiotherapy, no routine strategies of treatment adaptation have yet been established in clinic. Inter- and intrafraction organ motion (see above) and tumor shrinkage offer various opportunities to adjust the individual plans to these parameters. Most of the tumor regression occurs during external beam radiotherapy, with large interpatient variation in regression rates (Dimopoulos et al., 2009b). Depending on the timing of BT relative to external beam therapy, initial tumor volumes are reduced often by 80% and more at the time of BT (Hatano et al., 1999; Schmid et al., 2013). Several studies found tumor volume regression and reduction of signal intensity during treatment, as well as the amount of residual tumor after therapy, to be strong predictors of local control and disease-free survival (Mayr et al., 2010; Wang et al., 2010).

Excellent visualization of the tumor during treatment using repeated MRI is the prerequisite to performing adaptive treatments and shrinking CTV margins. Various scenarios of plan adaptation have been investigated: While reduced CTV-PTV margins of only 3 mm lead to a high risk for missing the tumor in 27% of patients, automated weekly MRI-based replanning (Stewart et al., 2010) and combinations of soft-tissue guidance with offline replanning (Oh et al. 2014) were found to ensure safe target coverage. Alternatively, a margin-of-the-day approach using CT-based plan libraries (Ahmad et al., 2013) has resulted in good target coverage without compromising normal tissue sparing. With repeated MRI imaging and adaptive replanning, which include deformable registration and dose accumulation, target coverage with small margins appears possible using a dosimetrically triggered method (Lim et al., 2014). So far, none of these strategies for EBT have been established in clinical routine because of challenges in deformable dose accumulation, the speed of adaptive offline and online replanning strategies, and limited labor/personnel resources. Adaptive strategies are expected to increase the therapeutic ratio by enabling delivery of additional boost doses to residual tumors, for example, pelvic LN (Ariga et al., 2013) or residual central tumors (Assenholt et al., 2014), or by selective sparing of high-dose normal tissue volumes. One of the requirements for adaptive planning is the ability to accumulate doses deformably between external beam fractions and BT sessions. Organ mobility, contouring uncertainties, and tumor and normal tissue position and volume changes, as well as significant noncontinuous position changes of the pelvic anatomy through insertion of applicators during BT, prohibit a straightforward voxel-by-voxel dose summation. Deformable dose summation of external beam and BT is the topic of ongoing research.

4.4.1.5 Potential for Normal Tissue Avoidance

MRI allows excellent soft-tissue visualization of normal tissue organs, such as the bladder, rectum, sigmoid, and vagina, and provides reliable information on the position of normal tissues on repeated MRI images during treatment. Repeated MRI imaging also allows for improved topographical correlation between normal tissue doses and observed treatment sequelae. In general, while high-dose normal tissue volumes D2cc exhibit variations over time, D2cc volumes and observed toxicities are well correlated for the bladder and rectum (Anderson et al., 2013; Jamema et al., 2013). Based on these observations, rectum D2cc doses below 75 Gy are recommended (Georg et al., 2011). IMRT planning with repeated MRI for image guidance has resulted in improved sparing of all normal tissue structures (Kerkhof et al., 2008). Reduced normal tissue toxicity was also seen, however, in non-MRI-guided IMRT and may be due to smaller margins, an overall reduction of bladder and rectum volumes during treatment, and dedicated protocols for bladder filling and rectum emptying for planning and delivery (Gandhi et al., 2013; Hasselle et al., 2011; Kerkhof et al., 2008; Mundt et al., 2003; Simpson et al., 2012).

4.4.2 MRI for Post-Therapy Assessment of Radiotherapy Effects

4.4.2.1 Follow-up Response Assessment Using MRI

MRI has been used to determine the need for additional surgical salvage in the presence of residual tumor early after radiochemotherapy and to detect tumor recurrence during post-treatment follow-up. MRI has excellent accuracy in assessing residual tumor (Hatano et al., 1999) with high negative predictive values (Vandecasteele et al., 2012). High false positive rates were observed particularly if MRI was obtained early in post-treatment with ongoing inflammatory changes (Vincens et al., 2008). The presence of residual tumor on MRI was validated against surgery, biopsy, or continued clinical follow-up. The diagnostic ability to detect residual disease was found to be superior in MRI compared to PET, with positive and negative predictive values greater than 50% (Vandecasteele et al., 2012; see Figure 4.4).

MRI has also been valuable in identifying tumor recurrence for central recurrences in the vagina and cervix as well as for tumor recurrences in the parametria and pelvic side

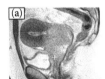 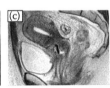 

Pretreatment MRI   Pretreatment PET-CT   Post-treatment MRI   Post-treatment PET-CT

FIGURE 4.4 Response assessment on PET-CT and MRI. Large cervix cancer with involvement of the uterine body (*) and vagina wall (+). (a) The tumor extent is particularly easy to see on T2-weighted MRI. (b) High radiotracer uptake on corresponding PET-CT with maximum SUV of 30. (c) Six months after radiochemotherapy decreased T2 signal and significant tumor volume reduction as a response to treatment. (d) Residual increased activity on PET scan (max SUV 6.5) likely as a sign of inflammatory changes versus residual tumor.

walls that are difficult to detect with clinical examination. The challenge in identifying tumor recurrence on MRI is to differentiate between tumor and post-therapy sequelae such as fibrosis, inflammation, and particularly necrosis. A longer interval between treatment and post-therapy assessment results in improved reliability to identify tumor recurrence correctly (Hricak et al., 1993). Knowledge of the topographical pattern of recurrence is helpful for determining tumor recurrence from other pathologies (Babar et al., 2007).

4.4.2.2 Identification of Treatment Sequelae in Normal Tissues

Radiotherapy with or without chemotherapy is known to cause a variety of acute and late side effects in healthy normal tissue. MRI has been investigated for its ability to visualize these changes, to determine dose and time dependence of changes, and to correlate imaging findings with clinical symptoms. Typical MRI signs of postradiation changes are increased wall thickness of the bladder and rectum, increased signal intensity on T2-weighted images due to acute inflammation, and decreased signal intensity on T1- and T2 weighted images for late fibrotic changes (Sugimura et al., 1990). In addition, soft-tissue changes of muscle, fat, and fasciae have been described. MRI has also been used to assess post-radiotherapy changes of the uterus, such as endometrial thickness and cervix length (Milgrom et al., 2013). MRI findings are positively correlated with clinical symptoms and show a clear dose dependence with doses greater than 45 Gy resulting in a marked increase in normal tissue changes (Sugimura et al., 1990).

MRI alone or combined with other tests also improves the diagnosis of bone and bone marrow changes that occur after external beam radiotherapy to the pelvis. A wide range in the rate of pelvic insufficiency fractures between 10% and nearly 50% has been described (Tokumaru et al., 2012). In addition, avascular necrosis and osteolysis are readily identified on post-radiotherapy MRIs (Kwon et al., 2008). Involution of active bone marrow into fatty bone marrow has also been diagnosed with MRI in 90% of patients 6–8 weeks after radiotherapy (Blomlie et al., 1995). Methods to test functional bone marrow sparing with IMRT, that is, reduced dose to active bone marrow, are under investigation (Liang et al., 2013). An issue with MRI for assessment of post-treatment changes pertains to its applicability across institutions. Most of the studies have been based on the experience of a single institution. Response assessment is challenging due to inconsistent imaging protocols, different scanners, and potentially disparate patient populations. More robust imaging features and techniques need to be developed to allow transfer between institutions and between scanners to help develop more quantitative imaging techniques.

4.4.3 Novel MRI Sequences and Their Application in Cervix Cancer

The availability of functional imaging modalities supports the concept of individualized treatment planning through investigation of biomarkers that are prognostic for patient outcome and predictive of therapeutic success. Functional imaging enables a more detailed tumor characterization than morphological imaging alone and may allow a personalized treatment approach through delivering an additional dose to radioresistant tumor volumes and/or a reduced dose to radiosensitive parts of the tumor. While FDG PET is a standard functional imaging modality for cervix cancer, the focus of this section will be on emerging imaging technologies. The combination of FDG PET with MRI, including fMRI may prove beneficial, however, as already indicated in this chapter.

Morphological MRI has become a standard imaging method for cervix cancer radiotherapy; however, fMRI sequences have been investigated more recently as biomarkers for radioresistivity. DW MRI is a noncontrast imaging methodology that measures water diffusion in tissue typically using apparent diffusion coefficients (ADCs) for quantification and comparison. ADC values are lower in a tumor than in normal tissue due to higher cellularity, and they increase as a result of therapy due to apoptosis and cell death, as well as inflammation and microvascular leakage (Padhani and Koh, 2011). Several small studies investigated DW MRI in radiotherapy of cervix cancer and observed that ADC change early during radiotherapy predicts tumor response on MRI and is associated with overall survival (Somoye et al., 2012). It was also observed that patients with high initial ADC values had worse outcome than patients with lower ADC, likely due to necrotic areas that are hypoxic and therefore radioresistant (Liu et al., 2009). FDG PET (SUV_{max}) and DW MRI (ADC_{min}) were found to be complementary prognostic factors of cervix cancer (Nakamura et al., 2012). DW MRI is not routinely used in radiotherapy now because of the lack of larger clinical validation of results and lack of integration into the radiotherapy workflow (Barwick et al., 2013).

Dynamic contrast-enhanced MRI (DCE MRI) is assumed to show vascular density and perfusion, which is thought to be correlated with hypoxia and radioresistance. Changes in signal intensity, indicating increasing tumor perfusion and volume change after 2–2.5 weeks of therapy, predict local tumor control and survival (Mayr et al., 2010). In a large study with repeated DCE MRI before and during therapy, functional risk volumes were generated for subvolumes with critically low DCE signal. Larger functional risk volumes predicted poor tumor control and disease-free survival before and during therapy. DCE MRI was a better predictor than anatomical volume change (Mayr et al., 2012).

New imaging methods under current investigation include blood oxygen level–dependent (BOLD) MRI (Kim et al., 2014) and MRS (Zietkowski et al., 2013). BOLD MRI assesses hypoxia, which is a common finding in cervix cancer and which has been shown to be a predictive factor for outcome. So far, only preliminary data are available to indicate the feasibility of BOLD imaging. Further study is required to confirm a relation with clinical results. Measurement of tissue metabolites, such as lipids and choline, using MRS in vivo has been challenging due to physiological organ motion in the pelvis, which limits MRS studies mostly to in vitro analysis. Further technological developments are necessary prior to determining clinical use.

4.5 SUMMARY

Multimodality imaging has been integrated into the radiotherapy process of many tumor sites, particularly during treatment planning where improved soft-tissue visualization and information on metabolic tumor activity obtained from diagnostic imaging support the delineation of the tumor target. Compared to a CT-only radiotherapy approach, multimodality imaging provides important advantages not only for radiotherapy planning but also for validation of the target position during treatment and assessment of the therapeutic response. The additional knowledge obtained through multimodality imaging of the tumor location, tumor extent, and functional information results in opportunities to increase dose conformality, reduce safety margins, and adapt the treatment to changes observed during treatment, thus promising improved tumor control and less side effects. Current research works to overcome challenges associated with multimodality imaging to provide reproducible image quality and quantitative measures, in particular for functional imaging modalities. In addition, robust registration methods are still under development to accommodate large deformations between images obtained at different times with different patient poses. Wider availability of multimodality imaging methods such as PET and MRI in radiotherapy departments is expected to improve their integration in all aspects of the treatment process and thereby deliver the information needed for the development of an individualized radiotherapy approach.

REFERENCES

Aerts, H. J. W. L., G. Bosmans, A. A. van Baardwijk, et al. 2008. Stability of 18F-deoxyglucose uptake locations within tumor during radiotherapy for NSCLC: A prospective study. *Int J Radiat Oncol Biol Phys* 71 (5): 1402–1407.

Aerts, H. J., A. A. van Baardwijk, S. F. Petit, et al. 2009. Identification of residual metabolic-active areas within individual NSCLC tumours using a pre-radiotherapy (18)fluorodeoxyglucose-PET-CT scan. *Radiother Oncol* 91 (3):386–392.

Ahmad, R., L. Bondar, P. Voet, et al. 2013. A margin-of-the-day online adaptive intensity-modulated radiotherapy strategy for cervical cancer provides superior treatment accuracy compared to clinically recommended margins: A dosimetric evaluation. *Acta Oncol* 52 (7): 1430–1436.

Albain, K. S., R. S. Swann, V. W. Rusch, et al. 2009. Radiotherapy plus chemotherapy with or without surgical resection for stage III non-small-cell lung cancer: A phase III randomised controlled trial. *Lancet* 374 (9687):379–386.

Anderson, C., G. Lowe, R. Wills, et al. 2013. Critical structure movement in cervix brachytherapy. *Radiother Oncol* 107 (1):39–45.

Ariga, T., T. Toita, G. Kasuya, et al. 2013. External beam boost irradiation for clinically positive pelvic nodes in patients with uterine cervical cancer. *J Radiat Research* 54 (4):690–696.

Arnesen, M. R., K. Bruheim, E. Malinen, and T. P. Hellebust. 2014. Spatial dosimetric sensitivity of contouring uncertainties in gynecological 3D-based brachytherapy. *Radiother Oncol* 113 (3):414–419.

Assenholt, M. S., A. Vestergaard, J. F. Kallehauge, et al. 2014. Proof of principle: Applicator-guided stereotactic IMRT boost in combination with 3D MRI-based brachytherapy in locally advanced cervical cancer. *Brachytherapy* 13(4):361–368.

Babar, S., A. Rockall, A. Goode, J. Shepherd, and R. Reznek. 2007. Magnetic resonance imaging appearances of recurrent cervical carcinoma. *Int J Gynecol Cancer* 17 (3):637–645.

Balik, S., E. Weiss, N. Jan, et al. 2013. Evaluation of 4-dimensional computed tomography to 4-dimensional cone-beam computed tomography deformable image registration for lung cancer adaptive radiation therapy. *Int J Radiat Oncol Biol Phys* 86 (2):372–379.

Balleyguier, C., E. Sala, T. Da Cunha, et al. 2010. Staging of uterine cervical cancer with MRI: Guidelines of the European Society of Urogenital Radiology. *Eur Radiol* 21 (5):1102–1110.

Barwick, T. D., A. Taylor, and A. Rockall. 2013. Functional imaging to predict tumor response in locally advanced cervical cancer. *Curr Oncol Rep* 15 (6):549–558.

Beadle, B. M., A. Jhingran, S. S. Yom, P. T. Ramirez, and P. J. Eifel. 2010. Patterns of regional recurrence after definitive radiotherapy for cervical cancer. *Int J Radiat Oncol Biol Phys* 76(5):1396–1403.

Belderbos J. S., L. Kepka, F. M. S. Kong, M. K. Martel, G. M. Videtic, and B. Jeremic. 2008. Report from the International Atomic Energy Agency (IAEA) consultants' meeting on elective nodal irradiation in lung cancer: Non-small-Cell lung cancer (NSCLC). *Int J Radiat Oncol Biol Phys* 72(2):335–342.

Blomlie, V., E. K. Rofstad, A. Skjønsberg, K. Tverå, and H. H. Lien. 1995. Female pelvic bone marrow: Serial MR imaging before, during, and after radiation therapy. *Radiology* 194(2):537–543.

Boer, P. De, J. A. Adam, M. R. Buist, et al. 2013. Role of MRI in detecting involvement of the uterine internal os in uterine cervical cancer: Systematic review of diagnostic test accuracy. *Eur J Radiol* 82(9):e422–e428.

Bradley, J. D., R. Paulus, R. Komaki, et al. 2015. Standard-dose versus high-dose conformal radiotherapy with concurrent and consolidation carboplatin plus paclitaxel with or without cetuximab for patients with stage IIIA or IIIB non-small-cell lung cancer (RTOG 0617): A randomised, two-by-two factorial phase 3 study. *Lancet Oncol* 16(2):187–199.

Bral, S., M. Duchateau, M. De Ridder, et al. 2009. Volumetric response analysis during chemoradiation as predictive tool for optimizing treatment strategy in locally advanced unresectable NSCLC. *Radiother Oncol* 91(3):438–442.

Brink, C., U. Bernchou, A. Bertelsen, O. Hansen, T. Schytte, and S. M. Bentzen. 2014. Locoregional control of non-small cell lung cancer in relation to automated early assessment of tumor regression on cone beam computed tomography. *Int J Radiat Oncol Biol Phys* 89(4):916–923.

Bunt, L. van De, I. M. Jürgenliemk-Schulz, G. De Kort, et al. 2008. Motion and deformation of the target volumes during IMRT for cervical cancer: What margins do we need? *Radiother Oncol* 88(2):233–240.

Chan, P., R. Dinniwell, M. A. Haider, et al. 2008. Inter- and intrafractional tumor and organ movement in patients with cervical cancer undergoing radiotherapy: A cinematic-MRI point-of-interest study. *Int J Radiat Oncol Biol Phys* 70(5):1507–1515.

Chang, J. H., D. L. Joon, B. T. Nguyen, et al. 2013. MRI scans significantly change target coverage decisions in radical radiotherapy for prostate cancer. *J Med Imaging Radiat Oncol* 58(2):237–243.

Chang, J. Y., S. Senan, M. A. Paul, et al. 2015. Stereotactic ablative radiotherapy versus lobectomy for operable stage I non-small-cell lung cancer: A pooled analysis of two randomised trials. *Lancet Oncol* 16(6):630–637.

Charra-Brunaud, C., V. Harter, M. Delannes, et al. 2012. Impact of 3D image-based PDR brachytherapy on outcome of patients treated for cervix carcinoma in France: Results of the French STIC Prospective Study. *Radiother Oncol* 103(3):305–313.

Christensen, G. E., B. Carlson, K. S. C. Chao, et al. 2001. Image based dose planning of intracavitary brachytherapy: Registration of serial-imaging studies using deformable anatomic templates. *Int J Radiat Oncol Biol Phys* 51(1):227–243.

Cuaron, J., M. Dunphy, and A. Rimner. 2013. Role of FDG-PET scans in staging, response assessment, and follow-up care for non-small cell lung cancer. *Front Oncol* 2:208.

D'Cunha, J., J. E. Herndon 2nd, D. L. Herzanet, et al. 2005. Poor correspondence between clinical and pathologic staging in stage 1 non-small cell lung cancer: Results from CALGB 9761, a prospective trial. *Lung Cancer* 48(2):241–246.

De Ruysscher, D., A. A. Baardwijk, J. Steevens, et al. 2012. Individualised isotoxic accelerated radiotherapy and chemotherapy are associated with improved long-term survival of patients with stage III NSCLC: A prospective population-based study. *Radiother Oncol* 102(2):228–233.

De Ruysscher, D., A. Houben, H. J. W. L. Aerts, et al. 2009. Increased 18F-deoxyglucose uptake in the lung during the first weeks of radiotherapy is correlated with subsequent radiation-induced lung toxicity (RILT): A prospective pilot study. *Radiother Oncol* 91(3):415–420.

Dimopoulos, J. C. A., S. Lang, C. Kirisits, et al. 2009a. Dose–volume histogram parameters and local tumor control in magnetic resonance image–guided cervical cancer brachytherapy. *Int J Radiat Oncol Biol Phys* 75(1):56–63.

Dimopoulos, J. C. A., G. Schirl, A. Baldinger, T. H. Helbich, and R. Pötter. 2009b. MRI assessment of cervical cancer for adaptive radiotherapy. *Strahlenther Onkol* 185(5):282–287.

Dimopoulos, J. C. A., P. Petrow, K. Tanderup, et al. 2012. Recommendations from Gynaecological (GYN) GEC-ESTRO Working Group (IV): Basic principles and parameters for MR imaging within the frame of image based adaptive cervix cancer brachytherapy. *Radiother Oncol* 103(1):113–122.

Ding, K., J. E. Bayouth, J. M. Buatti, G. E. Christensen, and J. M. Reinhardt. 2010. 4DCT-based measurement of changes in pulmonary function following a course of radiation therapy. *Medical Physics* 37(3):1261–1272.

Eisenhauer, E. A., A. P. Therasse, J. Bogaerts, et al. 2009. New response evaluation criteria in solid tumors: Revised RECIST guideline (version 1.1), *Eur J Cancer* 45:228–247.

Everitt S., A. Herschtal, J. Callahan, et al. 2010. High rates of tumor growth and disease progression detected on serial pretreatment fluorodeoxyglucose-positron emission tomography/computed tomography scans in radical radiotherapy candidates with nonsmall cell lung cancer. *Cancer* 116(21):5030–5037.

Feng, M., F.-M. M. Kong, M. Gross, S. Fernando, J. A. Hayman, and R. L. K. Ten Haken. 2009. Using fluorodeoxyglucose positron emission tomography to assess tumor volume during radiotherapy for non-dmall-cell lung cancer and its potential impact on adaptive dose escalation and normal tissue sparing. *Int J Radiat Oncol Biol Phys* 73(4):1228–1234.

Gandhi, A. K., D. N. Sharma, G. K. Rath, et al. 2013. Early clinical outcomes and toxicity of intensity modulated versus conventional pelvic radiation therapy for locally advanced cervix carcinoma: A prospective randomized study. *Int J Radiat Oncol Biol Phys* 87(3):542–548.

Georg, P., S. Lang, J. C. A. Dimopoulos, et al. 2011. Dose–volume histogram parameters and late side effects in magnetic resonance image–guided adaptive cervical cancer brachytherapy. *Int J Radiat Oncol Biol Phys* 79(2):356–362.

Gordon, J. J., E. Weiss, O. K. Abayomi, J. V. Siebers, and N. Dogan. 2011. The effect of uterine motion and uterine margins on target and normal tissue doses in intensity modulated radiation therapy of cervical cancer. *Phys Med Biol* 56(10):2887–2901.

Gould M. K., W. G. Kuschner, C. E. Rydzak, et al. 2003. Test performance of positron emission tomography and computed tomography for mediastinal staging in patients with non-small-cell lung cancer: A meta-analysis. *Ann Intern Med* 139 (11):879–892.

Guckenberger, M., J. Wilbert, A. Richter, and K. Baier. 2010. Potential of adaptive radiotherapy to escalate the radiation dose in combined radiochemotherapy for locally advanced non-small cell lung cancer. *Int J Radiat Oncol Biol Phys* 79(3):901–908.

Guerrero, T., K. Sanders, E. Castillo, et al. 2006. Dynamic ventilation imaging from four-dimensional computed tomography. *Phys Med Biol* 51(4):777–791.

Hanna G. G., J. R. van Sornsen de Koste, M. R. Dahele et al. 2012. Defining target volumes for stereotactic ablative radiotherapy of early-stage lung tumours: A comparison of three-dimensional 18F-fluorodeoxyglucose positron emission tomography and four-dimensional computed tomography. *Clin Oncol* 24(6):e71–e80.

Haack, S., S. K. Nielsen, J. C Lindegaard, J. Gelineck, and K. Tanderup. 2009. Applicator reconstruction in MRI 3D image-based dose planning of brachytherapy for cervical cancer. *Radiother Oncol* 91(2):187–193.

Haie-Meder, C., R. Pötter, E. Van Limbergen, et al. 2005. Recommendations from Gynaecological (GYN) GEC-ESTRO Working Group☆ (I): Concepts and terms in 3D image based 3D treatment planning in cervix cancer brachytherapy with emphasis on MRI assessment of GTV and CTV. *Radiother Oncology* 74 (3):235–245.

Hasselle, M. D., B. S. Rose, J. D. Kochanski, et al. 2011. Clinical outcomes of intensity-modulated pelvic radiation therapy for carcinoma of the cervix. *Int J Radiat Oncol Biol Phys* 80 (5):1436–1445.

Hatano, K., Y. Sekiya, H. Araki, et al. 1999. Evaluation of the therapeutic effect of radiotherapy on cervical cancer using magnetic resonance imaging. *Int J Radiat Oncol Biol Phys* 45 (3):639–644.

Hegazy, N., R. Pötter, C. Kirisits, et al. 2013. High-risk clinical target volume delineation in CT-guided cervical cancer brachytherapy: Impact of information from FIGO stage with or without systematic inclusion of 3D documentation of clinical gynecological examination. *Acta Oncol* 52 (7):1345–1352.

Hellebust, T. P., C. Kirisits, D. Berger, et al. 2010. Recommendations from Gynaecological (GYN) GEC-ESTRO Working Group: Considerations and pitfalls in commissioning and applicator reconstruction in 3D image-based treatment planning of cervix cancer brachytherapy. *Radiother Oncol* 96 (2):153–160.

Hellebust, T., K. Tanderup, C. Lervåg, et al. 2013. Dosimetric impact of interobserver variability in MRI-based delineation for cervical cancer brachytherapy. *Radiother Oncol* 107(1):13–19.

Hill, D. L. G., P. G. Batchelor, M. Holden, and D. J. Hawkes. 2001. Medical image registration. *Phys Med Biol* 46(3):R1–R45.

Hricak, H., P. S. Swift, Z. Campos, J. M. Quivey, V. Gildengorin, and H. Göranson. 1993. Irradiation of the cervix uteri: Value of unenhanced and contrast-enhanced MR imaging. *Radiology* 189(2):381–388.

Hricak, H. 2005. Role of imaging in pretreatment evaluation of early invasive cervical cancer: Results of the Intergroup Study American College of Radiology Imaging Network 6651-Gynecologic Oncology Group 183. *J Clin Oncol* 23(36):9329–9337.

Huang, K., S. Senthi, D. A. Palma, F. O. Spoelstra, et al. 2013. High-risk CT features for detection of local recurrence after stereotactic ablative radiotherapy for lung cancer. *Radiother Oncol* 109(1):51–57.

Hugo, G. D., K. Cao, C. Guy, et al. 2013. Measurement of local deformation due to lung tumor response to radiation therapy. In *The Fifth International Workshop on Pulmonary Image Analysis, Proc. of Medical Image Computing and Computer-Assisted Intervention.*, R. R. Beichel, M. de Bruijne, S. Kabus, A. Kiraly, A. Kitasaka, J. R. McClelland, E. van Rikxoort, and S. Rit (Eds.). City, ST: CreateSpace Independent Publishing Platform Nagoya, Japan.

Jadon, R., C. A. Pembroke, C. l. Hanna, N. Palaniappan, M. Evans, A.E. Cleves, and J. Staffurth. 2014. A systematic review of organ motion and image-guided strategies in external beam radiotherapy for cervical cancer. *Clin Oncol* 26(4):185–196.

Jamema, S. V., U. Mahantshetty, K. Tanderup, et al. 2013. Inter-application variation of dose and spatial location of D2cm3 volumes of OARs during MR image based cervix brachytherapy. *Radiother Oncol* 107(1):58–62.

Jansen, E. P. M., L. G. H. Dewit, M. Van Herk, and H. Bartelink. 2000. Target volumes in radiotherapy for high-grade malignant glioma of the brain. *Radiother Oncol* 56(2):151–156.

Johnson, H., M. M. McCormick, L. Ibanez, et al. 2015. *The ITK Software Guide: Introduction and Development Guidelines*. New York: Kitware.

Kalff, V., R. J. Hicks, M. P. MacManus, et al. 2001. Clinical impact of (18)F fluorodeoxyglucose positron emission tomography in patients with non-small-cell lung cancer: A prospective study. *J Clin Oncol* 19(1):111–118.

Kepka, L. and J. Socha. 2015. PET-CT use and the occurrence of elective nodal failure in involved field radiotherapy for non-small cell lung cancer: A systematic review. *Radiother Oncol* 115 (2):151–156.

Kerkhof, E. M., B. W. Raaymakers, U. A. Van Der Heide, L. Van De Bunt, I. M. Jürgenliemk-Schulz, and J. J. W. Lagendijk. 2008. Online MRI guidance for healthy tissue sparing in patients with cervical cancer: An IMRT planning study. *Radiother Oncol* 88 (2):241–249.

Kerkhof, E. M., R. W. Van Der Put, B. W. Raaymakers, U. A. Van Der Heide, I. M. Jürgenliemk-Schulz, and J. J. W. Lagendijk. 2009. Intrafraction motion in patients with cervical cancer: The benefit of soft tissue registration using MRI. *Radiother Oncol* 93 (1):115–121.

Kidd, E. A., B. A. Siegel, F. Dehdashti, and P. W. Grigsby. 2007. The standardized uptake value for F-18 fluorodeoxyglucose is a sensitive predictive biomarker for cervical cancer treatment response and survival. *Cancer* 110 (8):1738–1744.

Kim, C. K., S. Y. Park, B. K. Park, W. Park, and S. J. Huh. 2014. Blood oxygenation level-dependent MR imaging as a predictor of therapeutic response to concurrent chemoradiotherapy in cervical cancer: A preliminary experience. *Eur Radiol* 24 (7):1514–1520.

Kipritidis, J., S. Siva, M. S. Hofman, J.Callahan, R. J. Hicks, and P. J. Keall. 2014. Validating and improving CT ventilation imaging by correlating with ventilation 4D-PET/CT using 68Ga-labeled nanoparticles. *Med Phys* 41(1):011910. doi:10.1118/1.4856055.

Kirisits, C., M. J. Rivard, D. Baltas, et al. 2014. Review of clinical brachytherapy uncertainties: Analysis guidelines of GEC-ESTRO and the AAPM. *Radiother Oncol* 110 (1):199–212.

Kong, F. M., K. A. Frey, L. E. Quint, et al. 2007. A pilot study of [18F]fluorodeoxyglucose positron emission tomography scans during and after radiation-based therapy in patients with non small-cell lung cancer. *J Clin Oncol* 25 (21):3116–3123.

Kruis, M. F., J. B. van de Kamer, A. C. Houweling, J. J. Sonke, J. S. Belderbos, M. van Herk. 2013. PET motion compensation for radiation therapy using a CT-based mid-position motion model: Methodology and clinical evaluation. *Int J Radiat Oncol Biol Phys* 87 (2):394–400.

Kwon, J. W., S. J. Huh, Y. C. Yoon, et al. 2008. Pelvic bone complications after radiation therapy of uterine cervical cancer: Evaluation with MRI. *Am J Roentgenol* 191(4):987–994.

Liang, Y., M. Bydder, C. M. Yashar, et al. 2013. Prospective study of functional bone marrow-sparing intensity modulated radiation therapy with concurrent chemotherapy for pelvic malignancies. *Int J Radiat Oncol Biol Phys* 85 (2):406–414.

Lien, H. H., V. Blomlie, T. Iversen, C. Tropé, K. Sundfør, and V. M. Abeler. 1993. Clinical stage I carcinoma of the cervix. *Acta Radiol* 34 (2):130–132.

Lim, K., W. Small, L. Portelance, et al. 2011. Consensus guidelines for delineation of clinical target volume for intensity-modulated pelvic radiotherapy for the definitive treatment of cervix cancer. *Int J Radiat Oncol Biol Phys* 79 (2):348–355.

Lim, K., J. Stewart, V. Kelly, et al. 2014. Dosimetrically triggered adaptive intensity modulated radiation therapy for cervical cancer. *Int J Radiat Oncol Biol Phys* 90 (1):147–154.

Lin, W. U. C., Y. C. Hung, L. S. Yeh, C. H. Kao, R. F. Yen, and Y. Y. Shen. 2003. Usefulness of 18F-fluorodeoxyglucose positron emission tomography to detect para-aortic lymph nodal metastasis in advanced cervical cancer with negative computed tomography findings. *Gynecol Oncol* 89 (1):73–76.

Lindegaard, J. C., L. U. Fokdal, S. K. Nielsen, J. J. Christensen, and K. Tanderup. 2013. MRI-guided adaptive radiotherapy in locally advanced cervical cancer from a nordic perspective. *Acta Oncol* 52 (7):1510–1519.

Ling, C. C., J. Humm, S. Larson, et al. 2000. Towards multidimensional radiotherapy (MD-CRT): Biological imaging and biological conformality. *Int J Radiat Oncol Biol Phys* 47 (3):551–560.

Liu, Y., R. Bai, H. Sun, H. Liu, X. Zhao, and Y. Li. 2009. Diffusion-weighted imaging in predicting and monitoring the response of uterine cervical cancer to combined chemoradiation. *Clin Radiol* 64 (11):1067–1074.

Machtay, M., F. Duan, B. A. Siegel, et al. 2013. Prediction of survival by [18F]fluorodeoxyglucose positron emission tomography in patients with locally advanced non-small-cell lung cancer undergoing definitive chemoradiation therapy: Results of the ACRIN 6668/RTOG 0235 trial. *J Clin Oncol* 31 (30):3823–3830.

Mac Manus, M. P., Z. Ding, A. Hogg, et al. 2011. Association between pulmonary uptake of fluo-rodeoxyglucose detected by positron emission tomography scanning after radiation therapy for non-small-cell lung cancer and radiation pneumonitis. *Int J Radiat Oncol Biol Phys* 80 (5):1365–1371.

Mac Manus, M. P., S. Everitt S, M. Bayne, et al. 2013. The use of fused PET/CT images for patient selection and radical radiotherapy target volume definition in patients with non-small cell lung cancer: Results of a prospective study with mature survival data. *Radiother Oncol* 106(3):292–298.

Marks, L. B., D. P. Spencer, G. C. Bentel, et al. 1993. The utility of SPECT lung perfusion scans in minimizing and assessing the physiologic consequences of thoracic irradiation. *Int J Radiat Oncol Biol Phys* 26 (4):659–668.

Mayr, N. A., Z. Huang, J. Z. Wang, et al. 2012. Characterizing tumor heterogeneity with functional imaging and quantifying high-risk tumor volume for early prediction of treatment outcome: Cervical cancer as a model. *Int J Radiat Oncol Biol Phys* 83 (3):972–979.

Mayr, N. A., V. A. Magnotta, J. C. Ehrhardt, et al. 1996. Usefulness of tumor volumetry by magnetic resonance imaging in assessing response to radiation therapy in carcinoma of the uterine cervix. *Int J Radiat Oncol Biol Phys* 35 (5):915–924.

Mayr, N. A., J. Z. Wang, S. S. Lo, et al. 2010. Translating response during therapy into ultimate treatment outcome: A personalized 4-dimensional MRI tumor volumetric regression approach in cervical cancer. *Int J Radiat Oncol Biol Phys* 76 (3):719–727.

Mazeron, R., J. Gilmore, I. Dumas, et al. 2013. Adaptive 3D image-guided brachytherapy: A strong argument in the debate on systematic radical hysterectomy for locally advanced cervical cancer. *Oncologist* 18 (4):415–422.

Milgrom, S. A., H. A. Vargas, E. Sala, J. F. Kelvin, H. Hricak, and K. A. Goodman. 2013. Acute effects of pelvic irradiation on the adult uterus revealed by dynamic contrast-enhanced MRI. *BJR* 86 (1031):20130334.

Mirpour, S., J. C. Mhlanga, P. Logeswaran, G. Russo, G. Mercier, and R. M. Subramaniam. 2013. The role of PET/CT in the management of cervical cancer. *AJR* 201 (2):W192–W205.

Mitchell, D. G., B. Snyder, F. Coakley, et al. 2006. Early invasive cervical cancer: Tumor delineation by magnetic resonance imaging, computed tomography, and clinical examination, verified by pathologic results, in the ACRIN 6651/GOG 183 Intergroup Study. *J Clin Oncol* 24 (36):5687–5694.

Ménard, C., and U. van der Heide. 2014. Introduction: Systems for magnetic resonance image guided radiation therapy. *Sem Radiat Oncol* 24 (3):149–150.

Mundt, A. J., L. K. Mell, and J. C. Roeske. 2003. Preliminary analysis of chronic gastrointestinal toxicity in gynecology patients treated with intensity-modulated whole pelvic radiation therapy. *Int J Radiat Oncol Biol Phys* 56 (5):1354–1360.

Nakamura, K., I. Joja, C. Fukushima, et al. 2012. The preoperative SUVmax is superior to ADCmin of the primary tumour as a predictor of disease recurrence and survival in patients with endometrial cancer. *Eur J Nucl Med Mol Imaging* 40 (1):52–60.

Nestle, U., K. Walter, S. Schmidt, et al. 1999. 18F-deoxyglucose positron emission tomography (FDG-PET) for the planning of radiotherapy in lung cancer: High impact in patients with atelectasis. *Int J Radiat Oncol Biol Phys* 44 (3):593–597.

Nesvacil, N., K. Tanderup, T.P. Hellebust, et al. 2013. A multicentre comparison of the dosimetric impact of inter- and intra-fractional anatomical variations in fractionated cervix cancer brachytherapy. *Radiother Oncol* 107 (1):20–25.

Nomden, C. N., A. A. C. De Leeuw, J. M. Roesink, et al. 2013. Clinical outcome and dosimetric parameters of chemo-radiation including MRI guided adaptive brachytherapy with tandem-ovoid applicators for cervical cancer patients: A single institution experience. *Radiother Oncol* 107 (1):69–74.

Oh, S., J. Stewart, J. Moseley, et al. 2014. Hybrid adaptive radiotherapy with on-line MRI in cervix cancer IMRT. *Radiother Oncol* 110 (2):323–328.

Padhani, A. R., and D.-M. Koh. 2011. Diffusion MR imaging for monitoring of treatment response. *Magn Reson Imaging Clinics North America* 19 (1):181–209.

Paesmans, M., T. Berghmans, M. Dusart, et al. 2010. Primary tumor standardized uptake value measured on fluorodeoxyglucose positron emission tomography is of prognostic value for survival in non-small cell lung cancer: Update of a systematic review and meta-analysis by the European Lung Cancer Working Party for the International Association for the Study of Lung Cancer Staging Project. *J Thorac Oncol* 5 (5):612–619.

Park, H. K., K. Jeon, W. J. Koh, et al. 2010. Occult nodal metastasis in patients with non-small cell lung cancer at clinical stage IA by PET/CT. *Respirology* 15 (8):1179–1184.

Petit, S. F., W. J. van Elmpt, C. J. Oberije, et al. 2011. [(1)(8)F]fluorodeoxyglucose uptake patterns in lung before radiotherapy identify areas more susceptible to radiation-induced lung toxicity in non-small-cell lung cancer patients. *Int J Radiat Oncol Biol Phys* 2011 81 (3):698–705.

Petric, P., R. Hudej, P. Rogelj, et al. 2013. Uncertainties of target volume delineation in MRI guided adaptive brachytherapy of cervix cancer: A multi-institutional study. *Radiother Oncol* 107 (1):6–12.

Petric, P., R. Hudej,. Rogelj, et al. 2012. Comparison of 3D MRI with high sampling efficiency and 2D multiplanar MRI for contouring in cervix cancer brachytherapy. *Radiother Oncol* 46 (3):242–251.

Pötter, R., P. Georg, J. C. A. Dimopoulos, et al. 2011. Clinical outcome of protocol based image (MRI) guided adaptive brachytherapy combined with 3D conformal radiotherapy with or without chemotherapy in patients with locally advanced cervical cancer. *Radiother Oncol* 100 (1):116–123.

Reinhardt, J. M., K. Ding, K. Cao, G. E. Christensen, E. A. Hoffman, and S. V. Bodas. 2008. Registration-based estimates of local lung tissue expansion compared to xenon CT measures of specific ventilation. *Med Image Anal* 12 (6):752–763.

Rey, F., C. Chang, C. Mesina, N. Dixit, B.-K. K. Teo, and L. L. Lin. 2013. Dosimetric impact of interfraction catheter movement and organ motion on MRI/CT guided HDR interstitial brachytherapy for gynecologic cancer. *Radiother Oncol* 107 (1):112–116.

Reymen, B., J. van Loon, A. van Baardwijk, et al. 2013. Total gross tumor volume is an independent prognostic factor in patients treated with selective nodal irradiation for stage I to III small cell lung cancer. *Int J Radiat Oncol Biol Phys* 85 (5):1319–1324.

Rijkmans, E. C., R. A. Nout, I. H. H. M., et al. 2014. Improved survival of patients with cervical cancer treated with image-guided brachytherapy compared with conventional brachytherapy. *Gynecol Oncol* 135 (2):231–238.

Rizzo, S., G. Calareso, S. Maccagnoni, et al. 2014. Pre-operative MR evaluation of features that indicate the need of adjuvant therapies in early stage cervical cancer patients: A single-centre experience. *Eur J Radiol* 83 (5):858–864.

Roche, A., X. Pennec, M. Rudolph, et al. 2001. Generalized correlation ratio for rigid registration of 3D ultrasound with MR images. *IEEE TMI* 20 (10):25–31.

Rührnschopf, E.-P., and K. Klingenbeck. 2011. A general framework and review of scatter correction methods in X-ray cone-beam computerized tomography. Part 1: Scatter compensation approaches. *Med Phys* 38 (7):4296–4311.

Schmid, M. P., B. Mansmann, M. Federico, et al. 2013. Residual tumour volumes and grey zones after external beam radiotherapy (with or without chemotherapy) in cervical cancer patients. *Strahlenther Onkol* 189 (3):238–245.

Senthi, S., F. J. Lagerwaard, C. J. Haasbeek, B. J. Slotman, and S. Senan. 2012. Patterns of disease recurrence after stereotactic ablative radiotherapy for early stage non-small-cell lung cancer: A retrospective analysis. *Lancet Oncol* 13 (8):802–809.

Shirvani, S. M., J. Jiang, J.Y. Chang, et al. 2012. Comparative effectiveness of 5 treatment strategies for early-stage non-small cell lung cancer in the elderly. *Int J Radiat Oncol Biol Phys* 84 (5):1060–1070.

Siebers, J. V., and H. Zhong. 2008. An energy transfer method for 4D Monte Carlo dose calculation. *Med Phys* 35 (9):4096–4105.

Simpson, D. R., W. Y. Song, V. Moiseenko, et al. 2012. Normal tissue complication probability analysis of acute gastrointestinal toxicity in cervical cancer patients undergoing intensity modulated radiation therapy and concurrent cisplatin. *Int J Radiat Oncol Biol Phys* 83 (1):e81–e86.

Siva, S., J. Callahan, T. Kron, et al. 2014. A prospective observational study of gallium-68 ventilation and perfusion PET/CT during and after radiotherapy in patients with non-small cell lung cancer. *BMC Cancer* 14 (1):740.

Somoye, G., V. Harry, S. Semple, et al. 2012. Early diffusion weighted magnetic resonance imaging can predict survival in women with locally advanced cancer of the cervix treated with combined chemo-radiation. *Eur Radiol* 22 (11):2319–2327.

Sonke, J.J., and J. Belderbos. 2010. Adaptive radiotherapy for lung cancer. *Sem Radiat Oncol* 20(2):94–106.

Sotiras, A., C. Davatzikos, and N. Paragios. 2013. Deformable medical image registration: A survey. *IEEE TMI* 32 (7):1153–1190.

Sridhar, P., G. Mercier, J. Tan, M. T. Truong, B. Daly, R. M. Subramaniam. 2014. FDG PET metabolic tumor volume segmentation and pathologic volume of primary human solid tumors. *AJR* 202 (5):1114–1119.

Steenbakkers, R. J., J. C. Duppen, I. Fitton I, et al. 2006. Reduction of observer variation using matched CT-PET for lung cancer delineation: A three-dimensional analysis. *Int J Radiat Oncol Biol Phys* 64 (2):435–448.

Stewart, J., K. Lim, V. Kelly, et al. 2010. Automated weekly replanning for intensity-modulated radiotherapy of cervix cancer. *Int J Radiat Oncol Biol Phys* 78 (2):350–358.

Stiles, B. M., E. L. Servais, P. C. Lee, J. L. Port, S. Paul, N. K. Altorki. 2009. Point: Clinical stage IA non-small cell lung cancer determined by computed tomography and positron emission tomography is frequently not pathologic IA non-small cell lung cancer: The problem of understaging. *J Thorac Cardiovasc Surg* 137 (1):13–19.

Sugimura, K., B. M. Carrington, J. M. Quivey, and H. Hricak. 1990. Postirradiation changes in the pelvis: Assessment with MR imaging. *Radiology* 175 (3):805–813.

Takenaka, T., K. Yoshida, S. Tachiiri, et al. 2012. Comparison of dose-volume analysis between standard Manchester plan and magnetic resonance image-based plan of intracavitary brachytherapy for uterine cervical cancer. *Radiat Research* 53 (5):791–797.

Tanderup, K., S. K. Nielsen, G.-B. Nyvang, et al. 2010. From point A to the sculpted pear: MR image guidance significantly improves tumour dose and sparing of organs at risk in brachytherapy of cervical cancer. *Radiother Oncol* 94 (2):173–180.

Taylor, A., and M. E. B. Powell. 2008. An assessment of interfractional uterine and cervical motion: Implications for radiotherapy target volume definition in gynaecological cancer. *Radiother Oncol* 88 (2):250–257.

Thomas, L., B. Chacon, M. Kind, et al. 1997. Magnetic resonance imaging in the treatment planning of radiation therapy in carcinoma of the cervix treated with the four-field pelvic technique. *Int J Radiat Oncol Biol Phys* 37 (4):827–832.

Toita, T., T. Ohno, Y. Kaneyasu, et al. 2011. A consensus-based guideline defining clinical target volume for primary disease in external beam radiotherapy for intact uterine cervical cancer. *J J Clin Oncol* 41 (9):1119–1126.

Tokumaru, S. T. Toita, M. Oguchi, et al. 2012. Insufficiency fractures after pelvic radiation therapy for uterine cervical cancer: An analysis of subjects in a prospective multi-institutional trial, and cooperative ctudy of the Japan Radiation Oncology Group (JAROG) and Japanese Radiation Oncology Study Group (JROSG). *Int J Radiat Oncol Biol Phys* 84 (2):e195–e200.

Tustison, N., B. Avants, P. Cook, et al. 2010. N4ITK: Improved N3 bias correction. *IEEE TMI* 29 (6):1310–1320.

van Baardwijk, A., G. Bosmans, A. Dekker, et al. 2007. Time trends in the maximal uptake of FDG on PET scan during thoracic radiotherapy: A prospective study in locally advanced non-small cell lung cancer (NSCLC) patients. *Radiother Oncol* 82 (2):145–152.

van Elmpt, W., D. De Ruysscher, A. van der Salm A, et al. 2012. The PET-boost randomised phase II dose-escalation trial in non-small cell lung cancer. *Radiother Oncol* 104 (1):67–71.

van Meerbeeck, J. P., G. W. Kramer, P. E. van Schil PE, et al. 2007. Randomized controlled trial of resection versus radiotherapy after induction chemotherapy in stage IIIA-N2 non-small-cell lung cancer. *J Natl Cancer Inst* 99 (6):442–450.

Vandecasteele, K., L. Delrue, B. Lambert, et al. 2012. Value of magnetic resonance and 18FDG PET-CT in predicting tumor response and resectability of primary locally advanced cervical cancer after treatment with intensity-modulated arc therapy. *Int J Gynecol Cancer* 22 (4):630–637.

Vansteenkiste, J., D. De Ruysscher, W. E. E. Eberhardt, et al. 2013. Early and locally advanced non-small-cell lung cancer (NSCLC): ESMO Clinical Practice Guidelines for diagnosis, treatment and follow-up. *Ann Oncol* 24 (suppl 6):vi89–vi98.

Vidal Melo, M. F., D. Layfield, R. S. Harris. 2003. Quantification of regional ventilation-perfusion ratios with PET. *J Nucl Med* 44 (12):1982–1991.

Vincens, E., C. Balleyguier, A. Rey, et al. 2008. Accuracy of magnetic resonance imaging in predicting residual disease in patients treated for stage IB2/II cervical carcinoma with chemoradiation therapy. *Cancer* 113 (8):2158–2165.

Viswanathan, A. N., S. Beriwal, J.F. De Los Santos, et al. 2012b. American Brachytherapy Society Consensus Guidelines for Locally Advanced Carcinoma of the Cervix. Part II: High-dose-rate Brachytherapy. *Brachytherapy* 11 (1):47–52.

Viswanathan, A. N., and B. Thomadsen. 2012a. American Brachytherapy Society Consensus Guidelines for locally advanced carcinoma of the cervix. Part I: General principles. *Brachytherapy* 11 (1):33–46.

Viswanathan, A. N., B. Erickson, D. K. Gaffney, et al. 2014. Comparison and consensus guidelines for delineation of clinical target volume for CT- and MR-based brachytherapy in locally advanced cervical cancer. *Int J Radiat Oncol Biol Phys* 90 (2):320–328.

Wang, J. Z., N. A. Mayr, D. Zhang, et al. 2010. Sequential magnetic resonance imaging of cervical cancer. *Cancer* 116 (2):5093–5101.

Weiss, E., K. Eberlein, O. Pradier, H. Schmidberger, and C. F. Hess. 2002. The impact of patient positioning on the adequate coverage of the uterus in the primary irradiation of cervical carcinoma: A prospective analysis using magnetic resonance imaging. *RadiotherOncol* 63 (1):83–87.

Weiss, E., M. Fatyga, Y. Wu, N, et al. 2013. Dose escalation for locally advanced lung cancer using adaptive radiation therapy with simultaneous integrated volume-adapted boost. *Int J Radiat Oncol Biol Phys* 86 (3):414–419.

Weiss, E., S. Richter, T. Krauss, et al. 2003. Conformal radiotherapy planning of cervix carcinoma: Differences in the delineation of the clinical target volume. *RadiotherOncol* 67 (1):87–95.

Wells, W. M., P. Viola, H. Atsumi, S. Nakajima, and R. Kikinis. 1996. Multi-modal volume registration by maximization of mutual information. *Med Image Anal* 1 (1):35–51.

Weltens, C., J. Menten, M. Feron, et al. 2001. Interobserver variations in gross tumor volume delineation of brain tumors on computed tomography and impact of magnetic resonance imaging. *Radiother Oncol* 60 (1):49–59.

Wright, J. D., I. Deutsch, E. T. Wilde, et al. 2013. Uptake and outcomes of intensity-modulated radiation therapy for uterine cancer. *Gynecol Oncol* 130 (1):43–48.

Wu, D. H., N. A. Mayr, Y. Karatas, et al. 2005. Inter-observer variation in cervical cancer tumor delineation for image-based radiotherapy planning among and within different specialties. *J Appl Clin Med Phy* 6 (4):106–110.

Yang, J., J. Cai, H. Wang, et al. 2014. Four-dimensional magnetic resonance imaging using axial body area as respiratory surrogate: Initial patient results. *Int J Radiat Oncol Biol Phys* 88 (4):907–912.

Zahra, M. A., K. G. Hollingsworth, E. Sala, D.J. Lomas, and L. T. Tan. 2007. Dynamic contrast-enhanced MRI as a predictor of tumour response to radiotherapy. *Lancet Oncol* 8 (1):63–74.

Zhen, X., H. Chen, H. Yan, et al. 2015. A segmentation and point-matching enhanced efficient deformable image registration method for dose accumulation between HDR CT images. *Physics Med Biol* 60 (7):2981–3002.

Zietkowski, D., N. M. Desouza, R. L. Davidson, and G. S. Payne. 2013. Characterisation of mobile lipid resonances in tissue biopsies from patients with cervical cancer and correlation with cytoplasmic lipid droplets. *NMR Biomed* 26 (9):1096–1102.

Zunino, S., O. Rosato, S. Lucino, E. Jauregui, L. Rossi, and D. Venencia. 1999. Anatomic study of the pelvis in carcinoma of the uterine cervix as related to the box technique. *Int J Radiat Oncol Biol Phys* 44 (1):53–59.

Zwahlen, D., J. Jezioranski, P. Chan, et al. 2009. Magnetic resonance imaging-guided intracavitary brachytherapy for cancer of the cervix. *Int J Radiat Oncol Biol Phys* 74 (4):1157–1164.

CHAPTER 5

Advances in Computing Infrastructure

Yulong Yan, Alicia Yingling, and Steve Jiang

CONTENTS

5.1 INTRODUCTION

Radiation oncology is a profession that relies heavily on computation-based prediction to deliver the right amount of dose to the right place. Since its invention, computer technology has drastically influenced the practice of radiation oncology (Van Dyk, 2012). Initially, computer applications were developed to give faster and more accurate answers to dosimetric problems handled by manual methods (ICRU, 1976, 1987). With the advancement of technology, computation power has been further harvested by more advanced applications such as three-dimensional treatment planning systems (TPSs), which use three-dimensional anatomical imaging information to predict radiation accurate doses to the tumor and nearby organs. Although the display of three-dimensional dose distributions on human anatomy has been a breakthrough, developing better plans has been challenging. Instead of using conventional flat beams, intensity-modulated radiation therapy (IMRT) technology delivers modulated nonuniform beams to achieve higher dose conformity to the target while sparing nearby critical organs. At the core of this technology is inverse planning optimization, which uses computers to search for the optimal solution based on dosimetric objectives prescribed by a radiation oncologist.

Additional needs of more computer power stem from emergent new technologies for more accurate and efficient dose calculation methods, treatment of moving targets, stereotactic-body radiation therapy (SBRT), adaptive radiation therapy, and advanced online image guidance. A graphics processing unit (GPU) was designed to rapidly manipulate computer graphics and image processing to create realistic 3D graphics for the gaming industry. At present, GPU has become not only a graphics processing accelerator but also a supercomputer component to potentially benefit scientific computation needs. Since 2006, GPU parallel computing technologies have been gradually adopted by major vendors for routine use in the clinic (Jia and Jiang, 2016).

Computer-based information technology (IT) systems also play a critical role in the management of workflow and data in a radiation therapy clinic that is part of a multi-disciplinary healthcare complex. These IT systems include the treatment management systems (TMSs), hospital information systems (HISs), and radiological information systems (RISs) that need to be integrated to work best in the hospital setting. Committees have been established to supervise and regulate IT system integration. The initiative of Integrating the Healthcare Enterprise (IHE) was designed to integrate the information systems that support modern healthcare institutions (IHE, 2016). For radiation oncology, the IHE Radiation Oncology Technical Committee maintains the IHE Radiation Oncology Technical Framework (IHE-RO) (IHE Volume 1 & 2, 2014a,b).

As IT infrastructure grows to meet clinical needs, maintenance costs increase. Cloud computing emerges as a class of Internet-based technologies that harness shared resources on the internet to achieve networking and storage goals with minimum management effort.

In this chapter, we will introduce prevalent IT concepts and a common framework for professionals who design and manage IT infrastructure in radiation oncology clinics.

5.2 IT INFRASTRUCTURE MODEL

The infrastructure that supports information and computing solutions in radiation oncology clinics relies on personnel expertise and cutting-edge technology research. Because technology advances rapidly, users in the clinic need to become familiar with the ongoing changes and also understand the different technologies available. While IT infrastructures vary with design goals, their building blocks are consistent.

5.2.1 Infrastructure Building Blocks

The building blocks of an IT infrastructure include computing machines, desktops and servers, operating systems, storage and networking interfaces, endpoints, and data centers. Other devices such as handheld tablets, phablets, phones, and laptops are also key elements.

5.2.1.1 Operating Systems

Operating systems provide interfaces between the applications and the hardware. Prominent providers of operating systems in the market include Microsoft, Sun, Apple, Google, Android, and Linux. Operating systems are included in desktops, workstations, mobile platforms, and server components.

Mainstream operating systems provide more general-purpose capabilities and extensibility while maintaining a near real-time response and instantaneous user experience. Mainstream operating systems lay pathways for the applications and interconnects between systems that can be streamlined as well. The interconnectivity gives one the flexibility of designing system layout. Designs can call for homogenous or heterogeneous layouts. There are pros and cons for each layout. If time allows, it is recommended to remain homogeneous. The benefits of homogeneous operating system layout include common, out-of-the-box solutions and interfaces for networking, security, and administration, and a unified, streamlined training for users. When technologies and integrations require a heterogeneous design, which is often the case for large academic centers loaded with task-specific systems for research and education, the operator should plan accordingly for added technical and human resources. Heterogeneous solutions might require additional training and even expertise for end users, while the benefits can be greater customization and solutions tailored more closely to business processes. Interconnections between heterogeneous systems may require additional development or customization efforts.

Other embedded and real-time operating systems are highly specialized and provide single-purpose solutions to address niche purposes. They often drive mission-critical systems and require great redundancy and certainty of real-time behaviors. However, these systems will not be discussed in this chapter.

Technological proficiency with operating systems, interoperability, licensing costs, training, and application dependencies are all important components to review while planning the IT systems and their applications. Additionally, the current state and future roadmaps of any technology under consideration should be evaluated. As technology continues to evolve, the knowledge of the providers' needs is essential to avoid being forced into an expensive solution that may need to be added or replaced later.

5.2.1.2 Networking

Devices today depend highly on a connected infrastructure and can be linked by either wireless or wired interfaces. Wireless interfaces include Wi-Fi, cellular connections, Bluetooth, and other radio and infrared communication media. High-speed wired connections are needed for maximum performance devices, such as servers and high-end workstations. Wireless connections are most common for end-user laptops or handheld devices.

Wired and wireless networks must remain secure in terms of physical access and data transmission. Physical access is accomplished with appropriate routing of wires in secure areas and behind walls for optimal security. Wireless access points provide connectivity and require appropriate configurations to manage secure access and encryption. Currently, standard networking is based on ethernet connection using hardwired switches and cabling, and also controlled wireless access points. For current infrastructure needs, wireless technology at Institute of Electrical and Electronics Engineers (IEEE) 802.11ac standard is optimal for single-user systems. Gigabit Ethernet provides sufficient network bandwidth for multi-user systems.

5.2.1.3 Storage

Storage is available in different media and forms, and may be either fixed to a machine or portable for attaching and detaching on demand. Storage media provide for online use or offline standby, archiving, and recovery as requires. Solid-state devices are primarily used for fixed storage. Recent technology advancements and price points have made solid-state drives more appropriate given the increased performance and throughput, and smaller form factors. Older drives with spinning platters are still in use and may be discontinued in the near future. Initial models of solid-state drives have raised longevity concerns, although the same concerns apply to platter drives because the lifetime of the stored information is greater than that of the hardware. Storage for data archiving and offline requirements is available in tape medium, which provides substantial shelf life. Offsite storage services provide backup and storage through online transfers.

Recently, many clinics have switched to paperless systems, including the digital storage of patient treatment plans. Treatment plans may need to be reproduced for reporting purposes, retreatment, or peer review in clinical trials. For this reason, TPSs are equipped with backup and restore functions in their internal or proprietary formats. Because a plan may need to be reproduced from the raw data of a different TPS, plans may be stored in the standard Digital Imaging and Communications in Medicine (DICOM) format. A picture archiving and communications system specifically tailored for radiation therapy (RT PACS) is available to manage the increasing numbers of plans. The RT PACS provides users with DICOM services, such as storage and query/retrieve.

5.2.1.4 Virtualization

Virtualization is the creation of a virtual version of an operation system, server, storage device, or network resource. Virtualization started with a software tool that allows a piece of hardware to run multiple operating systems at the same time to avoid wasting expensive processing power. Advances in virtualization technologies have expanded and matured significantly in the past decade. Proprietary and open-source solutions provide excellent features and reliability to manage and control virtualized instances for both back-end server and front-end client requirements.

5.2.1.5 Servers

The primary objective of the IT infrastructure is to provide exceptional functionality and efficiency to end users. A collection of servers is the centerpiece of an IT infrastructure. Servers provide stability, consistency, efficiency, and other infrastructure components to protect the system by ensuring appropriate accessibility for end users. Hardware and software of servers should be configured so that they are readily available to users at a required uptime level. Servers can be categorized as devices that provide a consistent modular service or as a combination of services available within data centers or server rooms. Adequate proactive maintenance and upkeep are required to ensure availability of the configurations.

5.2.1.6 Data Centers

Infrastructure components providing the highest levels of service require secure physical access, consistent availability or uptime, climate control, redundancy, highly interconnected resource dependencies, and more. Data centers centralize these physical machines and components and provide multiple tenancy for many high-demand systems, although historically data centers were built and operated solely for a single organization. Data centers are tiered from 1 to 4 according to the ANSI/TIA-492 standard (American National Standards Institute/Telecommunications Industry Association). Increasing tiers require increasing levels of redundancy to provide uninterrupted service. A tier 1 data center allows for approximately 1,700 min of downtime per year, while the highest tier 4 reduces that downtime to 26 min. Tier 1 data centers and server rooms use modern modular racks to provide uniform installation and configuration. Some data centers also allow for co-locating hardware to be installed onsite with some limitations on size, heat generation, energy requirements, or interference levels. Data server designers should confirm with co-location providers that the hardware is compatible.

5.2.1.7 Client Devices

Client or end-user devices are growing and expanding rapidly. Widely used devices include tablets, ultraportable computing devices, high-power laptops, wearable technologies, phones, workstations, and personal desktops. Additional devices and innovative capabilities are becoming available as the Internet continues to push the boundaries of computing into ubiquitous infused objects such as wearables and sensor devices. Many of these devices use one of the major operating systems primarily and leverage most open standard technologies for common functionalities, such as networking, identity management, encryption, and remote access and management. Client devices can also be attached to more general-purpose workstation processors and storage. Personal devices may integrate scheduling and unify communications, but they may also open up vectors for additional security concerns for data and network exposure. Offerings are available to help to cordon applications, data, and access so that they are adequately controlled for sensitive information.

5.2.2 Cloud-Computing Models

Cloud computing is the on-demand delivery of computing power, database storage, applications, and other IT resources through a cloud services platform via the Internet. Cloud-computing models allow computing resource locations and their maintenance costs to be abstracted away from the user. A model allows computing, storage, and connectivity to be defined so that sufficient resources and bandwidth are available between resources and end users. Through this versatile arrangement, adequate security and functionality is ensured for all required activities.

5.2.2.1 Deployment Models

Cloud computing can improve computing efficiency. In addition to technical requirements, nontechnical aspects should also be considered in cloud computing. Data ownership and compliance with data usage, security, and access must be thought through at the start to ensure that data is adequately protected. Patient data have increasingly become targets for malicious capture. They require appropriate encryption, access control on primary data and copies, shards, backups, exports, and viewing. A review of all the designs should be confirmed with legal counsel.

5.2.2.1.1 Public Clouds   Hosting providers such as Amazon Web Services (AWS), Microsoft Azure, and Google offer computational and storage services that can be accessed via the Internet. They can be cordoned off as needed, but they provide full authorized access when required. Costs and performance can be scaled based on needs and constraints. Public clouds offer immediate value and in the long term remove the need to manage upgrades and security patches. In addition, public clouds allow common application development and implementations of turnkey solutions with some customization options.

5.2.2.1.2 Private Clouds   In many cases, private clouds are on-premise infrastructure-as-a-service (IaaS) solutions and offer the ability to add hardware to private data centers. Private clouds can be partitioned to allow specific applications and server access to resources as needed. Private clouds provide additional flexibility when specialized hardware needs to be included, but they still maintain scalable flexibility. In cases where large GPU clusters interact with other, more common back-end and front-end servers, a private cloud is as flexible as the public cloud solutions with specialized high-performance computing solutions. Private clouds require more complex installation and setup. Management of the patches and hardware add internal expense. Time and resources need to be reserved adequately.

5.2.2.1.3 Hybrid Cloud   Hybrid-cloud solutions allow hardware to reside in both local private and public data centers. Hybrid-cloud solutions have been used to address situations in which disaster recovery is required. Hybrid clouds require additional configuration and setup compared to private clouds. Synchronization, administration, and management need to be coordinated and verified between the private and public portions. Security also needs to be vetted to ensure that no additional attack vectors are introduced.

5.2.2.2 Service Models
Service models take various components of infrastructure, software, and other IT resources and offer them as services that can be interchanged and available through different encapsulated modules. The service model approach reduces tight coupling between systems and allows for greater efficiencies in upgrading improvised solutions. With new advances, these models are efficient, but in some cases they may impede progress. As internal rapid innovations occur, outside service providers often cannot keep the pace. The provision of the best service for the best price often requires a stable and optimized solution, which counters rapid innovation. As solutions become more specialized and are expected to encounter much faster rates of application or hardware changes, the level of the service model moves from software to platform, to infrastructure. Ironically, the need to reduce cost and maintain a more common solution would reverse the service model to move from infrastructure to platform, to software. For many common business needs, such as human resources, accounting, and finance, software services provide exceptional value.

5.2.2.2.1 Software-as-a-Service   New solution models are becoming available because the cloud offers greater flexibility, scalability, and performance at lower costs. It is a software licensing and delivery model in which software is licensed by subscription. Microsoft Office 365, Box, Google Apps, AWS, and DocuSign are a few examples of software-as-a-service (SaaS). Historically, the software service model has been offered through closed, black box solutions or licensed software, and it has been made available only recently through open source.

5.2.2.2.2 Platform-as-a-Service   Platform-as-a-service (PaaS) is a category of cloud-computing services. PaaS offers access to a platform that is ready and consistently configured to install custom solutions in most cases. By providing a common platform, many of the underlying interconnects allow the platforms to be connected and managed easily. In addition, a solid underlying grid of platforms is offered as leverage when custom computing solutions are needed. Microsoft Windows Azure Cloud Services is a typical example of a PaaS.

5.2.2.2.3 Infrastructure-as-a-Service   Infrastructure-as-a-service (IaaS) is a form of cloud computing. It offers hardware for users to install virtualized computing platforms or resources over the Internet. For example, Amazon S3 offers storage infrastructure over the Internet. Google Compute Engine offers flexible, self-managed virtual machines (VMs) hosted on Google's infrastructure. In cases where specialized operating systems and platforms need to be installed, IaaS can scale quickly and easily, without the user having to invest in hardware upfront or wait for it to be usable. When standard hardware requirements are defined as processors, memory, and storage, and there is no need for physical access (e.g., attaching dongles or specialized hardware to the machines), IaaS enables streamlining of the computing power online for rapid use. This is not a viable alternative when customized components, such as GPUs, sound cards, or other devices need to be connected or assembled with the machines.

5.2.2.2.4 X-as-a-Service   X-as-a-service (XaaS) , or "Anything" as a Service is a model that interconnects the "anything and everything else." In this model, defined levels of acceptable service can be compartmentalized and offered. XaaS models are often quite specialized but can alleviate management time and resources. For example, in cases where multiple platforms overlap and are highly connected, such as unified communications, XaaS provides a hybrid cloud solution with onsite components connected to cloud-based elements to offer email, phone, text, paging, and video chat through a managed service. With the rapid growth of connected elements and the Internet, these managed services are likely to become more standardized and similar to SaaS and PaaS. As new technologies are delineated into well-defined sets of behaviors and acceptable levels of service, markets will move from innovative XaaS designations to other service-level models. Examples include management-as-a-service (MaaS), communication-as-a-service (CaaS), and so on.

5.3 OPERATING SYSTEMS

Microsoft Windows, Linux/Unix, and Apple iOS, are the major operating systems.

5.3.1 Microsoft Operating Systems

Currently, Windows 10, 8, and 7 are the most widely used versions of Microsoft Windows. The use of older versions is progressively declining. The latest versions of Windows have high functionality for wireless connectivity and projecting display information. They also provide the most up-to-date support for using the latest hardware to its fullest capabilities.

5.3.2 Unix/Linux

Some of the most popular distributions of Linux are Ubuntu, Debian, Red Hat, and Open SUSE. They share the same Linux kernel with different layers of functionality. Many aspects should be considered when selecting a version, including enterprise management, user and application requirements, support, and licensing.

5.3.3 Apple Macintosh

The desktop operating system OSX from Apple (version 10.7) includes the Mac OS X Desktop and Mac OS X Server combined into one. This UNIX-based operating system shares features with the Microsoft operating systems. The enterprise solutions for identity management are recommended only for small organizations.

5.3.4 Operating Systems for Smartphones

Android, iOS, Windows, and Blackberry are the most popular operating systems for smartphones. Android and iOS have shown a strong growth pattern for the midterm future because they offer most popular applications on both platforms. All Android data and accounts are associated with Google and are available for easy export. Data synchronization via cloud services provides seamless offerings between any devices utilizing Google services. Apple ties much of the data into the device and requires physical connection in many cases to synchronize, back up, and restore data.

5.4 NETWORKING

Connected computing provides excellent features and performance when adequately secured. The network-connecting devices are both physical and broadcast signals, along with virtual networks and multiple layers of security. Advances in wireless technologies have made networking ubiquitous and have enabled many innovations in collecting and processing data, providing results through several networked displays and interfaces.

5.4.1 Local Area Network (LAN) Infrastructure

The infrastructure of local networks consists of the hardware for wired and wireless connections, and switches to route network traffic between endpoint devices. Copper and fiber-optic cables and the ports that provide physical connections are the common visible aspects of the local area network. The wireless components may either be seen as mounted

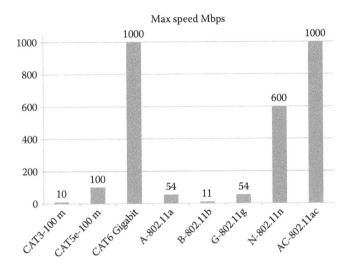

FIGURE 5.1   Network type and speed capacity.

antennas or hidden. Local area network management of access to network resources requires appropriate configuration, maintenance, monitoring, and protections for physical and on-network transmissions.

5.4.2 Wireless Local Area Network

Wireless connectivity is continuously improving to provide greater range and transmission speeds with reduced energy consumption. Common transmission standards are named according to industry conventions. Popular implementations include 802.11a, 802.11b/g/n, and 802.11.ac with most devices supporting AC or 802.11ac. The bandwidth of AC at 1,300 Mbps is three times greater than the bandwidth of N at 450 Mbps (Figure 5.1).

5.4.3 Firewalls

Firewalls provide access control to network resources and can be placed on or between every device. Their design should allow minimum required access. Firewalls offer increasing levels of restricted access on the network to resources. Some resources may be accessible only between the servers, while some servers may provide partial access to ports on an internal network, further reducing external accessibility to the outside Internet.

5.5 STORAGE

In most cases, inputs and outputs to computing machines must be stored at different levels of permanence. Storage is often inversely correlated to the speed at which that data can be accessed. As the speed of access increases, online storage is faster and offline storage is slower; fast storage has the least stability for long-term accessibility.

5.5.1 Tapes

Modern tape media are effective and efficient means for backup and recovery. Tape native capacities that are currently in use range from 100 GB to 10 TB. Historically, tapes were exposed and threaded on machines. Modern tape devices are housed within cartridges, allowing for quicker mounting and unmounting as well as more autonomous solutions, such as robotic arm loaders. In theory, the shelf life of tapes is on average a few decades, but in practice the data is accessible for up to a decade only. Tapes often outlast burned disc-based offline storage, which is good for only a couple of years.

5.5.2 Disks

Disk storage in enclosed formats has been advancing steadily. Devices are more affordable and effective because of increased data densities (and reduced sizes) and increased throughput due to greater data volume under a read head. Because applications on workstations use disk storage extensively, performance can improve noticeably in terms of application response. Modern solid-state storage has the misnomer of a solid-state disk (SSD), but in reality it consists of solid-state, nonvolatile storage memory in the same form factor as historic disk drives.

Open or unenclosed disk storage such as CD, DVD, and Blu-ray continue to grow in capacity, with increasing densities and improved transfer rates. Mass-produced burning techniques generate disks with shelf lives as high as a century. When unburned with data, an unenclosed disk medium has a shorter lifespan until it is burned. The latest Blu-ray disks called M-DISCs will allegedly last up to 1,000 years, setting an optical archival longevity record (Figure 5.2).

5.5.3 RAID

Redundant array of independent disks (RAID) is a technique of combining multiple physical disk drive components into a single logical unit for the purposes of data redundancy

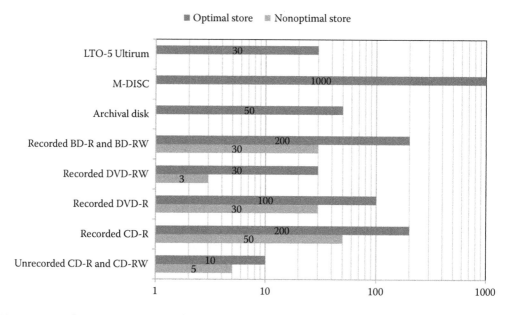

FIGURE 5.2 Common storage media storage length.

and performance improvement. RAID levels and configurations combine various methods of repeating devices on multiple devices and striping data across devices, or placing one part of a sequence on a drive and the next on another device, in succession. Striping is intended to improve performance. Copies on multiple devices allow one device to fail but data loss is still avoided, assuming the failed device is replaced and data is restored to it before another failure occurs. Common configurations consist of RAID 0, RAID 1, RAID 1 + 0 (10), RAID 5, RAID 6. Just a bunch of disks) (JBOD) is similar but not specifically RAID. JBOD allows larger storage volumes to be created from multiple irregular online storage devices (Delmar, 2003; Layton, 2011).

5.5.4 SAN

Storage area network (SAN) components are connected directly to servers. They provide the highest level of throughput between processing and storage. This solution is useful in cases where large data volumes need to be quickly accessible to single processing clusters.

5.5.5 NAS

Network attached storage (NAS) components provide large storage solutions connected to a network. They allow multiple data users to access the data effectively. NAS often utilizes RAID technologies to ensure maximum performance and redundancy with uninterrupted access.

5.5.6 Cloud Storage

Cloud storage allows data to be replicated on remote machines so that they are accessible through other devices and interfaces. Popular providers include Google Drive; Microsoft's OneDrive; and Apple's iCloud, Dropbox, JustCloud, and Box; and so on. Cloud storage offers a few benefits compared to local storage: (1) employees can work from anywhere, (2) no server maintenance is needed, (3) data is easily moved and synchronized, (4) some of them have built-in version controls, and (5) they include a viable option for a disaster recovery plan.

5.5.7 Tiered Storage

Storage of high performance is expensive. Tiered storage is available commercially. It allows users to balance between cost and performance. Slower, less expensive storage is made available for low-priority systems, while faster, more expensive storage is provided to high-priority systems.

5.6 VIRTUALIZATION

Virtualization is the creation of a virtual version of something, such as an operation system, a server, a storage device, or network resources. It offers different solutions by eliminating the need for the physical element(s) of the interaction through an abstract layer that simulates or redirects the action to a given allocation with expected behavior. Performance of a virtualized component depends on how the abstraction is implemented. In the cases of virtual memory, virtual networks, and virtual processors, performance is reduced. In the cases of virtualized storage and virtualized hardware, the ability to bring on additional underlying hardware resources rapidly increases performance. The downsides are often outweighed by the benefits of greater scalability and flexibility of resource allocation.

5.6.1 Virtual Machines

A VM is a software representation of a computer and is often compartmentalized in a virtual server or host system. It allows many VMs to be configured on a single physical hardware platform. Many commercial virtualization providers offer both onsite and cloud instances. VirtualBox, VM Ware, and Virtual PC are popular for local installations. Some IaaS providers, such as Microsoft's Azure and AWS, also provide services on the virtualization market. They offer cloud-based virtual servers to run custom solutions.

5.6.2 Virtual Memory

Virtual memory is a technique that uses slower storage (e.g., hard drives) to access and map memory for operating systems. Many operating systems have a swap area where memory can be stored on a drive. Ideally this memory would not be used to avoid any performance problems. The speed of memory is between 2 and 20 GB/s, and the fastest SSDs are still under 550 MB/s. Since many operations occur within memory, speed difference can be noticeable for a very large number of operations. When systems exhaust their physical memory and are forced to use their virtual memory frequently, the machine may become unresponsive, spoiling the user experience.

5.6.3 Virtual Storage

Storage can be virtualized so that the underlying physical storage can be switched out, relocated, and upgraded with minimal down time. Storage area network (SAN) is one form of virtualized storage. Aspects of virtual storage may be transparent, such as in band solutions, or they may require additional requestor/client awareness and software when out of band.

5.6.4 Virtual Networking

Virtual networking provides an interface between the underlying physical layers, such as wires and switches, and the connection points. Common aspects of virtualization include local area networks to provide connectivity between either virtual or physical machines, often without any actual network or wires, but as a set of hosted VMs. Virtual private networks (VPNs) allow devices to connect across networks and map endpoints for secure access as if they are connected on a single physical network.

5.6.5 Virtualization Performance

Performance can be measured by evaluating user efficiency and computational efficiency. In cases such as virtualized hardware, reduced performance of the applications running on this setup is observed as the host virtualization system distributes available resources between guests. VMs can be started and stopped so that functions such as testing or load balancing can be performed in real time. In some cases, virtualization performs better with a highly tuned configuration.

5.7 SERVERS

Servers provide the backbone of an infrastructure. One may organize servers into multiple tiers, creating a secure onion-like structure to protect the most critical elements. Appropriately authenticated and authorized users and accounts have access to these tiers, allowing minimum required access to data and resources. Servers with more dynamic configurations are replacing monolithic ones because virtualization has advanced and because the incoming global data stream has grown in volume, indicating the need to distribute storage, networking, and processing.

5.7.1 Server Components

Servers include hardware processors, memory, and motherboards that provide interfaces, which in turn connect primarily with storage and networking. Server components are hardened and expected to be under load at all times. They are designed for additional redundancy, management capabilities, and optimal efficiency. Server components provide greater durability by removing future expandability in the interfaces. Currently, a more streamlined infrastructure is being developed and the latest information can be found at the OpenCompute website: http://www.opencompute.org/.

5.7.1.1 Processors

Processors, also known as central processing units (CPUs), handle all the basic system instructions in a computer. Over the decades, they have evolved into microprocessors to perform the basic operations of fetching data, conduct operations from memory into chip registers, operate using the arithmetic logic unit (ALU), and store the results into registers. Other processors found in physical machines offer specialized functionality, such as controlling data mappings in storage controllers for various RAID implementations, or GPUs, which provide high-speed calculations across many parallel cores. CPUs now offer multiple cores in a single chip and are functionally independent. In processors, high-speed memory is expensive and used only during the calculation and result display stages. Then the results are returned to the main system memory, which is slower than the chip memory by several magnitudes.

5.7.1.2 Memory

Memory or random access memory (RAM) allows quick data access. Data in memory is available only while the machine is powered on; it is lost when the machine is powered off. Data must be copied to a long-term storage medium, such as solid-state or hard disk. When a machine is turned on, information is copied from these storage media into memory and brings the operating systems, applications, and data into memory for quick processing. High-speed memory on the CPUs and cores is often referred to as L1, L2, and L3 memory caches.

Some systems need high-speed storage and high-speed data buses for large multiple-terabyte datasets. Other systems need subterabyte common data loads and gigabytes worth of fast memory to optimize the hardware and allow the processes to be performed on the server.

5.7.1.3 Interfaces

Computing hardware has internal and external interfaces. Common external interfaces include USB, Thunderbolt, and 110/220-volt cords. Internal interfaces include serial advanced technology attachment (SATA), serial attached small computer system interface (SAS), small computer system interface (SCSI), and fiber channel for many storage drives, which connect through the motherboard buses and cabling. Trade-offs between cost and performance among different interfaces are continuously improving. Other aspects, such as the controller cards and support of RAID capabilities, can offset the limitations in some storage interfaces.

5.7.2 Server Performance

Tailoring performance to the task is an important need for many servers. Networking servers, firewalls, load balancers, web servers, database servers, file storage servers, and computational performance present different needs, and require prioritizing certain components and limiting the number of channels for certain devices. The server should be designed and tailored to the task at hand. With current virtualization trends, the ability to provide scalability by increasing the nodes is ideal, while many options are available to partition additional processing cores, memory, and storage. These approaches are not seamless, but they can be managed to minimize noticeable impacts for system users.

5.7.3 Server Availability

Server availability requires a balance between cost of increased redundancy to minimize down time and what is adequate for operational requirements. Ideally, systems would be available at all times and at full capacity, but the need to address security changes with patches, upgrades, and functional improvements requires a system to be offline, even if briefly. Although these downtimes are scheduled and often coordinated to minimize disruption, some situations may require unplanned downtime. Downtime periods are more frequently due to security exposure or logical errors than to hardware or system instability.

5.7.4 Server Operating Systems

The most popular server operating systems are Linux, Windows, and UNIX-based servers. Popularity percentages are approximately 37% for Linux, 33% for Windows, and 30% for UNIX (http://w3techs.com/technologies/overview/operating_system/all). Linux is robust, cost-effective, and globally improved by peer contributors as an open-source solution, which allows many bugs and security issues to be detected and fixed faster than they can be for other server operating systems.

5.8 DATA CENTERS

Data centers provide stable facilities for housing servers and key interconnects in networking backbones. They are constructed and maintained to provide the greatest redundancy and to ensure constant uptime and access to the computing resources. Data centers are geographically diverse. Larger organizations commonly have fully redundant resources not only within a data center but also across other data centers. Despite the highest levels of protection offered by data centers, natural calamities such as floods and earthquakes can reduce their operability.

5.8.1 Traditional Data Centers

Traditional data centers provide in-house accessibility and control for large organizations with substantial processing needs. Over the past decades, data centers have grown in capacity, often requiring increasing levels of cooling, energy, and space. Traditional data centers provide physical cloud solutions for organizations that have very high security concerns or provide services to others.

5.8.2 Cloud Data Centers

Cloud data centers are designed primarily or exclusively for dynamic computing resources that can be partitioned out to customers leasing the service. The underlying actual physical hardware is abstracted away from the operating systems, and the hardware is managed independently of the service users. The hardware is most commonly assigned to users tiered based on the number of the cores, amount of storage, type of storage, and amount of memory for each compute node.

5.8.3 Costs of Data Centers

The costs of data centers include power, climate control, maintenance, security, fault tolerance and disaster planning, and replacing equipment for both failure and upgrades. Besides the costs to service subscribers, commoditized service costs are substantial. However, through large-scale volume service with homogenized equipment, where the circumstances allow leveraging these data centers, costs are shared across subscribers and offer significant savings to end users.

5.9 PERIPHERAL AND MOBILE DEVICES

Additional devices are commonly used within the infrastructure systems. Examples include medical devices and systems that have devices tracking subjects' (devices' or peoples') locations in real time. Many of the peripherals are used by end users to interface with machines and, in many cases, for machines to autonomously detect, monitor, and offer guidance for operations and treatments. Mobile devices are as capable as many larger form-factor computing devices, and should be treated as though they have as much risk and opportunity as any additional computing resource.

5.9.1 Printing and Scanning

Current-generation printing technologies are precise and fast, with larger sizes up to 11 in. × 17 in. Although many of the printing and scanning requirements are still in place to support legacy implementations, digital and paperless solutions are becoming more available. Previous cost analyses that determined the need to move from paper to digital media did not include the costs of time in using paper with an increasing volume of patients. However, it is now evident that the time and resources needed for workflows that use paper are more expensive than digital solutions. A major advantage of a paperless solution is the ability to sign authorization forms and other documents digitally from anywhere with adequate network connections.

Similar to the conversion from printing on paper to recording on digital media, films used traditionally for pretreatment and treatment quality assurance and calibration during radiation therapy have been replaced with electronic measuring devices.

5.9.2 Laptops and Tablets

Laptops and tablets allow users to remain connected to the electronic information infrastructure without being tethered to physical power or network connections. The onboard batteries provide all-day power and common popular networking technologies.

5.9.3 Bring Your Own Device

Personal cell phones and tablets allow users to balance professional and personal schedules. These bring your own device (BYOD) situations bring improved efficiency and user experience. Risks associated with data access in case a device is lost need to be evaluated and addressed to prevent any unauthorized use. Data encryption is often required for the device if access to patient data is needed. A shared agreement may be needed so that the data can be wiped remotely to avoid any unwanted access or liability exposure.

5.9.4 Real-Time Location Systems

Real time location systems (RTLSs) have an increasing presence in medical practice. Efforts to improve patient outcomes and workflow practices and efficiency may benefit from leveraging these technologies. The systems incorporate various locator tags offered in active and passive response varieties. The latest transceivers can provide improved accuracy for tagged items. Some operational systems include tagging assets for inventory management and tracking. Others include equipment used in procedures to ensure that the appropriate equipment is in use for any given treatment. Also, some systems allow tracking the response to accidents and ensure that patients are scheduled and examined as quickly as possible. Radio frequency identification (RFID) is a component of real-time location systems. Many other components are active in systems such as Global Positioning System (GPS), Wi-Fi, ZigBee, RFID, ultrasonic, and infrared to provide identification and location information. They provide inputs, such as scheduling and notification to both patients and clinicians, that can be integrated into various systems from hospital information systems.

5.10 CLOUD COMPUTING

Computing technologies have advanced to the point that resources are often virtualized. This allows the technologies to be commoditized and offered in standard tiers of computing power, speed, and storage to meet various requirements. Cloud computing is not a solution for all computing requirements, but it offers new and exciting alternatives to the traditional onsite computing infrastructure.

5.10.1 Mobility Needs

Increasing requirements to access information from multiple locations and different devices with different operating systems, form factors, capabilities, and interfaces emphasize the need to advance cloud computing. Mobile devices are used widely in healthcare, and data acquisition and responsiveness to these signals is essential. Cloud computing is a useful tool to scale up and out to handle demand and growth.

As access to information becomes more distributed, every endpoint needs to be protected. Authentication through more than a single factor should be considered standard practice. Authentication of multiple form factors along with encryption and strong password policies increases protection against unintended data exposure.

5.10.2 Cloud Architecture

Cloud architecture provides great variability and configurability. As the need for cloud applications grows, implementations and technologies can be scaled to handle growth while also providing protection when technological limits are reached and changes must occur behind the scenes. Performance, capability, and application functionalities are expanded by adding to the web-based interfaces. Multiple formats in display and information collection are enabled by decoupling back-end processing from the front-end presentation and user interactions.

5.10.3 Provide Services through Clouds

Providing services through clouds streamlines many aspects of application use, management, and growth. Users have the ability to interact with other users in a highly integrated manner. Insights can be tagged, highlighted, notated, and shared with colleagues quickly within a unified system to improve review and auditability. Additional integration provides displays, notifications, and other system information to patients and clinicians.

5.10.4 Big Data

Big data are large datasets that can be aggregated and analyzed in real time. In many cases, big data may be analyzed in overlapping and correlating events across many different domains. Examples include associating polls with social media streams mined for keywords, or natural-language processing systems that determine favorable or unfavorable reactions to certain events. Other examples can also include image processing to determine if certain Internet memes and images convey reactions toward events. Also, social media streams can be used to capture many audio- and videorecordings and understand how people and events are related. In the medical field, relationships and causality are being investigated on larger scales with larger datasets, such as genomic data. Big-data applications often process huge amounts of data, with sizes in the range of hundreds of terabytes and petabytes. Innovative techniques, tools, and technologies are needed to scale data across hundreds and thousands of nodes and process them efficiently to provide results rapidly.

5.10.5 Managing Cloud Resources

Cloud resources reduce the level of direct management because many of the efforts are handled by the cloud service provider. Knowledge of data volumes and relations to business growth and use rates help determine when increased levels of computing performance or storage are required. While scaling up and scaling out are available from cloud providers, technical understanding of needed resources, use, and expectations is still required. Depending on the system design on the cloud platforms, adding resources may be seamless and may reflect a higher cost for increased use. Additional resources may also result in slower throughput and a diminished end-user experience.

5.11 HIGH-PERFORMANCE COMPUTING

High-performance computing is the designation given to high-level capacity computation machines. Over the past half century, these machines have evolved from a few processors to hundreds of thousands of processors providing quadrillions of floating point operations (FLOPS), measured as peta-FLOPS. These machines are used for the most computationally expensive and intensive efforts, such as cryptographic exposure and breaching, and simulations of the most complex systems, including initial moments of the universe and nuclear-related behaviors. In the medical field, high-performance computing is used both in biological and treatment simulations.

5.11.1 Clusters

Clusters interconnect computational resources to increase either parallelized computing or redundancy. With highly parallelized clusters, the computation workload is distributed among the nodes in the cluster. A hybrid approach may be needed to connect meshes of highly interlinked nodes with other meshes across the Internet.

In the case of redundancy, clusters act as readily available machines that share sufficient state and data information. If an active node fails, a mirror node takes over, allowing activities to continue with minimal disruption to end users.

5.11.2 GPU-Based High-Performance Computing

Recent developments in radiotherapy therapy demand high computation powers to solve challenging problems in a timely fashion in the clinical environment. The GPU has been introduced to radiotherapy as an emerging high-performance computing platform. It is particularly attractive due to its high computational power, small size, and low cost for facility deployment and maintenance (Jia and Jiang, 2016). GPU-based technologies make feasible many precision radiotherapy procedures, such as online adaptive treatment planning and delivery using Monte Carlo simulation that demands high-performance computing.

5.12 SECURITY STRATEGIES

Security strategies require data to be protected through adequate authentication and authorization processes. Appropriate segmentation within the infrastructure by tiers and applications is needed for proper security. A single solution or layer of protection against unwanted access is currently inadequate. Access needs to be safeguarded between each layer, such as between a database and an application server. The application client, the end-user interface, and the end-users' endpoint such as a mobile device, also need to be secured. The transmission of data between these layers should be encrypted and, when possible, the network infrastructure should be limited to those channels specifically.

5.12.1 Authorized Access Strategies

Authorized access strategies are the methods by which access to resources and the actions on those resources are defined. Some resources may allow only new information to be added; others allow the ability to read the data, access metadata, access the actual data, and prevent all access to the resource. Resources can include machines; shared files; devices such as printers; or other, more complex equipment.

5.12.1.1 Microsoft Active Directory

The Microsoft Active Directory has now become the Microsoft Active Directory Domain Services; it allows for enterprise management of security. Microsoft products are known for their highly integrated controls in the Active Directory security model that enables streamlined deployment, and also development, updating, and management of resources from centralized servers. It offers tools for trust-based authentication and authorization of resources. Despite being proprietary, Microsoft Active Directory offers a set of mature Application Programming Interfaces (APIs) for third-party applications to extend the capabilities and interfaces for managing network security and reporting. The increase of third-party options coupled with the ability to extend program functionality has allowed the Microsoft Active Directory to keep up with requirements and changes with both cloud and local implementations.

5.12.1.2 Novell Networking

The latest enterprise security management solution from Novell Networking is the Novell Open Enterprise Server, which adds daemons to Linux installations to provide common central management for authentication and authorization across managed machines. Novell legacy solutions were market leaders in the early 1990s. Competition and emerging open-source technologies have reduced Novell dominance. Current offerings support heterogeneous environments where open-source, Apple, and Microsoft system security are required.

5.12.2 Antivirus Protection

Protection of information resources is a vital aspect of ensuring that the infrastructure is ready for use. Such protection applies to small installations as well as the largest enterprises. Because of the increasing connected nature of data exchanges and user interfaces, both internally and externally, it is necessary to verify and update security software to protect against virus and malware threats.

5.12.3 Data Backup Plans

Data backup plans are crucial components to address any unforeseen failures. Backups should be able to restore data and eliminate risk to any business. Backups are conducted daily and taken offsite nightly to avoid excess data loss. To prevent data loss on a local machine, other backups run in real time and move the data offsite through the cloud. Through this system, data could be restored within seconds of the failure.

5.12.4 Data Breach Prevention

Data breaches should be prevented by implementing computer practices that ensure password complexity, changing passwords frequently, and multiform authentication. In addition to explicit prevention, implicit measures such as active auditing of access and usage are also advised.

5.13 PLANNING BUSINESS CONTINUITY AND DISASTER RECOVERY

Contingency plans and resources to ensure continuity and recovery should be developed to respond adequately to disasters. Key elements of the plan should include prioritizing activities, logistics sequencing, and redundant alternatives.

5.13.1 Planning Hardware Updates

Systems should plan for hardware and firmware updates. Updating operations may require resetting the system or bringing it to a level of operation that cannot be accessed by other systems and users. Downtimes can be scheduled outside business hours to minimize any work flow disruption. If systems need to remain online, redundant clustering techniques allow one system to remain in service while others are updated. Then the system is switched over to the updated one, and the out-of-service cluster is updated. Common processes ensure that adequate service levels are maintained for the system.

5.13.2 Planning Software Updates

Software updates are required to maintain security, improve functionality, and reduce unintended effects of certain behaviors. Updates can involve the operating system, application, scripting component, or even data used by the software. Despite recent advances, some updates may still require a system to be removed from service temporarily. As updates increasingly involve the application layer, updates can lower uptime. Because application updates may be incompatible with older versions, system data and functionality should always be tested after upgrading and connecting with other systems.

5.13.3 Backup

Backups are essential to protect systems and users. Their frequency depends on data sensitivity; media on which backups reside; and whether the backups are kept onsite, offsite, or both. Backups should be tested regularly by restoring the schedules and verifying the data to ensure proper function during emergency situations.

5.13.4 Disaster Recovery Plan

The first step of a disaster recovery plan is to identify the business-critical systems. The plan is to determine the needs of those systems and bring them to a fully operational state. Examples include hardware specifications or standby hardware ready to bring fully redundant hardware online in remote locations with access to backup data. Dependent systems need to be brought online in a certain order, with clearly defined technical needs and business priorities. The plans also identify end-user steps and alternatives to ensure that the operations of a functional organization are resumed to maintain business continuity.

5.14 SYSTEM INTEGRATION

A hospital environment has many systems. As systems continue to evolve, the need for them to be interconnected is growing. Integration between systems occurs through both physical wired and wireless means. Also, the logistical layers of data formats and hardware signaling are interconnected. Optimal integrations leverage many communication standards to allow systems to quickly identify and consume data, often with much self-described metadata that allow systems to consume data sources dynamically and more readily.

5.14.1 Hospital Information Systems

Hospital information systems provide a comprehensive set of applications designed to manage all financial, legal, and medical operations of a hospital. Most systems operate with central servers and distributed clients across different clinics and administrative locations. The primary goal of the system is to provide a centralized system to secure health records and to improve efficiency and the quality of care.

5.14.2 Information Systems for Radiation Therapy

Within a hospital, radiation therapy (RT) requires extensive IT support. Computer programs such as TPSs, treatment management systems (TMSs), treatment delivery systems (TDSs), Radiation Oncology-specific Electronic Medical Records (ROEMR), and image-viewing systems (e.g., picture archiving and communication systems [PACSs]) are used routinely (Siochi and Balter, 2009). As we move toward personalized precision medicine, radiation therapy is adopting emerging new technologies. As a result, planning, delivery, and management information systems are becoming more complex. Also, as the amount of data increases, comprehensive information management schemes are more and more a prerequisite to correct treatment delivery. The information system provides the backbone for communication, documentation, and quality control within a radiation therapy clinic. In academic centers, clinical data is constantly mined by various systems or applications to improve patient care. In many cases, a physical or electronic barrier is placed between the information systems in the clinic and those in the research areas to allow the smooth operation of clinical procedures and to meet Health Insurance Portability and Accountability Act (HIPAA) regulations. A typical diagram of an RT information system is illustrated in Figure 5.3.

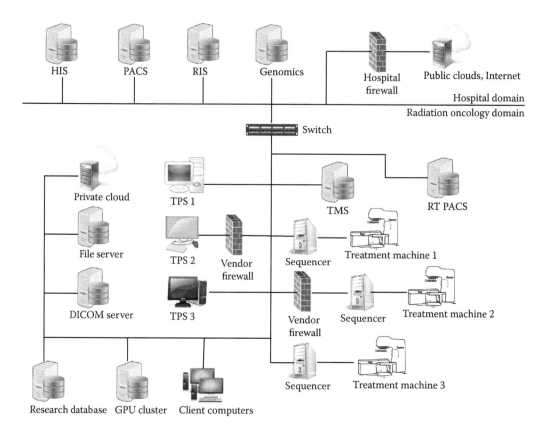

FIGURE 5.3   An RT information system.

5.14.3 Communication Standards

Methods integrating information between electronic systems are continuously evolving to allow multiple data providers to generate a format that can be used consistently. DICOM and Health Level Seven International (HL7) are the most relevant to daily clinical practice.

5.14.3.1 DICOM

DICOM was devised by the National Electrical Manufacturers Association (NEMA) in collaboration with the American College of Radiology (ACR) to facilitate the interoperability of systems in a multi-vendor environment. DICOM covers most image formats in medicine, including messaging and communication protocols between imaging modalities and systems. DICOM is the backbone of the Picture Archiving and Communication System (PACS).

One of the extensions of DICOM was devised for RT in 1997 and is known as DICOM-RT. DICOM-RT uses five extended information object definitions (IODs), namely, RT image, RT structure set, RT plan, RT dose, and RT record to model clinical activities in radiation therapy. A PACS tailored to fit RT clinics is called RT PACS. DICOM-RT accommodates the different clinical settings, a wide range of modalities, and intermodality interactions.

To address the interoperability issue typical of the RT multi-vendor environment, the Integrating the Healthcare Enterprise—Radiation Oncology (IHE-RO) initiative was developed.

IHE-RO is an American Society for Therapeutic Radiology and Oncology (ASTRO) initiative that improves system-to-system connections in radiation oncology through the coordinated use of established standards. The IHE-RO task force addresses any connectivity or communication problems that may arise with software or equipment. The goal is to improve patient care and reduce medical errors by refining operability and eliminating incompatibilities.

DICOM standards are evolving rapidly to accommodate emerging new technologies. They are revised multiple times a year by RT experts worldwide.

5.14.3.2 HL7

HL7 is the healthcare data format for interoperability and is named after the seventh level in the open systems interconnection (OSI) model. It is the standard of choice to convey nonimage information in hospital information systems.

5.14.4 Programming Strategies

Strategies for development efforts are essential to ensure long-term success in leveraging infrastructure. Modern development practices include distributed version control repositories, integrated development environments (IDEs), automated deployment, and testing.

Design patterns are object-oriented, language-agnostic best practices that reduce cost of change and improve maintainability of software codes. Advanced concepts require greater development acuity and are strongly encouraged for software development efforts expected to grow and add functionality over time. Examples of development include scripting additional functionality in applications, programming automated exchange between systems for improved workflow efficiency and accuracy, and modeling new simulations of treatment plans. The complexity of programming or software engineering efforts should be matched appropriately to adequate levels of modularity, encapsulation, and isolation.

5.15 TELEPHONY TECHNOLOGIES

Telephony has expanded and evolved over the years. Communication is being unified so an originating seeker of an individual or party can focus on the content or question to send. New technology routes different types of messages by tracking activities in various interfaces. The format of the messages can be translated between audio and text (or vice versa) so that the medium of the message is less of a concern for both sender and receiver. Translation into different languages is also readily available, though it needs to be improved. Technology has also facilitated message routing through texting, web application interface, email, and machine speech audio. Smart integrated systems allow notifications and reminders to be delivered on mobile devices through ringer muting or vibration with minimal disturbance to unintended recipients.

5.15.1 Internet Protocol Phone Solutions

Internet Protocol (IP) phones offer the ability to be reached virtually through the Internet. Some may require network access through VPNs, but they allow a greater range of flexibility to users who need to be reached easily while changing locations frequently. Some IP phones also allow users to check e-mail quickly and integrate voice mails into e-mails (or vice versa).

5.15.2 Telephony Integration

Unified communications systems are integrated tools that help people communicate with others across multiple channels. Services include voice connecting, texting, paging, and video systems. Present systems allow users to eliminate or minimize the effort spent reaching others.

5.16 IT STAFFING

Staffing an IT department depends on the goals and objectives of the organization. A thorough understanding of the organization culture helps to identify the appropriate candidates. Innovation-centered organizations generally look for people who can learn new technologies quickly and solve problems effectively. Candidate credentials should be aligned with the organizations' objectives based on their specific needs. IT staff should be recruited carefully according to current policies, procedures, and resources.

5.16.1 Desktop Support

Desktop support requires careful planning and requires excellent technical and organizational skills, knowledge of security systems, and the ability to troubleshoot. For small organizations, outsourced support can provide expertise and affordable rates.

5.16.2 System Administration

System administration ensures that one or more systems are running appropriately and efficiently. System administration oversees user management, backups, and scheduling software and hardware upgrades as well as security patches. Expertise is required to integrate new systems into the infrastructure. System administrators are involved in upgrading and updating custom development solutions. Best practices should be followed to ensure that resources are secured and that accounts remain functional.

5.16.3 Database Administration

Database administration oversees data monitoring, maintenance, and management. Database backups must restore systems to near failure points quickly and consistently while large volumes of transactions are occurring in the system. Databases need to be maintained and secured adequately, and they must be accessible to other systems. Database administration tasks may also be extended to managing mirror databases, publishing, and subscribing to datasets. The number of users and applications accessing the database also contributes to the workload of the role.

5.16.4 Security and Compliance Officers

Security and compliance officers oversee security policies and procedures. Policies need to be documented to provide adequate guidance and consistency for developers, administrators, and users. Officers ensure that mismatches between the desired levels of security are mitigated through appropriate remediation plans and policy updates. The systems should be audited regularly to avoid data breaches and to ensure appropriate levels of access.

5.17 CONCLUSION

As an interdisciplinary profession, radiation oncology features highly complex equipment that generates ionizing radiation to treat cancer patients. Multidisciplinary expertise is needed to commission, maintain, and operate these machines. In addition to the physical equipment, information technology systems are a crucial component of a functioning radiation oncology clinic. The IT system represents the central nerve system of the "living" radiation therapy clinic that handles memory (database), knowledge (machine learning), communication (data transfer, scheduling, and worklist), and impulses (digital monitoring and safety interlocks). In the future, advanced information technologies will (1) improve treatment safety, (2) improve treatment quality, (3) improve treatment efficiency, (4) provide decision support, (5) improve patient experience, and (6) enhance education and research and development

REFERENCES

Graves, M. 2005. *Data Recovery and Fault Tolerance: The Complete Guide to Networking and Network+*. Delmar Cengage Learning, Clifton Park, NY.

IHE. 2014a. Radiation oncology (RO) technical framework, Volume 1 10 IHE RO TF-1 profiles.

IHE. 2014b. Radiation oncology (RO) technical framework, Volume 2 10 IHE RO TF-2 Transactions.

IHE. 2016. IT infrastructure (ITI) technical framework, Volume 1 (ITI TF-1) Integration profiles.

International Commission on Radiation Units and Measurements (ICRU). 1987. Report 42: Use of computers in external beam radiotherapy procedures with high-energy photons and electrons. Bethesda, MD.

International Commission on Radiation Units and Measurements (ICRU). 1976. Report 24: Determination of absorbed dose in a patient irradiated by beams of X or gamma rays in radiotherapy procedures. Washington, DC.

Jia, X., and S. B. Jiang. 2016. *Graphics Processing Unit-Based High Performance Computing in Radiation Therapy*. CRC Press, Boca Raton, FL.

Layton, Jeffrey B. 2011. Intro to nested-RAID: RAID-01 and RAID-10, Linux-Mag.com. *Linux Magazine*.

Siochi, R. A., P. Balter, and C.D. Bloch. 2009. Information technology resource management in radiation oncology. *Journal of Applied Clinical Medical Physics*. 10:16–35.

Van Dyk, J. 2012. Commissioning and quality assurance. In *Treatment Planning in Radiation Oncology*, (Eds.) F. M. Khan, and B. J. Gerbi, 144–168. Philadelphia: PA: Lippincott, Williams and Wilkins.

CHAPTER 6

Advances in Inverse Planning Algorithm and Strategy

Masoud Zarepisheh, Baris Ungun, Ruijiang Li,
Yinyu Ye, Stephen Boyd, and Lei Xing

CONTENTS

6.1 INTRODUCTION

The landscape of radiation therapy (RT) has changed dramatically in the past two decades due to the development of dynamic multileaf collimators (MLCs). Today, a number of intensity modulation techniques are available to deliver high doses of radiation to the tumor while sparing the surrounding critical structures. These techniques include fixed gantry intensity-modulated radiation therapy (IMRT) (Ezzell et al., 2003; Bortfeld, 2006) (typically, with five to ten beams), volumetric modulated arc therapy (VMAT) (Yu and Tang, 2011) (typically with one to two arcs), and tomotherapy (Mackie et al., 1993). Many inverse planning optimization algorithms have been developed by researchers to support these techniques.

Despite significant advances in inverse planning and optimization algorithms, treatment planning is still a patient-specific and time-consuming trial-and-error process. A planner needs to specify a set of goals along with their importance weights based on past experience. Then the planner runs the optimization and tweaks the objectives and weights until a satisfactory plan is found. A significant effort has been dedicated recently to automate or facilitate the planning process using either prior knowledge or some advanced optimization techniques, and Section 6.2 of this chapter is devoted to this subject.

Optimization algorithms are grouped into two main categories: direct machine parameter optimization (DMPO) (also known as direct aperture optimization [DAO]) and beamlet-based optimization (BBO). In BBO, each beamlet intensity is optimized as an independent variable, and then leaf sequencing is carried out to segment the optimal beam profile. BBO does not account for deliverability, so the optimal beam profiles can be complex and far from deliverable. DMPO, on the other hand, includes hardware constraints and optimizes aperture shapes directly, but its formulation suffers from nonconvexity; consequently its implementations are computationally expensive. In Section 6.3, compressed sensing (CS) inverse planning is introduced; it could be considered an intermediate approach that incorporates the interplay between delivery and planning and enjoys the convexity of BBO.

The last few years have witnessed significant advances in linac technology and therapeutic dose delivery methods. A new generation of digital linacs has become available that offers unique features such as programmable motion of station parameters and high dose-rate flattening filter free (FFF) beams. Current treatment planning techniques (IMRT and VMAT) are designed for traditional machines and do not leverage the power of the new digital linacs. Station parameter optimized radiation therapy (SPORT) (Li and Xing, 2011; Xing et al., 2013; Li and Xing, 2013; Xing and Li, 2014; Zarepisheh et al., 2015) and 4π non-coplanar delivery (Dong et al. 2013) have been introduced recently to deliver highly conformal radiation dose distributions efficiently by fully utilizing the technical capability of emerging digital linacs (Wang et al. 2012; Fahimian et al., 2013; Xing and Li, 2014; Xing et al., 2014). Section 6.4 of this chapter is dedicated to SPORT. First, the main idea behind SPORT and some heuristic implementations and delivery strategies are explained. Finally, a more rigorous, mathematical optimization framework, with the fundamental station point parameters as the decision variables, is derived (a station control point describes the state of delivery system including linac configurations such as beam energy, aperture shape and weight, gantry/collimator angle, and auxiliaries such as the couch).

List of Abbreviations

Acronym	Expansion
RT	Radiation therapy
MLC	Multileaf collimator
IMRT	Intensity modulated radiation therapy
VMAT	Volumetric modulated arc therapy
DMPO	Direct machine parameter optimization
DAO	Direct aperture optimization
BBO	Beamlet-based optimization
FFF	Flattening filter free
SPORT	Station parameter optimized radiation therapy
MCO	Multiple criteria optimization
DVH	Dose volume histogram
PO	Prioritized optimization
CS	Compressed sensing
TV	Total variation
TFOCS	Templates for First-Order Conic Solvers
ADMM	Alternating direction method of multipliers
BEVD	Beam eye's view dosimetric
OAR	Organ at risk

Notation Glossary

Symbol	Description	Symbol	Description
X	Aperture shape (leaves' positions)	i	MLC row index
X^A/X^B	Left/right leaves' positions	j	MLC column index
MU	Cumulative monitor units	s	station index
θ	Beam angle	k	Voxel index
d	Maximum delivered dose without exceeding the prescription	N_k^T	Total number of tumor voxels
		N_i	Total number of MLC rows
p	Prescription dose	N_j	Total number of MLC columns
p^T	Target prescription dose	N_k	Total number of voxels
D	Dose deposition matrix	N_s	Total number stations
z	Delivered dose		
x	Beamlet intensity		
w	Importance factor		
F	Objective function		
η	Aperture intensity		
$\Phi(X_b,\theta_b)$	Delivered dose from station b		

6.2 AUTOMATED AND FACILITATED TREATMENT PLANNING TECHNIQUES

Currently, inverse treatment planning is a patient-specific, time-consuming, and resource-intensive task, with plan quality heavily dependent on the institution and the planners' skill and experience. In this section, a brief review of some recent advances to facilitate this process is presented.

6.2.1 Multiple Criteria Optimization: Pareto Surface Navigation

The main challenge in treatment planning is to find clinically acceptable trade-offs between the conflicting objectives of (1) delivering enough radiation to the tumor and (2) not overdosing the surrounding tissues. The conventional approach turns the multiobjective problem into a single-objective problem by minimizing the weighted sum of the objectives, and the weights are tuned in a trial-and-error fashion. The multiple criteria optimization (MCO) approach (Kufer et al., 2000; Zhu and Xing, 2009; Craft et al., 2012) suggests generating a pool of potentially good plans in advance, known as Pareto optimal plans, and then allowing the planner to navigate among them and select the best one.

A plan is called Pareto optimal if it cannot be improved in any objective without worsening the others. Figure 6.1 explains the Pareto optimality concept in a schematic way for a very simple case where there are only two objectives: (1) Planning Target Volume (PTV)-minimum dose that needs to be maximized, and (2) Organ at Risk (OAR)-maximum dose that needs to be minimized. The shaded area represents the set of all feasible (achievable) plans. Plan A inside the feasible region is not Pareto optimal because there exists plan B with the same PTV-Min dose and better OAR-Max dose. In this situation, plan B dominates plan A. However, Plan B is Pareto optimal because one cannot improve one of the objectives without worsening the other; in other words, there is no plan inside the feasible region that dominates plan B. The set of all Pareto optimal plans is called Pareto surface.

MCO simplifies the treatment planning by eliminating the non-Pareto optimal plans and generating a discrete representation of the Pareto surface. How many Pareto plans are needed to represent the Pareto surface and how these plans can be generated have been the topics of active research recently (Craft et al. 2005; Craft, 2006, 2010).

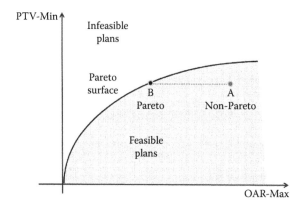

FIGURE 6.1 Plan A is not Pareto optimal because there is another feasible plan, plan B, with the same PTV-Min objective and better OAR-Max objective. Plan B is Pareto optimal because none of the objectives can be improved without worsening the other.

6.2.2 Knowledge-Based Treatment Planning

The main idea of knowledge-based treatment planning is to extract information from plans made for previously treated patients with similar medical conditions in order to facilitate the treatment of a new patient. This information could be, for instance, beam angles (Pugachev and Xing, 2001; Zhang et al., 2011; Kim et al., 2015), objective function weights (Boutilier et al., 2015), or dose-volume histogram (DVH) points (Wu et al., 2011; Zarepisheh et al., 2014a). Of these, DVH points have been the most popular for researchers who have been trying to extract information using patient geometric similarity metrics.

Once a new patient comes in, one can mine a dataset of treated patients and find achievable DVH points. These points, which are hereafter referred to as reference DVHs for simplicity, can serve as good initial goals in the conventional treatment planning process, or one can take advantage of sophisticated optimization algorithms to reproduce them on the new patient's geometry (Cotrutz and Xing, 2003a; Zarepisheh et al., 2014a). Reference DVHs may be either inside or outside the feasible region of the new patient; using optimization algorithms, one can project them onto the Pareto surface.

6.2.3 Prioritized Optimization

Prioritized optimization (PO), sometimes referred to as lexicographic optimization, reduces the need for trial-and-error fine-tuning by prioritizing the planning goals and optimizing them in order (Jee et al. 2007; Wilkens et al., 2007; Long et al., 2012). PO is a stepwise technique where the highest priority goal (e.g., tumor coverage) is optimized first. At each iteration step, the previous objectives are turned into constraints and a new goal is optimized. It has been shown empirically that the objective function at each iteration improves significantly if one slightly relaxes the constraints introduced by the previous objectives. Figure 6.2 illustrates how the PO may look for a prostate case. On the first step, tumor coverage is maximized while respecting maximum dose constraints on the other structures. Then the tumor coverage achievement of the first step is turned into a constraint for the second step, and the rectum dose is minimized. Finally, at the last step, the bladder dose is minimized subject to the constraints of the previous steps.

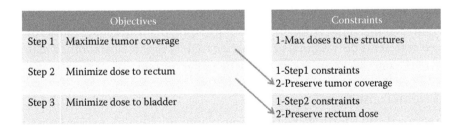

Objectives		Constraints
Step 1	Maximize tumor coverage	1-Max doses to the structures
Step 2	Minimize dose to rectum	1-Step1 constraints 2-Preserve tumor coverage
Step 3	Minimize dose to bladder	1-Step2 constraints 2-Preserve rectum dose

FIGURE 6.2 A schematic of the prioritized optimization workflow. Tumor coverage is optimized first and then turns into a constraint for the second step, where the rectum dose is minimized. Bladder dose is minimized at the last step subject to the constraints of the previous steps.

6.3 COMPRESSED SENSING INVERSE PLANNING

In the traditional beamlet-based algorithms (BBO) (Xia et al., 2005; Breedveld et al., 2006; Salari et al., 2012; Zarepisheh et al., 2014a), each beamlet intensity is optimized as an independent and continuous variable. Because the physical constraints of the dose delivery hardware are not included in the optimization, the optimized beamlet intensity map has a high complexity, and the number of apertures for dose delivery is usually large after leaf sequencing. A large number of apertures reduces treatment efficiency and also reduces treatment accuracy due to increased patient motion during beam delivery and the use of irregularly shaped apertures.

On the other hand, DMPO methods (also known as [DAO]) (Shepard et al., 2002; Cotrutz and Xing, 2003b; Romeijn et al., 2005; Otto, 2008; Zarepisheh et al., 2015) have been proposed to eliminate the need for leaf sequencing and can include hardware constraints in the optimization. While DMPO is a better approach in principle, solving the resultant nonconvex optimization problem is challenging, and computational efficiency relies on developing advanced specialized optimization techniques (using a generic nonconvex optimization technique like simulated annealing or pattern search is usually computationally prohibitive). Before presenting the recently developed DMPO method, the CS inverse planning idea is introduced that could be considered as an intermediate approach between BBO and DMPO. CS inverse planning enjoys the convexity and simplicity of BBO on one hand, and it takes into account the interplay between deliverability and planning on the other hand, and it can be employed for either IMRT or SPORT.

Instead of directly including the nonconvex deliverability constraints in the optimization, CS inverse planning proposes an efficient method to achieve a global optimal solution in a sparsified space of fluence maps in which the physical constraints are implied. The derivation is based on the fact that beamlet intensity maps, which can be delivered using a small number of segments, must be piecewise constant, and hence their spatial derivatives must be sparse. The proposed algorithm can be regarded as an application of CS methods from the field of signal processing. Briefly, CS is a technique for acquiring and reconstructing signals that are known to be sparse or compressible. By analogy, the fluence maps are treated as the "signal"; the assumed sparsity of the derivative of the fluence maps makes CS a viable solution to the treatment planning problem. It should be noted that many attempts have been made to reduce fluence map complexity by using various data-smoothing techniques. These algorithms smooth the edges and help delete spiky behaviors of fluence maps. However, the overall shapes of the final fluence maps remain the same; thus, the solution so obtained represents only a small perturbation to the original unsmoothed plan, and the reduction of the number of segments is usually rather limited.

6.3.1 L-1 Norm Formulation

The conventional BBO for inverse treatment planning is based on the linear relationship between the delivered dose distribution on the patient, z, and the intensity of the beamlets, x:

$$z = Dx$$

where z is a vectorized dose distribution for a three-dimensional volume, and the beamlet intensity x is a vector that consists of row-wise concatenations of beamlet intensities for all beams. Matrix D, called the *dose deposition matrix* (also known as dose influence matrix or kernel matrix), relates the beamlet intensity to the delivered dose. Each column of the matrix D is a beamlet kernel, corresponding to the dose distribution achieved by one beamlet at unit intensity. The beamlet kernels are precomputed based on the CT images of the patient, the treatment machine settings, and the beam geometry.

For computational efficiency, a convex function is usually used as the objective function in the optimization. If the objective function $\varphi_1(x)$ is defined to be the square of the l_2-norm of the difference between the delivered dose and the target dose, the treatment planning problem can be expressed as:

$$minimize\, \varphi_1(x) = \sum_k w_k (D^k x - p_k)^2 \tag{6.1}$$

$$subject\ to: x \geq 0$$

where:

w_k is the relative importance factor of each voxel, each column of the matrix
D^k is the beamlet kernel corresponding to the kth voxel
p_k is the prescribed dose.

A piecewise constant fluence map, or a fluence map with a small number of signal levels, is desirable for reducing the number of apertures. The total variation (TV) regularization used in many optimization problems produces an optimal solution that is piecewise constant (Kolehmainen et al., 2006; Block et al., 2007). Inspired by these facts, an additional term, $\varphi_2(x)$, is included in the new optimization problem, and it corresponds to the TV (or l_1-norm) of the beam intensity x:

$$\Phi_2(x) = \sum_{s=1}^{N_s} \sum_{i=2}^{N_i} \sum_{j=2}^{N_j} \left| x_{i,j,s} - x_{i-1,j,s} \right| + \left| x_{i,j,s} - x_{i,j-1,s} \right| \tag{6.2}$$

where:

i and j are MLC row and column indices, and the beamlet intensity
x is parameterized by these variables
N_i, N_j, and N_s are the total number of rows, columns, and stations

Now, the optimization problem can be reformulated as follows:

$$minimize\, \varphi_1(x) + \beta \Phi_2(x) \tag{6.3}$$

$$subject\ to: x \geq 0$$

where β is the penalty parameter associated with the TV term. The TV term calculates the sum of absolute values of the discrete spatial derivatives, and the penalties drive the derivatives toward zero and force the optimized beam intensity to be close to piecewise constant.

Although not exactly equivalent, the TV regularization term in the optimization implies the aperture constraint on the intensity maps. The resulting optimized intensity maps are more readily achievable using apertures, at the price of slightly degraded dose distributions compared to that of the beamlet-based plans. This trade-off can be adjusted by using different values of β empirically (Zhu et al., 2008). An alternative approach is to reformulate Problem (6.3) as a multi-objective optimization problem and navigate the resultant Pareto frontier to find a solution with desirable trade-offs among the objectives (Zhu and Xing, 2009) (see Section 6.3.2).

The proposed approach has been evaluated using two clinical cases (Zhu et al., 2008), a prostate case and a head-and-neck (HN) case. Under the condition that the clinical acceptance criteria of the treatment plan are satisfied, for the prostate patient, the total number of apertures for five fields is reduced from 61 using the Eclipse planning system to 35 using the proposed algorithm; for the head-and-neck patient, the total number of segments for seven fields is reduced from 107 to 28. Figure 6.3 shows one example of the optimized beamlet intensities for the second incident field of the prostate case, optimized once with quadratic smoothing (i.e., replacing the l_1-norm in Problem (6.2) with a l_2-norm) and again with TV regularization. The quadratic smoothing reduces the complexity of the intensity map to some extent, but it also smooths the sharp edges. The TV regularization further reduces the field complexity by preserving the edges, and the resulting intensity map is close to a piecewise constant function. Figure 6.4 shows the DVHs of the PTV and the sensitive structures for $N_s = 35$ stations using the beamlet-based method without regularization, using the quadratic smoothing and using the TV regularization (a leaf sequencing algorithm has been modified to yield exactly 35 apertures [Zhu et al., 2008]). It is clear that TV regularization achieves a better plan on the conformity of the PTV dose distribution when a small number of segments are used.

6.3.2 Pareto Frontier Navigation

Problem (6.3) is inherently a multiobjective optimization problem with two competing objectives, φ_1 and φ_2. A common approach to deal with multiobjective problems is to find a good discrete representation of the Pareto frontier (e.g., evenly distributed points on the Pareto frontier). Each point on the frontier corresponds to a candidate solution embodying different trade-offs, and the user ultimately picks a Pareto solution with a trade-off that matches his or her goals. For a given β, Problem (6.3) yields a point on the Pareto frontier (Figure 6.5a) and different Pareto points can be generated by varying β; however, these points are not necessarily evenly distributed on the Pareto frontier for evenly distributed weights. Finding a good representation of a Pareto frontier is a well-studied problem for which many algorithms have been developed (Craft et al., 2006). Zhu and Xing (2009) explained how this task can be accomplished by solving Problem (6.4) for different value of parameter ϵ.

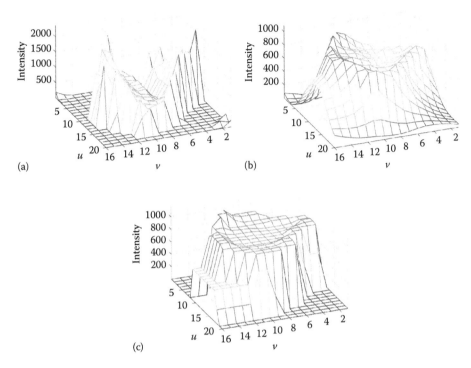

FIGURE 6.3 Optimized beam intensity maps with no regularization, with quadratic smoothing and with total variation regularization. The data are from the second field of the prostate study. (a) With no regularization ($\beta = 0$ in the objective function [3]). (b) With quadratic smoothing. (c) With total variation regularization. (From Zhu, L. et al., *Phys. Med. Biol.*, 53, 6653–6672, 2008.)

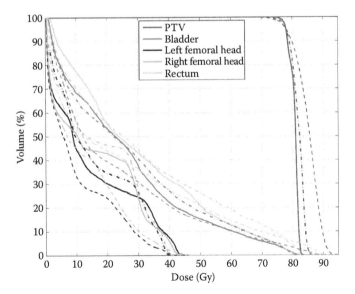

FIGURE 6.4 DVHs of the prostate plans using the beamlet-based method without regularization (dashed lines), using quadratic smoothing (dotted-dashed lines), and using total variation regularization (solid lines). All plans use the same number of segments ($N_s = 35$). (From Zhu, L. et al., *Phys. Med. Biol.*, 53, 6653–6672, 2008.)

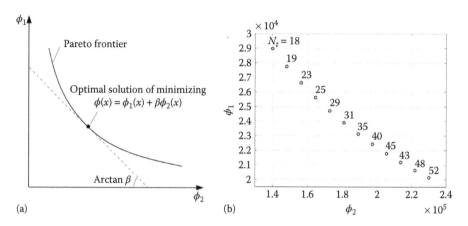

FIGURE 6.5 (a) The optimal point on the Pareto frontier if the optimization is solved using a regularization-based method (Problem [6.3]). (b) The calculated Pareto frontiers of the prostate plans. The derived number of apertures (N_s) corresponding to each data point is marked on the plot. (From Zhu, L. and Xing, L., *Med. Phys.*, 36, 1895–1905, 2009.)

$$minimize\ \varphi_2(x) \qquad\qquad (6.4)$$

$$subject\ to: x \geq 0, \phi_1(x) \leq \epsilon$$

Problem (6.4) imposes an upper limit constraint ϵ on the first objective function, and the minimization is carried out only on the second objective function. (The reader is referred to the original paper to see how different values for ϵ are selected.) Figure 6.5b illustrates the calculated Pareto frontier of the plans for the prostate patient. The number of apertures (N_s) corresponding to each Pareto efficient point after applying a leaf-sequencing algorithm is also marked on the plot. In general, a small (large) φ_2 value on the Pareto frontier achieves a small (large) number of apertures, while the dose distribution is degraded (improved), as indicated by the increase (decrease) in the φ_1 value. However, because the l_1-norm objective in the algorithm only implies the uniformity constraint of the apertures while the connectivity constraint is enforced by the subsequent leaf sequencing procedure, the above relationship is not exactly monotonic. As shown in Figure 6.5b, in some local areas (for instance where N_s varies from 35 to 45 to 43), a larger φ_2 value may yield a smaller number of apertures.

6.3.3 Reweighted l_1-Norm Formulation

Theoretically, l_0-norm, which minimizes the number of nonzero elements, provides the ideal solution for sparse signal reconstruction. In practice, however, solving the l_0-norm problem is computationally intractable because minimizing the number of the nonzero elements is a mixed-integer optimization problem (Boyd and Vandenberghe, 2004) with exponential runtime complexity. As earlier applied, l_1-norm has been widely used as an approximation of l_0-norm. However, Candès et al. (2008) demonstrated that the reweighted l_1-norm provides a better approximation of l_0-norm and can further enhance the sparsity. Thus, it makes sense to employ the reweighted l_1-norm in the application to further simplify the fluence map and enhance the delivery efficiency.

The reweighted l_1-norm method consists of solving a sequence of weighted l_1-norm problems. The weight assigned to each element for each iteration is computed from the value of the previous solution and is inversely related to the magnitude of the corresponding element. If x^k denotes the solution from iteration k of Problem (6.3), then x^{k+1} is obtained by solving Problem (6.3) with $\phi_2(x)$ replaced by $\phi_2^k(x)$ defined as follows:

$$\Phi_2^k(x) = \sum_{s=1}^{N_s} \sum_{i=2}^{N_i} \sum_{j=2}^{N_j} w_{i,j,s}^k \left(\left| x_{i,j,s}^k - x_{i-1,j,s}^k \right| + \left| x_{i,j,s}^k - x_{i,j-1,s}^k \right| \right) \tag{6.5}$$

$$\left(w_{i,j,s}^1 = 1, w_{i,j,s}^k = \frac{1}{\left| x_{i,j,s}^k - x_{i-1,j,s}^k \right| + \left| x_{i,j,s}^k - x_{i,j-1,s}^k \right| + \delta^k} (k \geq 2) \right)$$

where δ^k is a small positive number. Iteratively solving Problem (6.3) with the above $\phi_2^k(x)$ can be seen as an approximation to l_0-norm problem. As δ^k approaches zero, the weighted l_1-norm exactly recovers the l_0-norm. In practice, however, δ^k should be greater than 0 to prevent divergence when the corresponding elements go to zero. Candès et al. (2008) recommended $\delta^k \ll \max\left(\left| x_{i,j,s}^k - x_{i-1,j,s}^k \right| + \left| x_{i,j,s}^k - x_{i,j-1,s}^k \right| \right)$.

The reweighted l_1-norm technique has been evaluated using five clinical cases. A conformation number (CN) (Riet et al., 1997; Oozeer et al., 2000; Paddick, 2000), shown in Problem (6.6), consisting of PTV coverage (> 95%) (CN1) and the healthy tissue protection (CN2), has been used to assess the quality of the plans.

$$\text{CN} = \frac{V_{\tau,ref}}{V_\tau} \cdot \frac{V_{\tau,ref}}{V_{ref}} \cong (\text{CN1}) \cdot (\text{CN2}) \tag{6.6}$$

where V_τ is the volume of PTV, $V_{\tau,ref}$ represents the target volume receiving the dose greater than or equal to the reference (prescribed) dose, and V_{ref} is the total volume receiving the dose greater than or equal to the reference dose. CN1 is set to be 0.95 for all plans while trying to see the variations of CN; a higher CN implies better target dose conformity with healthy tissue protection. Table 6.1 summarizes the results for five clinical datasets. It has been observed that to attain a similar CN, the number of apertures of the plans obtained with the reweighted l_1-norm method was reduced by 10–15 apertures relative to the plans made using conventional l_1-norm, and reduced by 30–35 apertures compared with the quadratic smoothing-based plans.

TABLE 6.1 Comparison of the Plans Acquired by Beamlet-Based Optimization using L_1-Norm, Reweighted L_1-Norm, and Quadratic Smoothing Regularizations for Five Clinical Datasets in Dose Conformity to the Target (CN), and Number of Apertures: The Proposed Method Enables the Attainment of a Similar CN with Many Fewer Apertures

Patient Case	Criterion	Algorithms		
		Quadratic Smoothing	L_1-Norm	Reweighted L_1-Norm
Prostate-1	Apertures	70	50	40
Prostate-2	CN	0.8629	0.8671	0.8632
Lung	Apertures	70	50	40
HN1	CN	0.8753	0.8779	0.8731
HN-2	Apertures	65	50	35
	CN	0.8843	0.8854	0.8870
	Apertures	70	50	35
	CN	0.8739	0.8719	0.8710
	Apertures	95	80	65
	CN (PTV66)	0.7521	0.7522	0.7549
	CN (PTV54)	0.7286	0.7728	0.7878

Source: Kim, H. et al., *Med. Phys.*, 40, 071719, 2013.

6.3.4 Solving the Large-Scale L_1-Norm Problems

Solving Problem (6.4) could be challenging, especially because of the increased number of variables in SPORT and also the presence of the L_1-norm regularization. Here, an open-source software called Templates for First-Order Conic Solvers (TFOCS) is briefly introduced, as well as a parallel-friendly optimization technique known as alternating direction method of multipliers (ADMM), both of which are able to handle l_1-norm problems efficiently. The use of clustering techniques to reduce the size of the planning problem is also introduced.

6.3.4.1 Templates for First-Order Conic Solvers

The huge size of the problem makes first-order optimization techniques a more viable option compared to the second-order methods (e.g., Newton's method, interior point method), which need to form and store the large Hessian matrix. Among the several l_1-solvers that have been developed, the newly released and open-source l_1-solver TFOCS (Becker et al., 2011) is proposed here. TFOCS can easily handle problems in the form of Problem (6.3) or Problem (6.4), and is based on the dual-variable updates, which acquires the primal solutions by updating the dual variables at each step in iterative solutions. This helps overcome the problem caused by the possibly ill-conditioned dose matrices in SPORT/IMRT fluence-map optimization. The computational experiments (Kim et al.; 2012a, 2012b) using two clinical cases (an HN case and a prostate case) revealed that, compared with the second-order algorithms, the computational speed using TFOCS is increased by a factor of 4 to 6, and at the same time the memory requirement is reduced by a factor of 3 to 4.

6.3.4.2 Distributed Optimization Using Accelerated ADMM

An alternative approach to solving the l_1-norm optimization problems efficiently, especially Problem (6.3), is to employ the optimization technique called ADMM (Boyd, 2010).

ADMM has recently received a lot of attention, especially because of its capability to harness the power of parallel and distributed computing environments. Some techniques to accelerate the original ADMM have been recently introduced, and it has been demonstrated that the accelerated ADMM can solve BBO problems very efficiently (Zarepisheh et al., 2017). Because ADMM is a parallel-friendly optimization algorithm, the accelerated ADMM could be a method of choice for solving the large-scale l_1-norm problems arising in SPORT. It seems to be an especially compelling technique because most computations today are moving toward the cloud.

6.3.4.3 Approximate Optimization Using Clustering Techniques

For large SPORT planning problems, the computational burden of the increased number of optimization parameters can be compensated for by reducing the effective number of voxels in the dose deposition matrix. Solving an optimization using this reduced-size matrix can be much faster, but its solution will necessarily be an approximation of the solution to the full problem.

Past work in this domain has explored downsampling the resolution of the voxel grid, either in geometrically regular blocks or in irregular—but geometrically contiguous—blocks with grouping based on the numerical content of the voxels' corresponding rows in the dose deposition matrix (Scherrer et al., 2005; Scherrer and Küfer, 2008). Those studies introduce methods of adaptively refining the clusters to improve the quality of the approximate solution. The work of Martin et al. (2007) on stochastic IMRT optimization using voxel sampling describes another variation on this concept: They assign voxels to cluster regions, then implement a stochastic gradient descent that relies on dose calculations for, and objective evaluations at, a small number of voxels that are randomly sampled from each region on each iteration. As expected, the solution returned by this algorithm converges to the solution to the full problem.

A related form of voxel clustering is proposed here: clustering based entirely on the numerical content of the dose deposition matrix D while omitting the requirement for geometric contiguity of the clustered voxels (co-clustered voxels need to be belong to the same organ in order to avoid generating ambiguities in treatment objectives). The benefit of this approach is that the purely numerical clustering, compared to geometrically restricted clustering, yields approximate problems that are "closer" to the full problem, while still benefiting from reduced computational burden, and therefore yield solutions that are closer to the solution of the full problem. In this approach, the dose deposition matrix D is first created and then an approximate factorization of the matrix is provided:

$$D \approx UD_1$$

Here, D_1 is a compressed dose deposition matrix with k rows (corresponding to the number of clusters) and the same number of columns as D (i.e., the same beams/beamlets are under consideration). The matrix U encodes the clustering information: It consists of one "1" per row and zeros elsewhere. If $U_{i\kappa} = 1$, the ith voxel is assigned to the κth cluster. The goal here is to compress the matrix in a way that preserves maximal fidelity to its numerical content.

In other words, the entries of the product UD_1 should be "close" to the entries of D for some metric or a loss function, l. D_1 can be obtained by solving

$$D_1 = \mathbf{argmin} \sum_{i,j} l(D_{ij} - (UD_1)_{ij})$$

$$\mathbf{subject\ to}\ U_{i\kappa} \in \{0,1\},$$

$$U^T 1 = 1$$

If one chooses $l(\cdot) = 1/2(\cdot)^{\wedge}2$, this becomes equivalent to assembling D_1 by performing k-means clustering on the rows of D. However, the reader is by no means limited to considering k-means clustering; other choices of clustering methods may yield even higher-fidelity reduced matrices for the same extent of compression.

Given the size of the dose deposition matrix, clustering its rows may be a computational challenge in its own right; however, clustering algorithms like k-means are readily parallelizable, making the clustering task well suited to implementation in distributed computation settings.

Depending on computational specifics, the solution obtained by solving the approximate problem may be used instead of solving the full problem; alternately, one may use this solution as a guess to start optimization of the full problem. By chaining together a series of increasingly large (i.e., less compressed) approximate matrices, one can implement a multigrid-like method in which one sequentially solves problems of different sizes and obtain solutions of increasing fidelity.

6.4 SPORT

6.4.1 Bridging the Gap between IMRT and VMAT

SPORT has been introduced to bridge the gap between IMRT and VMAT by utilizing the new capabilities of digital linacs (Li and Xing, 2011) (Figure 6.6). Given a limited number of beam angles (typically five to ten beams), IMRT plans often suffer from insufficient angular sampling, which prevents them from achieving desirable spatial dose distributions for complicated clinical cases. Increasing the number of beams prolongs treatment delivery and makes practicality an issue. On the other hand, single-arc VMAT plans fail to provide the desired beam intensity modulation in some or all directions. The use of multiple arcs does not address the need for modulation of each individual beam and so may still result in under- or overmodulation in some or all directions. SPORT aims to bridge the gap between IMRT and VMAT by optimizing the interplay between intensity modulation and beam sampling.

Figure 6.6 compares SPORT with IMRT and VMAT in a schematic way. Intuitively, SPORT plans can be viewed as an IMRT plans with a higher angular sampling rate and less beam modulation or, alternatively, as a VMAT plan with fewer angular samples and more modulation per beam. In other words, SPORT picks optimal beam angles, and for each specific angle, it modulates the beam as needed. As shown in Figure 6.6, SPORT explores a large area of uncharted territory in terms of the number of beams and levels of intensity modulation, potentially leading to a significant improvement in plan quality.

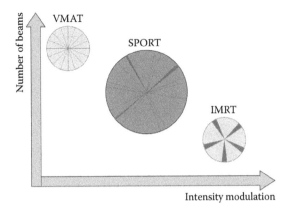

FIGURE 6.6   Schematic plot of SPORT, conventional IMRT, and VMAT in terms of the number of beams and level of intensity modulation. SPORT optimizes the interplay between intensity modulation and angular sampling, and explores the uncharted territory of decision space left out by IMRT and VMAT. (From Li, R., and Xing, L., *Med. Phys.*, 38, 4912–4919, 2011.)

The first feasibility study in 2011 (Li and Xing, 2011) provided insight into the interplay between angular sampling rates and intrabeam modulation. It was proposed to increase the angular sampling rate in IMRT and to compensate for the increase in delivery time caused by adding beams and concomitantly eliminate the dispensable segments of the incident beams. (At the time, the new modality was coined "dense angularly sampled and sparse intensity modulated RT" [DASSIM-RT]). Three clinical cases—a HN case, a pancreas case, and a lung case—were used in that study to evaluate the new scheme. For each patient, a SPORT plan was compared (with 15 beams and 5 intensity levels) against an IMRT plan (with 7 beams and 10 intensity levels) and a 2-arc VMAT plan in terms of the following: (1) conformality index (CI) (Oozeer et al., 2000), (2) dose-volume histogram (DVH), and (3) equivalent uniform dose (EUD) (Niemierko, 1997; Wu et al., 2002). All the plans turned out to have similar target coverage, whereas the SPORT plan had better organ sparing than IMRT and VMAT in all three patients.

From the deliverability perspective, digital linacs made SPORT efficient by (1) concatenating the stations so that they can be delivered in sequence automatically (i.e., auto-field sequencing) and (2) using high dose-rate beams. Furthermore, eliminating dispensable segments also shortened delivery time. It was estimated that, using the step-and-shoot technique conventionally used in IMRT cases, the SPORT plans would take about 4.5, 4.4, and 4.2 min for the HN, pancreas, and lung case, respectively, similar to the delivery times of the corresponding IMRT plans with 7 beams and 10 intensity levels (more details can be found in the Li and Xing's [2011] original paper).

6.4.2 Heuristic Implementation and Rotational arc SPORT

SPORT can also be delivered in a rotational arc. Analogous to VMAT and/or sliding window IMRT beam delivery, radiation is on when going from one station to the next. A novel scheme of a rotational arc SPORT has been implemented, and it has been shown that the need for multiple arcs in conventional VMAT can be eliminated by distributing the station points nonuniformly in the same coplanar sampling space (Li and Xing, 2013). Conventional VMAT such as Varian RapidArc™ discretizes the angular space into equally spaced station points (178 stations) during planning and then optimizes the apertures and weights of the stations. The aperture at an angle between two stations is obtained through interpolation. This approach tacitly ignores the differences in the levels of intensity modulation required at different angles. Unless the discretization is exceedingly fine (which is computationally prohibitive and unavoidably oversamples some angles), the approach is incapable of providing sufficient modulation for some or all directions for many disease sites. As such, multiple arcs are often required to produce a clinically acceptable treatment plan.

In the one-arc SPORT implementation, the angular sampling rate (i.e., the number of stations per unit angle) is modulated according to the need of the individual angles for intensity modulation. The key components of this approach are the means to identify individual angles that need intensity modulation and a method that provides the necessary intensity modulation for those angles. In the next two subsections, the reader will become familiar with two heuristic approaches. A more rigorous optimization approach will be introduced in Section 6.4.3.

6.4.2.1 Demand-Metric-Based Technique

In the initial study in 2013 (Li and Xing, 2013), for ease of computation, a heuristic approach was followed and the treatment was planned in an adaptive fashion after starting from a conventional 1-arc VMAT plan with equally spaced stations. The angles that need higher intensity modulation were identified with the help of a "demand metric", defined as:

$$\sum_{\substack{k=-K \\ k \neq 0}}^{K} \sum_{i=1}^{60} \left(\left| X_s^A(i) - X_{s+k}^A(i) \right| + \left| X_s^B(i) - X_{s+k}^B(i) \right| \right) \left| \frac{MU_s - MU_{s+k}}{\theta_s - \theta_{s+k}} \right|.$$

Here, $X_s^A(i)$ and $X_s^B(i)$ are the positions of the ith MLC leaves in banks A (left leaves) and B (right leaves) for station s, MU is the cumulative monitor units at a station, and θ is the station's gantry angle. Intuitively, the metric is the geometric modulation weighted by the segmental MU per gantry angle. The neighboring $2K$ stations are used to calculate the demand metric of a point. ($K = 10$ is used in the study.) To boost the level of modulation for an angle with a high "demand metric" value, a number of stations around the point are added. The rationale here is that, if the aperture or MU changes a lot at an angle in the original 1-arc VMAT, it suggests that more apertures or stations are needed in the vicinity of the angle to meet the angular demand for more intensity modulation. The original and the added stations are then optimized again. A single-arc sequence is then constructed for delivery using the TrueBeam™.

The aforementioned technique has been applied to four previously treated clinical cases: two HN cases, one prostate case, and one liver case. The SPORT plans are compared with 1-arc and 2-arc VMAT plans in terms of DVH, dose distribution, MU, and EUD. Here, only comparison in terms of DVH is reported, and the reader is referred to the original paper for the remaining comparisons. Figure 6.7 depicts the DVHs of three different treatment plans for the four clinical cases. For HN case 1 (Figure 6.7a), SPORT clearly provides improvements relative to both 1-arc and 2-arc VMAT in terms of target coverage and critical structure sparing. For the more complicated HN case 2 (Figure 6.7b), SPORT provides similar target coverage while achieving better critical structure sparing, particularly for the left parotid, spinal cord, and chiasm. In the liver and prostate cases (Figure 6.2c and d), SPORT far outperforms 1-arc VMAT in terms of critical structure sparing and achieves a similar or better plan quality compared to 2-arc VMAT. For instance, the technique significantly improves the dose to the right kidney. It is worth mentioning that, compared to the 2-arc VMAT, the single arc SPORT decreases the MUs by 20%–30%, depending on the case. This reduction not only speeds up dose delivery and improves clinical workflow, it also decreases the total-body radiation dose and the risk of secondary cancers (Williams and Hounsell, 2001; Hall, 2006; Hardcastle et al., 2007).

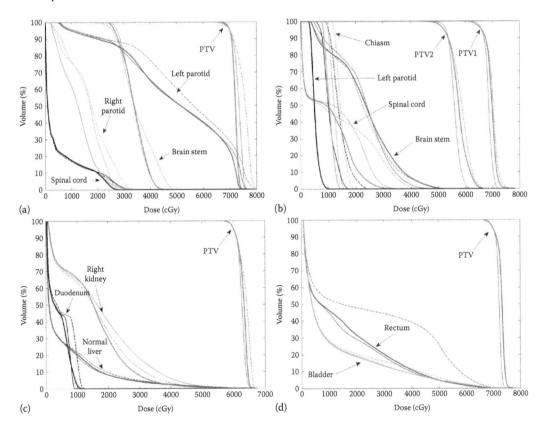

FIGURE 6.7 DVHs of three different treatment plans: 1-arc VMAT (dashed line), 2-arc VMAT (solid thin line), and SPORT (solid thick line) for (a) HN case 1, (b) HN case 2, (c) liver, and (d) prostate. (From Li, R., and Xing, L., *Med. Phys.*, 38, 4912–4919, 2011.)

6.4.2.2 Prior-Knowledge-Based Technique

Another recent heuristic approach (Kim et al., 2015) identifies the need of an individual angle for intensity modulation using prior geometric and dosimetric knowledge. In this approach, a metric named beam eye's view dosimetrics (BEVD), introduced in 2001 (Pugachev and Xing, 2001), is employed for a priori ranking of the angular space. The general consideration in constructing the BEVD is that a beam needs more modulation if it can deliver more dose to the target without exceeding the dose tolerance of the OARs located on the path of the beam. For computational purposes, a beam portal is divided into a grid of beamlets. Each beamlet crossing the target is assigned the maximum intensity that could be used without exceeding the dose tolerances of the OARs. A forward dose calculation using this maximum beam intensity profile is performed, and the score of the given beam direction is calculated according to

$$
\frac{1}{N_k^T} \sum_{k \in Target} \left(\frac{d_\theta^k}{p^T} \right)^2,
$$

where:

d_θ^k is the maximum dose delivered to the voxel k from beam angle θ without exceeding the dose tolerance to all structures involved

N_k^T is the number of voxels in the target

p^T is the target prescription dose

The BEVD score function goes beyond the simple geometric BEV concept in three-dimensional conformal radiotherapy (3D CRT) and captures the main features of a planner's judgment about the quality of a beam. The benefit of using BEVD over the heuristic presented in Section 6.4.2.1 is that it neither has to optimize the equally spaced conventional VMAT nor needs to implement the re-optimization with the chosen angles. The effectiveness of this approach has been shown in three clinical cases, and the reader is referred to the original paper for more information.

6.4.3 Direct Machine Parameter Optimization

As mentioned earlier, in BBO, plan deliverability is totally decoupled from dose optimization, and CS inverse planning has been introduced to alleviate this issue. However, CS inverse planning still yields suboptimal plans because it does not optimize the machine parameters (e.g., apertures, beam angles) directly, and leaf sequencing is still needed to segment the optimal beam profile after optimization. The beam angles are fixed in conventional CS inverse planning, which makes the beam angle optimization a difficult and nontrivial task.

To identify an optimal SPORT plan, all the machine parameters need to be optimized directly and simultaneously. It leads to a difficult optimization problem, especially because the dose distribution is a nonconvex function of the machine parameters. Many optimization algorithms have been developed to optimize beam angles (Pugachev and Xing, 2002; Craft, 2007; Bertsimas et al., 2013; Yarmand and Craft, 2013) or aperture shapes (Cotrutz and Xing, 2003b; Romeijn et al., 2005; Otto, 2008; Men et al., 2010; Cassioli and Unkelbach, 2013; Papp and Unkelbach, 2013) in IMRT or VMAT. In this section, a recently developed optimization algorithm that optimizes the machine parameters directly and simultaneously is presented.

6.4.3.1 Optimization Model

A set of stations is sought, characterized by corresponding beam angles θ_s, aperture shapes X_s, and aperture intensities η_s ($s = 1, \ldots, N_s$), to minimize a predefined metric function F of the dose distribution (F is typically a quadratic function like ϕ_1 in Problem (6.1), but it could be any clinically meaningful function [Zarepisheh et al., 2014b]). The dose distribution z is related to the decision variables (X_s, θ_s, η_s) as described in the following paragraphs.

Let $D_{i,j}^k(\theta_s)$ denote the dose delivered to voxel k by beamlet (i, j) (beamlet in row i and column j) of unit intensity from beam angle θ_s. The dose at voxel k from the station s of unit intensity, $\Phi^k(X_s, \theta_s)$, is then obtained by summing the contributions from all open beamlets at that station. Figure 6.8 illustrates this process using an imaginary beam discretized into 7×7 beamlets. Shaded squares labeled "0" show the beamlets blocked by leaves, while the remaining squares labeled "1" represent the open beamlets. $X_s^A(i)$ and $X_s^B(i)$ represent the position of the left and right leaves, respectively, of the ith row. The accumulated dose from all stations is obtained by multiplying $\Phi^k(X_s, \theta_s)$ by the intensity of the station, η_s, and then summing up over all the stations.

Now the optimization problem can be formulated as

$$\min\{F(z) \mid z^k = \sum_{s=1:n}\Phi^k\left(X_s,\theta_s\right)\eta_s,\ \eta_s \geq 0,\ k \in \{1, \ldots, N_k\}, (X_s, \theta_s) \in \Omega \times B,\ n \leq N_s\} \quad (6.7)$$

where

B and Ω are the set of allowed beam angles and aperture shapes, respectively.

$D_{i,j}^k(\theta_s)$: Dose delivered to voxel k by beamlet (i, j) of unit intensity from beam angle θ_s.

	1	2	3	4	5	6	7	
$X_S^A(1) = 1$	0	1	1	1	1	0	0	$X_S^B(1) = 6$
$X_S^A(2) = 1$	0	1	1	1	1	1	0	$X_S^B(2) = 7$
$X_S^A(3) = 0$	1	1	1	1	1	1	1	$X_S^B(3) = 8$
$X_S^A(4) = 0$	1	1	0	0	0	0	0	$X_S^B(4) = 3$
$X_S^A(5) = 1$	0	1	1	1	1	1	0	$X_S^B(5) = 7$
$X_S^A(6) = 5$	0	0	0	0	0	1	0	$X_S^B(6) = 7$
$X_S^A(7) = 7$	0	0	0	0	0	0	0	$X_S^B(7) = 8$

FIGURE 6.8 Calculation of $\Phi(X_s, \theta_s)$ as a function of the leaf position and the beam angle. The dose contribution of each beamlet to each voxel is precalculated by $D_{i,j}^k(\theta_s)$, so $\Phi^k(X_s, \theta_s)$ can be obtained by summing on $D_{i,j}^k(\theta_s)$ over all open beamlets at station (X_s, θ_s). Open beamlets are labeled as 1 in this figure. (From Zarepisheh, M. et al., *Med. Phys.*, 42, 1012–1022, 2015.)

$$\Phi^k(X_s, \theta_s) = \sum_{i=1}^{7} \sum_{j=X_s^A(i)+1}^{X_s^B(i)-1} D_{i,j}^k(\theta_s)$$

$$\Phi(X_s, \theta_s) \; D_{i,j}^k(\theta_s) \; \Phi^k(X_s, \theta_s) \; D_{i,j}^k(\theta_s) \; (X_s, \theta_s)$$

6.4.3.2 Integrated Algorithm

The difficulty in Problem (6.7) lies in the lack of standard implicit and explicit representations of $\Phi(X_s, \theta_s)$ as a function of X_s and θ_s. On the other hand, once $\Phi(X_s, \theta_s)$ is known, one can easily find the optimal intensity by solving Problem (6.7) with the set of η_s as the only set of the variables. Therefore, it makes sense to split Problem (6.7) into a two-layer problem where the aperture shapes and beam angles are optimized in an outer layer, as in Problem (6.8), while aperture intensities are optimized in an inner layer, as in Problem (6.9). The two problems are given as:

$$\min \{G(X, \theta) \,|\, (X, \theta) = ((X_1, \theta_1), (X_2, \theta_2), \dots, (X_n, \theta_n)), (X_s, \theta_s) \in \Omega \times B \; \forall b, \; n \le N_s \} \quad (6.8)$$

$$G(X, \theta) = \min \{F(z) \,|\, z^k = \sum_{s=1:n} \Phi^k(X_s, \theta_s) \eta_s, k \in \{1, \dots, N_k\}, \; \eta \ge 0, \; n \le N_s \} \quad (6.9)$$

It can be readily shown that an optimal solution of Problems (6.8) and (6.9) would also be optimal for Problem (6.7). While, in principle, Problem (6.8) can be solved only by using computationally expensive techniques such as pattern search, column generation and the subgradient methods are employed to speed up the optimization process. Column generation finds the number of stations required to produce a satisfactory SPORT plan. In subsections, the subgradient method, column generation, and pattern search are briefly introduced.

6.4.3.2.1 Column Generation   Column generation is a sequential optimization technique specialized to handle problems with a large number of variables. Its objective is to find the most beneficial variable(s) at each iteration by solving an extra optimization problem, usually referred to as the "pricing" problem, and then add that variable to the pool of existing variables. This technique has been employed in IMRT and VMAT to optimize the aperture shapes (Romeijn et al., 2005; Men et al. 2010; Peng et al., 2012) and can be adopted here for SPORT. At each iteration, the pricing problem searches through all plausible aperture shapes from all angles to find the "best" station and adds that to the pool of selected stations if it provides a sufficient improvement in the objective function (e.g., the relative improvement in the objective function is more than 1%).

The best station is the station with the highest rate of improvement in the objective of Problem (6.8) with respect to the intensity of the station. For a beamlet (i, j) at beam angle θ_s, the rate of change in the objective function with respect to the intensity of the beamlet is obtained by using the chain rule:

$$\pi_{i,j}(\theta_s) = \sum_{k=1}^{N_k} \frac{\partial F}{\partial Z_k} D_{i,j}^k(\theta_s) \qquad (6.10)$$

Now, the rate of change in the objective function with respect to the intensity of a station (X_s, θ_s) can be obtained by adding together the results of Problem (6.10) for all open beamlets:

$$\sum_i \sum_{j=X_s^A(i)+1}^{X_s^B(i)-1} \pi_{i,j}(\theta_s)$$

Problem (6.11) represents the pricing problem.

$$\min\left\{ \sum_i \sum_{j=X_s^A(i)+1}^{X_s^B(i)-1} \pi_{i,j}(\theta_s), (X_s, \theta_s) \in (\Omega, B) \right\} \qquad (6.11)$$

The pricing problem is an easy problem to solve. It can be solved efficiently by using the enumeration techniques introduced in Romeijn et al. (2005).

6.4.3.2.2 The Subgradient Method   While column generation provides a good initial solution quickly and also helps identify how many stations are required, it does not guarantee the optimality of the solution. So, the subgradient method (the generalized version of the gradient method) and pattern search are carried out to improve the current solution. Let $(X, \theta) = ((X_1, \theta_1), (X_2, \theta_2), \ldots, (X_n, \theta_n))$ denote the current solution. According to the gradient descent method, this solution needs to be updated using the gradient of the objective function, as follows:

$$(X, \theta) \leftarrow (X, \theta) - \lambda(\nabla G_X(X, \theta), \nabla G_\theta(X, \theta)) \qquad (6.12)$$

where λ is obtained using a backtracking search method to ensure improvement in the objective function. Function G is right and left differentiable, allowing us to define an improving subgradient direction g to take over the gradient in Problem (6.12) by using the fact that the objective function improves if one increases a variable with the negative right derivative, or if one decreases a variable with the positive left derivative:

$$g_i = \begin{cases} \partial_i^+ G, \partial_i^+ G < 0 \\ \partial_i^- G, \partial_i^- G > 0 \\ 0, otherwise \end{cases}$$

The subgradient g can be approximated efficiently by assuming that the optimal value of the variable η does not change in the neighborhood of the current solution (X, θ). See the original paper (Zarepisheh et al., 2015) for more details.

While the subgradient method improves the results of column generation, it is not enough to guarantee optimality because of (1) the use of an approximate subgradient direction, (2) the presence of the integer decision variables (leaf positions and beam angles, whose values are solved as continuous variables and rounded after each iteration), (3) the nonconvexity of the problem, and (4) the lack of global search in the subgradient method. While a comprehensive local and global search such as a pattern search is required to guarantee optimality, the subgradient method is computationally much more efficient and hence is integrated with a pattern search in the implementation. The derivative information obtained in the subgradient method is also used in the pattern search, improving the computational efficiency of the pattern search.

6.4.3.2.3 Pattern Search   Pattern search seeks an improving direction in the *positive basis* to improve the current solution, and it concludes that the solution is optimal if there is no improving direction (Conn et al. 2009; Rocha et al., 2013). The positive basis here is defined as set of vectors $V = \{v_1, v_2, \ldots, v_m\}$, which represents possible gantry and MLC motions: It contains two directions for each leaf (moving to the right and left) as well as two directions corresponding to each delivery angle (moving clockwise and counterclockwise). Pattern search has two search strategies known as search step and poll step. In the poll step, one aims to improve the current solution by moving toward one of the directions in V (poll directions); in the search step, one examines a finite set of directions where each is a non-negative and integer combination of the poll directions. A random combination of the poll directions can be used to construct directions for the search step. Given the construction of the positive basis, the poll step aims to improve the current solution by moving an individual leaf (toward the right or left) or changing a beam angle (clockwise or counterclockwise), and the search step aims to make an improvement by changing multiple leaves and/or beam angles.

For improved performance, one can reorder the poll directions and check the directions that are more likely to improve the solution by incorporating the derivative information obtained in the subgradient method. As pointed out before, the problem is nonconvex, and both local and global searches are required to ensure global optimality. The local search is performed by taking the step size equal to 1 in both search and poll steps (it is the smallest possible step size in the problem). As for the global search, the step size should be randomly selected between 2 and the maximum possible value.

6.4.3.2.4 Flowchart   Figure 6.9 illustrates the algorithm in a flowchart. Column generation adds the most beneficial stations sequentially until the plan quality improvement saturates with any further increase in the number of station points. Column generation provides an initial solution and also finds the number of stations needed to produce a satisfactory SPORT plan. At this stage, the subgradient method is applied to improve the selected stations locally by reshaping the apertures and updating the beam angles toward a descent subgradient direction. The algorithm continues to improve the selected stations locally and globally by a pattern search algorithm to explore the part of search space not reachable by the subgradient method. The algorithm terminates when neither gradient method nor pattern search improves the current solution sufficiently; alternatively, when there is a room in the plan for extra stations, one may call column generation again to check whether new stations can improve the objective.

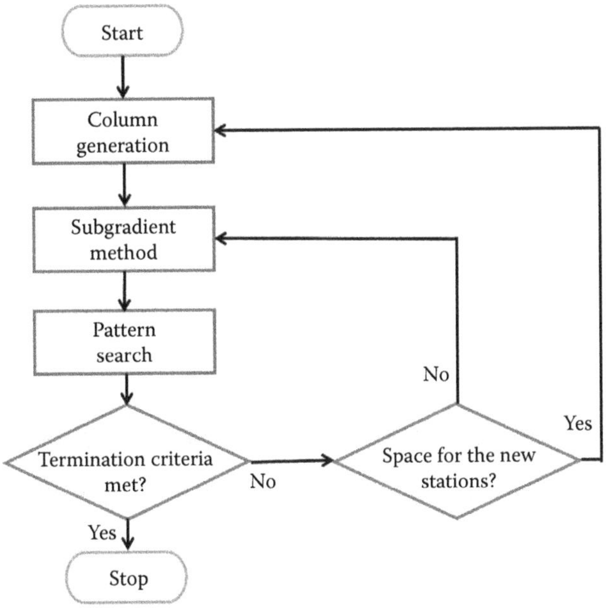

FIGURE 6.9   Flowchart of the SPORT optimization algorithm. Column generation adds the most beneficial stations sequentially, provides a good initial solution, and also identifies the number of stations needed. The subgradient method and pattern search are carried out afterward to refine the stations generated by the column generation. (From Zarepisheh, M. et al., *Med. Phys.*, 42, 1012–1022, 2015.)

6.4.3.3 Results

The integrated optimization algorithm has been tested on two previously treated clinical cases (an HN case and a prostate case). Figure 6.10 demonstrates the convergence behavior of the algorithm by depicting the objective function value as a function of time. The solid, dashed, and dotted lines are used to illustrate the contributions of three optimization techniques, respectively. For the HN case (Figure 6.10a), first, column generation runs and finds an initial solution that uses 24 stations, with a corresponding objective function value of approximately 264 (adding more aperture would improve the objective function less than 1% at this point). Then the subgradient method and pattern search run to improve the 24 selected stations, reducing the objective function by 13%, to approximately 230. Next, column generation runs again and finds three additional beneficial stations, and the algorithm continues similarly thereafter, resulting in 36 optimized stations when the algorithm terminates. For the prostate case (Figure 6.10b), the algorithm finds 34 stations. As shown in Figure 6.10, the column generation takes the least time, whereas pattern search is the slowest process. Figure 6.11 illustrates the distribution of the optimal beam angles that are nonuniformly distributed.

Figure 6.12 compares the new SPORT plan with an IMRT plan for the same case, which represents the ideal situation for a given seven fixed-gantry beam configuration without any constraint on beam intensity modulation. An improvement can be observed in critical structure sparing as well as tumor coverage by delivering the SPORT plan.

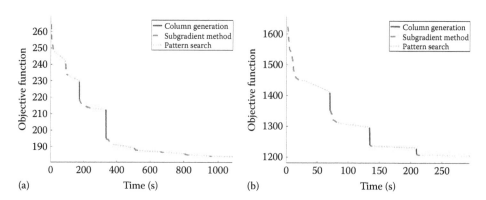

FIGURE 6.10 The convergence behavior of the algorithm for (a) the head and neck and (b) the prostate case. (From Zarepisheh, M. et al., *Med. Phys.*, 42, 1012–1022, 2015.)

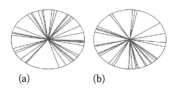

FIGURE 6.11 The optimal beam angle distributions for (a) the head and neck case and (b) the prostate case. (From Zarepisheh, M. et al., *Med. Phys.*, 42, 1012–1022, 2015.)

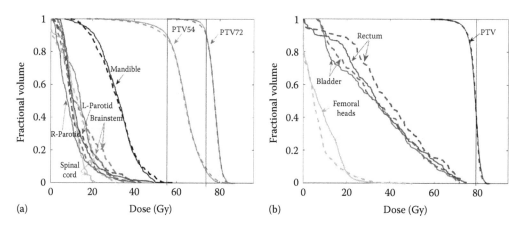

FIGURE 6.12 Comparison of the SPORT plan (solid line) with the ideal IMRT plan (dashed line) for (a) the head and neck and (b) the prostate case. (From Zarepisheh, M. et al., *Med. Phys.*, 42, 1012–1022, 2015.)

6.5 CONCLUSION

The conventional treatment planning paradigm consists of a trial-and-error technique to adjust the objective function types and parameters, which makes the treatment planning a time-consuming and expensive task. Research efforts are being made toward the more automated process with less user intervention. This chapter briefly reviewed three techniques: (1) MCO, (2) knowledge-based treatment planning, and (3) PO. Although these techniques need much less fine-tuning compared to conventional treatment planning, they are all computationally expensive, and researchers are trying to speed up these techniques either by developing better algorithms or by utilizing the capabilities of emerging hardware technologies.

SPORT has been introduced as a new treatment modality to improve the plan quality by taking advantage of the capabilities of the new generation of digital linacs. In practice, SPORT can be implemented and realized in different ways depending on the clinical goal and/or predefined solution space. For newly available digital linacs, such as TrueBeam™ (Varian Medical Systems, Palo Alto, California), SPORT delivery can be made efficient by the high dose-rate beams and by concatenating the beams so that they can be delivered sequentially in an automated fashion and without an operator's intervention. On the optimization side, SPORT is supported by two optimization frameworks: (1) DMPO, which optimizes the station parameters (machine parameters) directly using an integrated algorithm, and (2) CS inverse planning, which incorporates the delivery constraints indirectly into the BBO using appropriate regularization terms.

Future research is needed to address the following issues:

1. More automated and efficient treatment planning process.

2. Enhancing the automated and SPORT optimization by leveraging the power of the modern distributed computing environments (e.g., Graphic Processing Unit (GPU), cloud) and employing parallel-friendly optimization techniques like ADMM.

3. The clinical implementation of SPORT with collimator modulation, noncoplanar, and/or even nonisocentric beams.

4. Incorporating the physical constraints of the delivery system in SPORT optimization.

We wish to thank grant supports from NIH (5R01 CA176553 and 1R01 EB016777) and Varian Medical Systems.

REFERENCES

Becker, S. R., E.l J. Candès, and M. C. Grant. 2011. Templates for convex cone problems with applications to sparse signal recovery. *Mathematical Programming Computation* 3 (3): 165–218. doi:10.1007/s12532-011-0029-5.

Bertsimas, D., V. Cacchiani, D. Craft, and O. Nohadani. 2013. A hybrid approach to beam angle optimization in intensity-modulated radiation therapy. *Computers & Operations Research* 40 (9): 2187–2197. doi:10.1016/j.cor.2012.06.009.

Block, K. T., M. Uecker, and J. Frahm. 2007. Undersampled radial MRI with multiple coils: Iterative image reconstruction using a total variation constraint. *Magnetic Resonance in Medicine* 57 (6): 1086–1098. doi:10.1002/mrm.21236.

Bortfeld, T. 2006. IMRT: A review and preview. *Physics in Medicine and Biology* 51 (July): R363–R379. doi:10.1088/0031-9155/51/13/R21.

Boutilier, J. J., T. Lee, T. Craig, M. B. Sharpe, and T. C. Y. Chan. 2015. Models for predicting objective function weights in prostate cancer IMRT. *Medical Physics* 42 (4): 1586–1595. doi:10.1118/1.4914140.

Boyd, S. 2010. Distributed optimization and statistical learning via the alternating direction method of multipliers. *Foundations and Trends® in Machine Learning* 3 (1): 1–122. doi:10.1561/2200000016.

Boyd, S., and L. Vandenberghe. 2004.*Convex Optimization*. City, ST: Cambridge University Press, New York.

Breedveld, S., P. R. M. Storchi, M. Keijzer, and B. J. M. Heijmen. 2006. Fast, multiple optimizations of quadratic dose objective functions in IMRT. *Physics in Medicine and Biology* 51 (July): 3569–3579. doi:10.1088/0031-9155/51/14/019.

Candès, E. J., M. B. Wakin, and S. P. Boyd. 2008. Enhancing sparsity by reweighted ℓ 1 minimization. *Journal of Fourier Analysis and Applications* 14 (5–6): 877–905. doi:10.1007/s00041-008-9045-x.

Cassioli, A., and J. Unkelbach. 2013. Aperture shape optimization for IMRT treatment planning. *Physics in Medicine and Biology* 58 (2): 301. doi:10.1088/0031-9155/58/2/301.

Conn, A. R., K. Scheinberg, and L. N. Vicente. 2009. *Introduction to Derivative-Free Optimization*. Philadelphia, PA: Society for Industrial and Applied Mathematics/Mathematical Programming Society.

Cotrutz, C., and L. Xing. 2003a. IMRT dose shaping with regionally variable penalty scheme. *Medical Physics* 30: 544–551. doi:10.1118/1.1556610.

Cotrutz, C., and L. Xing. 2003b. Segment-based dose optimization using a genetic algorithm. *Physics in Medicine and Biology* 48 (18): 2987–2998. doi:10.1088/0031-9155/48/18/303.

Craft, D., T. Halabi, and T. Bortfeld. 2005. Exploration of tradeoffs in intensity-modulated radiotherapy. *Physics in Medicine and Biology* 50: 5857.

Craft, D. 2007. Local beam angle optimization with linear programming and gradient search. *Physics in Medicine and Biology* 52 (7): N127–N135. doi:10.1088/0031-9155/52/7/N02.

Craft, D. 2010. Calculating and controlling the error of discrete representations of Pareto surfaces in convex multi-criteria optimization. *Physica Medica* 26 (4): 184–191. doi:10.1016/j.ejmp.2009.11.005.

Craft, D. L., T. F. Halabi, H. A. Shih, and T. R. Bortfeld. 2006. Approximating convex Pareto surfaces in multiobjective radiotherapy planning. *Medical Physics* 33 (9): 3399. doi:10.1118/1.2335486.

Craft, D. L., T. S. Hong, H. A. Shih, and T. R. Bortfeld. 2012. Improved planning time and plan quality through multicriteria optimization for intensity-modulated radiotherapy. *International Journal of Radiation Oncology*Biology*Physics* 82 (1): e83–e90. doi:10.1016/j.ijrobp.2010.12.007.

Dong, P., P. Lee, D. Ruan, T. Long, E. Romeijn, Y. Yang, D. Low, P. Kupelian, and K. Sheng. 2013. 4π non-coplanar liver SBRT: A novel delivery technique. *International Journal of Radiation Oncology*Biology*Physics* 85 (5): 1360–1366. doi:10.1016/j.ijrobp.2012.09.028.

Ezzell, G. A., J. M. Galvin, D. Low, J. R. Palta, I. Rosen, M. B. Sharpe, P. Xia, Y. Xiao, L. Xing, and C. X. Yu. 2003. Guidance document on delivery, treatment planning, and clinical implementation of IMRT: Report of the IMRT Subcommittee of the AAPM Radiation Therapy Committee. *Medical Physics* 30 (8): 2089–2115. doi:10.1118/1.1591194.

Fahimian, B., V. Yu, K. Horst, L. Xing, and D. Hristov. 2013. Trajectory modulated prone breast irradiation: A LINAC-based technique combining intensity modulated delivery and motion of the couch. *Radiotherapy and Oncology* 109 (3): 475–481. doi:10.1016/j.radonc.2013.10.031.

Hall, E. J. 2006. Intensity-modulated radiation therapy, protons, and the risk of second cancers. *International Journal of Radiation Oncology*Biology*Physics* 65 (1): 1–7. doi:10.1016/j.ijrobp.2006.01.027.

Hardcastle, N., P. Metcalfe, A. Ceylan, and M. J. Williams. 2007. Multileaf collimator end leaf leakage: Implications for wide-field IMRT. *Physics in Medicine and Biology* 52 (21): N493. doi:10.1088/0031-9155/52/21/N01.

Jee, K. W., D. L. McShan, and B. A. Fraass. 2007. Lexicographic ordering: Intuitive multicriteria optimization for IMRT. *Physics in Medicine and Biology* 52 (7): 1845–1861. doi:10.1088/0031-9155/52/7/006.

Kim, H., S. Becker, R. Lee, S. Lee, S. Shin, E. Candès, L. Xing, and R. Li. 2013. Improving IMRT delivery efficiency with reweighted L1-minimization for inverse planning. *Medical Physics* 40 (7): 071719. doi:10.1118/1.4811100.

Kim, H., R. Li, R. Lee, T. Goldstein, S. Boyd, E. Candes, and L. Xing. 2012a. Dose optimization with first-order total-variation minimization for dense angularly sampled and sparse intensity modulated radiation therapy (DASSIM-RT). *Medical Physics* 39 (7): 4316–4327. doi:10.1118/1.4729717.

Kim, H., R. Li, R. Lee, and L. Xing. 2015. Beam's-eye-view dosimetrics (BEVD) guided rotational station parameter optimized radiation therapy (SPORT) planning based on reweighted total-variation minimization. *Physics in Medicine and Biology* 60 (5): N71. doi:10.1088/0031-9155/60/5/N71.

Kim, H., T. S. Suh, R. Lee, L. Xing, and R. Li. 2012b. Efficient IMRT inverse planning with a new L1-solver: Template for first-order conic solver. *Physics in Medicine and Biology* 57 (13): 4139. doi:10.1088/0031-9155/57/13/4139.

Kolehmainen, V., A. Vanne, S. Siltanen, S. Jarvenpaa, J. P. Kaipio, M. Lassas, and M. Kalke. 2006. Parallelized Bayesian inversion for three-dimensional dental X-ray imaging. *IEEE Transactions on Medical Imaging* 25 (2): 218–228. doi:10.1109/TMI.2005.862662.

Kufer, K. H, H. W. Hamacher, and T. R. Bortfeld. 2000. A multicriteria optimization approach for inverse radiotherapy planning. In *Proceedings of the 13th ICCR (Heidelberg, 2000)*, T. R. Bortfeld and W. Schlegel (Eds.). Berlin, Germany: Springer, pp. 26–29.

Li, R., and L. Xing. 2011. Bridging the gap between IMRT and VMAT: Dense sngularly sampled and sparse intensity modulated radiation therapy. *Medical Physics* 38 (9): 4912–4919. doi:10.1118/1.3618736.

Li, R., and L. Xing. 2013. An adaptive planning strategy for station parameter optimized radiation therapy (SPORT): Segmentally boosted VMAT. *Medical Physics* 40 (5): 050701.

Long, T., M. Matuszak, M. Feng, B. A. Fraass, R. K. Ten Haken, and H. E. Romeijn. 2012. Sensitivity analysis for lexicographic ordering in radiation therapy treatment planning. *Medical Physics* 39 (6): 3445–3455. doi:10.1118/1.4720218.

Mackie, T. R., T. Holmes, S. Swerdloff, P. Reckwerdt, J. O. Deasy, J. Yang, B. Paliwal, and T. Kinsella. 1993. Tomotherapy: A new concept for the delivery of dynamic conformal radiotherapy. *Medical Physics* 20 (6): 1709–1719. doi:10.1118/1.596958.

Martin, B. C., T. R. Bortfeld, and D. A. Castañon. 2007. Accelerating IMRT optimization by voxel xampling. *Physics in Medicine and Biology* 52 (24): 7211.

Men, C., X. Jia, and S. B. Jiang. 2010. GPU-based ultra-fast direct aperture optimization for online adaptive radiation rherapy. *Physics in Medicine and Biology* 55 (15): 4309. doi:10.1088/0031-9155/55/15/008.

Niemierko, A. 1997. Reporting and analyzing dose Distributions: A concept of equivalent uniform dose. *Medical Physics* 24 (1): 103–110. doi:10.1118/1.598063.

Oozeer, R., B. Chauvet, R. Garcia, C. Berger, C. Felix-Faure, and F. Reboul. 2000. Evaluation of three-dimensional conformai radiation therapy dose distributions: Conformity index. *Cancer/Radiothérapie* 4 (3): 207–216. doi:10.1016/S1278-3218(00)89096-4.

Otto, K. 2008. Volumetric modulated arc therapy: IMRT in a single gantry arc. *Medical Physics* 35 (1): 310–317. doi:10.1118/1.2818738.

Paddick, I. 2000. A simple scoring ratio to index the conformity of radiosurgical treatment plans. *Journal of Neurosurgery* 93 (supplement 3): 219–222. doi:10.3171/jns.2000.93.supplement.

Papp, D., and J. Unkelbach. 2013. Direct leaf trajectory optimization for volumetric modulated arc therapy planning with sliding window delivery. *Medical Physics* 41 (1): 011701. doi:10.1118/1.4835435.

Peng, F., X. Jia, X. Gu, M. A. Epelman, H. E. Romeijn, and S. B. Jiang. 2012. A new column-generation-based algorithm for VMAT treatment plan optimization. *Physics in Medicine and Biology* 57 (14): 4569–4588. doi:10.1088/0031-9155/57/14/4569.

Pugachev, A., and L. Xing. 2001. Pseudo beam's-eye–view as applied to beam orientation selection in intensity-modulated radiation therapy. *International Journal of Radiation Oncology*Biology*Physics* 51 (5): 1361–1370. doi:10.1016/S0360-3016(01)01736-9.

Pugachev, A., and L. Xing. 2002. Incorporating prior knowledge into beam orientaton optimization in IMRT. *International Journal of Radiation Oncology*Biology*Physics* 54 (5): 1565–1574. doi:10.1016/S0360-3016(02)03917-2.

Riet, A. V., A. C. A. Mak, M. A. Moerland, L. H. Elders, and W. van der Zee. 1997. A conformation number to quantify the degree of conformality in brachytherapy and external beam irradiation: Application to the prostate. *International Journal of Radiation Oncology*Biology*Physics* 37 (3): 731–736. doi:10.1016/S0360-3016(96)00601-3.

Rocha, H., J. M. Dias, B. C. Ferreira, and M. C. Lopes. 2013. Beam angle optimization for intensity-modulated radiation therapy using a guided pattern search method. *Physics in Medicine and Biology* 58 (9): 2939–2953. doi:10.1088/0031-9155/58/9/2939.

Romeijn, H. E., R. K. Ahuja, J. F. Dempsey, and A. Kumar. 2005. A column generation approach to radiation therapy treatment planning using aperture modulation. *SIAM Journal on Optimization* 15 (3): 838–862. doi:10.1137/040606612.

Salari, E., J. Wala, and D. Craft. 2012. Exploring trade-offs between VMAT dose quality and delivery efficiency using a network optimization approach. *Physics in Medicine and Biology* 57 (17): 5587–5600. doi:10.1088/0031-9155/57/17/5587.

Scherrer, A., and K. H. Küfer. 2008. Accelerated IMRT plan optimization using the adaptive clustering method. *Linear Algebra and Its Applications*, Special Issue on Linear and Nonlinear Models and Algorithms in Intensity-Modulated Radiation Therapy, 428 (5–6): 1250–1271. doi:10.1016/j.laa.2007.03.025.

Scherrer, A., K.-H. Küfer, T. Bortfeld, M. Monz, and F. Alonso. 2005. IMRT planning on adaptive volume structures—A decisive reduction in computational complexity. *Physics in Medicine and Biology* 50 (9): 2033.

Shepard, D. M., M. A. Earl, X. A. Li, S. Naqvi, and C. Yu. 2002. Direct aperture optimization: A turnkey solution for step-and-shoot IMRT. *Medical Physics* 29 (6): 1007–1018. doi:10.1118/1.1477415.

Wang, L., K. N. Kielar, E. Mok, A. Hsu, S. Dieterich, and L. Xing. 2012. An end-to-end examination of geometric accuracy of IGRT using a new digital accelerator equipped with onboard imaging system. *Physics in Medicine and Biology* 57 (3): 757. doi:10.1088/0031-9155/57/3/757.

Wilkens, J. J., J. R. Alaly, K. Zakarian, W. L. Thorstad, and J. O. Deasy. 2007. IMRT treatment planning based on prioritizing prescription goals. *Physics in Medicine and Biology* 52 (6): 1675–1692. doi:10.1088/0031-9155/52/6/009.

Williams, P. C., and A. R. Hounsell. 2001. X-ray leakage considerations for IMRT. *The British Journal of Radiology* 74 (877): 98–100. doi:10.1259/bjr.74.877.740098.

Wu, B., F. Ricchetti, G. Sanguineti, M. Kazhdan, P. Simari, R. Jacques, R. Taylor, and T. McNutt. 2011. Data-driven approach to generating achievable dose–volume histogram objectives in intensity-modulated radiotherapy planning. *International Journal of Radiation Oncology*Biology*Physics* 79 (4): 1241–1247. doi:10.1016/j.ijrobp.2010.05.026.

Wu, Q., R. Mohan, A. Niemierko, and R. Schmidt-Ullrich. 2002. Optimization of intensity-modulated radiotherapy plans based on the equivalent uniform dose. *International Journal of Radiation Oncology*Biology*Physics* 52 (1): 224–235. doi:10.1016/S0360-3016(01)02585-8.

Xia, P., N. Yu, L. Xing, X. Sun, and L. J. Verhey. 2005. Investigation of using a power function as a cost function in inverse planning optimization. *Medical Physics* 32: 920. doi:10.1118/1.1872552.

Xing, L., J. Wong, and R. Li. 2014. WE-D-BRD-01: Innovation in radiation therapy delivery: Advanced digital linac features. *Medical Physics* 41 (6): 494–495. doi:10.1118/1.4889398.

Xing, L., and R. Li. 2014. Inverse planning in the age of digital LINACs: Station parameter optimized radiation therapy (SPORT). *Journal of Physics: Conference Series* 489 (1): 012065. doi:10.1088/1742-6596/489/1/012065.

Xing, L., M. H. Phillips, and C. G. Orton. 2013. DASSIM-RT is likely to become the method of choice over conventional IMRT and VMAT for delivery of highly conformal radiotherapy. *Medical Physics* 40 (2): 020601. doi:10.1118/1.4773025.

Yarmand, H., and D. Craft. 2013. Two effective heuristics for beam angle optimization in radiation therapy. *ArXiv e-Print 1305.4959*. http://arxiv.org/abs/1305.4959.

Yu, C. X., and G. Tang. 2011. Intensity-modulated arc therapy: Principles, technologies and clinical implementation. *Physics in Medicine and Biology* 56 (5): R31–R54. doi:10.1088/0031-9155/56/5/R01.

Zarepisheh, M., R. Li, Y. Ye, and L. Xing. 2015. Simultaneous beam sampling and aperture shape optimization for SPORT. *Medical Physics* 42 (2): 1012–1022. doi:10.1118/1.4906253.

Zarepisheh, M., T. Long, N. Li, Z. Tian, H. E. Romeijn, X. Jia, and S. B. Jiang. 2014a. A DVH-guided IMRT optimization algorithm for automatic treatment planning and adaptive radiotherapy replanning. *Medical Physics* 41 (6): 061711. doi:10.1118/1.4875700.

Zarepisheh, M., A. F. Uribe-Sanchez, N. Li, X. Jia, and S. B. Jiang. 2014b. A multicriteria framework with voxel-dependent parameters for radiotherapy treatment plan optimization. *Medical Physics* 41 (4): 041705. doi:10.1118/1.4866886.

Zarepisheh, M., L. Xing, and Y. Ye. 2017. A computation study on an integrated alternating direction method of multipliers for large scale optimization. *Optimization Letters*, February, 1–13. doi:10.1007/s11590-017-1116-y.

Zhang, X., X. Li, E. M. Quan, X. Pan, and Y. Li. 2011. A methodology for automatic intensity-modulated radiation treatment planning for lung cancer. *Physics in Medicine and Biology* 56 (13): 3873–3893. doi:10.1088/0031-9155/56/13/009.

Zhu, L., L. Lee, Y. Ma, Y. Ye, R. Mazzeo, and L. Xing. 2008. Using total-variation regularization for intensity modulated radiation therapy inverse planning with field-specific numbers of segments. *Physics in Medicine and Biology* 53 (December): 6653–6672. doi:10.1088/0031-9155/53/23/002.

Zhu, L., and L. Xing. 2009. Search for IMRT inverse plans with piecewise constant fluence maps using compressed sensing techniques. *Medical Physics* 36 (5): 1895–1905. doi:10.1118/1.3110163.

CHAPTER 7

Advances in Patient Setup and Target Localization

Lei Ren and Fang-Fang Yin

CONTENTS

7.1 INTRODUCTION

Patient setup and localization in radiation therapy is increasingly important as more precise delivery with high fractional dose is becoming widely accepted in clinical practices. Target localization accuracy is closely related to patient setup and immobilization. Without proper patient setup, precision radiation therapy will be suboptimal regardless of the use of any sophisticated localization technique. In this chapter, both current approaches and future developments of techniques used in patient setup and localization will be discussed.

7.2 PRINCIPLES OF PATIENT SETUP AND TARGET LOCALIZATION

The treatment outcome of radiation therapy depends heavily on the accuracy of the radiation delivery (Zelefsky et al., 2012; Soike et al., 2013). Accurate radiation delivery ensures adequate dose delivery to the target while minimizing radiation dose delivered to the healthy tissues surrounding the target. Many steps contribute to the success of accurate radiation delivery, including patient setup and target localization. The principle of patient setup is to ensure that patients are positioned in a convenient, comfortable, and repeatable position so that both patients' interfraction and intrafraction positioning variations can be minimized throughout the entire treatment course. Proper setup of the patient is also critical for designing and delivering radiation with optimal geometry, especially for certain types of machines, to avoid issues with mechanical clearance. Target localization is another important aspect of radiotherapy and is designed to align the target location with the treatment beam either before or during the treatment. The principle of target localization is to use an appropriate methodology such as an imaging technique to infer the treatment target coordinate precisely with respect to the machine (radiation beam) coordinate as determined during the process of treatment planning. Accurate localization not only ensures adequate dose coverage of the target but also allows reduction of the margin required from the clinical target volume (CTV) to the planning target volume (PTV) to minimize radiation dose to the surrounding healthy tissues.

The techniques selected for patient setup and target localization should be based on the specific patient, specific treatment site, specific technique, and the specific type of radiotherapy machine used for the patient treatment. The radiotherapy treatment schemes can typically be classified into two major categories: conventional fractionated treatments (10–35 fractions with low dose per fraction) and hypofractionated stereotactic radiosurgery (SRS) or stereotactic body radiation therapy (SBRT) treatments (1–5 fractions with high dose per fraction). Compared to fractionated treatments, SRS/SBRT delivers much higher fractional dose and is much less tolerable to treatment errors, especially localization errors in any single fraction. Thus, the demands for better patient immobilization, setup reproducibility, and target localization accuracy are much higher for SRS/SBRT.

Overall, multiple factors such as efficiency, accuracy and imaging dose (when X-ray imaging is used for localization) need to be considered for each patient when selecting proper setup and localization techniques, which will affect the entire planning and treatment process. The uncertainties associated with the selected techniques determine the target margin needed in the treatment planning process, which in turn will affect the dose that can be delivered to the target and will be delivered to the surrounding healthy tissues.

7.3 APPROACHES FOR PATIENT SETUP AND IMMOBILIZATION

The approaches for patient setup and immobilization depend very much on factors such as treatment site and patient condition, so they will be discussed for each specific treatment site. The primary goal for immobilization and patient setup is to minimize patient and organ movement during treatment. Different devices could be used for this purpose. We will describe some devices commonly used in the clinic.

7.3.1 Central Nervous System Cancers

Central nervous system (CNS) consists of the brain and spinal cord. Patients with brain cancers are typically set up in a supine position (occasionally with prone position for some special cases) and immobilized by either a frameless mask or a frame-based system. The frameless system is used only to immobilize the patient, while the frame-based system often functions for both immobilization and establishing the relationship between the patient and machine coordination systems. The conventional frameless mask is made out of a plastic product that hardens and is molded to hold the patient's head in position. The mask is customized noninvasively for individual patients and can be used over multiple treatment fractions as long as the patient anatomy is not substantially changed during the course of treatment. It is critical to ensure that the mask conforms to the patient head/body tightly so that patient motion is minimized within the mask. The frame-based system immobilizes the patient's head by attaching a rigid stereotactic frame to the patient's skull with screws penetrating the skin. This method allows a high level of precision in immobilizing the patient and is used for single fraction of SRS treatment. However, it is an invasive procedure and requires completion of the patient treatment on the same day. For treatment of spinal tumors, it is preferable to set up patients in a supine position. For lesions in the C-spine, a long frameless mask is preferred because it immobilizes the patient from the head down to the shoulder region to ensure minimal motion of the C-spine region. For lesions in the T–S spinal regions, patients can be immobilized using either an alpha cradle or a vacuum cushion that conforms to the patient's body. One additional consideration is to avoid radiation beams penetrating through the patients' arms such as by positioning the arms up over the head.

7.3.2 Head-and-Neck Region Cancers

Head-and-neck cancer patients are usually set up in a supine position and immobilized using a head-and-neck mask. Patient weight loss or tumor shrinkage over the treatment course may require making a new mask. One difference from brain cancer setup is that a longer mask system is favored for head-and-neck cancer patients to immobilize the shoulder region.

7.3.3 Breast Cancers

Patients can be set up in either the supine position or the prone position for breast cancer treatments. In the supine position, patients can be set up using the breast board with arms raised over the head, head turned toward the contralateral breast, and chin extended. A matched supraclavicular (SCV) field and/or a medial internal mammary field can be added to cover lymph nodes superior and/or medial to the breast fields. Breath-hold technique may be used for treating the left breast to minimize heart dose. In the prone position, patients can be set up using a prone breast board. The board contains a movable insert at the level of the breast so that the treated breast is allowed to hang below the board while the healthy breast rests on the insert. The prone position can also be used when treating breast tumors without any nodal (internal mammary or SCV) involvement. Studies showed that prone setup reduces the amount of lung and heart volumes irradiated, especially for left breast cancer patients (Lymberis et al., 2012) In general, the supine position is relatively easier to set up than the prone position, while the prone position has less respiratory motion than the supine position for free breathing treatments.

7.3.4 Thorax and Abdominal Region Cancers

The patient is usually set up in the supine position, with arms up and head and legs in a neutral and comfortable position. The patient can be immobilized in a wing board for regular fractionated treatments and in an alpha cradle or vacuum cushion for better immobilization for SBRT treatments. The alpha cradle consists of an exterior polystyrene cradle that can be filled with a foaming agent to form a mold that fits the patient's body shape. The advantage of the alpha cradle is its ability to be formed to different sizes and orientations. A cradle can be made to immobilize the entire body or just the treated area. However, the cradle can be used only for one patient. The vacuum cushion system is a rectangular plastic bag (available in varying sizes) filled with pellets. When the patient is positioned in the desired position on the bag, a vacuum is applied to remove the air in the bag, leaving the pellets to form a rigid mold around the patient. (Fuss et al., 2004; Han et al., 2010) The vacuum cushion is easier to make than an alpha cradle and can be reused for different patients. It is important to select an adequate size for the vacuum cushion to fit the patient and to prevent it from leaking through the treatment course. Abdominal compressors are sometimes used in conjunction with the vacuum cushion for some patients for the purpose of reducing organ motion. However, abdominal compression is not always achievable (Eccles et al., 2011).

Different respiratory motion management techniques have been developed for patient setup to minimize the effects of respiratory motion for treatment planning and delivery in the thorax and abdominal region. Three types of motion management techniques are typically employed: free breathing (FB), breath-hold (BH), and respiratory gated treatment. FB treatments allow patients to breathe freely, and the respiratory motion of the target is accounted for by enlarging the PTV volume to encompass the range of target motion. The FB technique is easy to implement and generally faster to treat because radiation can be delivered continuously without any interruption. However, the PTV volume treated in FB is larger than that treated with the other motion management techniques, which can lead to unwanted radiation dose to a larger volume of the surrounding healthy tissues. The BH technique requires the patient to hold the breath at a certain level during radiation delivery to minimize target motion. The PTV volume in BH treatments is substantially reduced compared to that in FB treatments because of to the reduction of target motion, which in turn reduces the dose to healthy tissue and allows potential dose escalation to the target. Because the BH technique requires the patient to hold the breath for around 20 s or longer, it may not be feasible for patients with limited lung function. Respiratory gated treatments allow the patient to breathe freely and allows delivery of the radiation only when patient breathing is within a certain gating window. The gating window is typically selected around the end-expiration phase due to the quiescence of the phase. The window width is usually set around 30%–50% of the breathing cycle to balance between delivery efficiency and residual target motion within the gating window. Compared to FB treatments, respiratory gating effectively reduces target motion during radiation delivery to reduce the PTV volume. Compared to BH treatments, respiratory gating allows patients to breathe freely and is therefore feasible for patients who are incapable of holding their breath. As the start of each delivery, the gating window is controlled based on the prediction of the respiratory phases from the breathing signals acquired on-the-fly during the treatment and requires the patient to breathe regularly and repeatedly. Irregularities in patient breathing may introduce dosimetric and geometric errors in the gated delivery. To improve the reproducibility of BH level or respiratory cycles during gating, an audiovisual biofeedback system (AVBFS) has been introduced to give patients direct feedback signal of their breathing curve to guide their breathing (Park et al., 2011) An optical camera or a spirometer are the two devices most commonly used to monitor patient breathing. Optical imaging-based respiration monitoring showed advantages over spirometer-based monitoring in terms of patient compliance and system compactness (Nelson et al., 2005). AVBFS can potentially reduce treatment errors and improve delivery efficiency of BH and respiratory gating, which is especially critical when these motion management techniques are employed for SBRT treatments.

7.3.5 Pelvic Region Cancers

The patient can be immobilized in an alpha-cradle or vacuum bag in a supine position with arms on chest and legs straight with a foot stabilizer. The immobilization device needs to be long enough to cover the pelvic region and legs and thus minimize patient motion and setup deviations.

7.3.6 Extremities

Immobilization techniques for extremities are most variable, including the use of special immobilization devices and special postures to avoid irradiation to other parts of the body. In some occasions, the feet are positioned toward the gantry. In general, immobilization and patient setup are designed specifically for the individual patient, the treatment site, and the treatment machine. One should also factor in the method of verification and imaging because some types of positioning and immobilization may prohibit the use of certain imaging techniques due to mechanical clearance issues.

7.4 APPROACHES FOR TARGET LOCALIZATION

Target localization is critical for ensuring accurate delivery of the radiation dose to the tumor with minimal dose to the surrounding healthy tissues. The primary goal for target localization is to establish a relationship of patient geometry with machine geometry as well as to provide methods of managing target motion. Different target localization methodologies and techniques are discussed in detail in the following subsections.

7.4.1 X-Ray Imaging Techniques

X-ray imaging systems are the most commonly available techniques in the clinic. They include two-dimensional (2D) megavoltage (MV) or kilovoltage (kV) imaging and three-dimensional (3D) cone-beam computed tomography (CBCT) imaging.

7.4.1.1 Onboard 2D Imaging

2D-MV images can be acquired from the treatment beams using an onboard flat panel electronic portal imager device (EPID), while 2D-kV images can be acquired using an onboard kV X-ray source and a flat panel detector (FPD) that are mounted either in the room or on the treatment machine gantry. The 2D imaging techniques can acquire images in real time with low imaging dose to the patient. Due to the lack of volumetric information, however, 2D imaging is mostly used for localization of rigid body bony anatomy or implanted markers and has limited applications in localizing soft tissues. Implanted metal markers may be used to highlight the treatment target in the 2D X-ray images, especially for the treatment of abdominal and pelvic lesions. However, it involves an invasive procedure for marker implantation. Types of markers, number of markers, and location of markers also have an impact on the accuracy of localization using implanted markers. Typically, 2D images are acquired in pairs sequentially or simultaneously for accurate target (marker) localization.

7.4.1.2 Onboard 3D CBCT

The conspicuousness of anatomy in 2D X-ray images inherently suffers when multilayered anatomy is projected onto a single image plane. Three-dimensional imaging techniques can substantially improve anatomical visibility by reconstructing tomographic images of patient anatomy and performing soft-tissue matching on a daily basis. Current gantry-based kV imaging systems are generally capable of acquiring volumetric image data that can be reconstructed as CBCT. Whereas traditional fan-beam CT systems contain only a limited number of detector rows, CBCT is implemented with a rectangular FPD, which usually contains several hundred or a few thousand rows. The X-ray source is also opened in the third dimension (like a cone shape rather than a fan shape) to irradiate the rectangular detector. As such, CBCT acquires a full 3D volumetric image during one single rotation of the source-detector tandem with no translation of the patient supporting device (i.e., helical CT). The detector size determines the body volume to be imaged. Commercial CBCT systems generally allow for the reconstruction of a large volume (e.g., 25 × 25 × 17 cm) with submillimeter voxels, enabling all three dimensions to be viewed in high resolution. Resultant CBCT localization volumes can be registered directly with planning CT image data for highly accurate patient positioning (Feldkamp et al., 1984; Jaffray and Siewerdsen, 2000; Jaffray, 2005; Letourneau et al., 2005; Oldham et al., 2005)

In radiation therapy applications, CBCT is typically implemented by mounting a kV X-ray tube and an FPD on the treatment gantry along the axis orthogonal to the treatment beam axis, as shown in Figure 7.1. Full rotation of the gantry takes roughly one minute, so one cone-beam projection image is acquired through a period of multiple breathing cycles. A bow-tie shaped filter and an antiscatter collimator are usually applied to improve image quality. A filtered back projection algorithm is typically used to reconstruct CBCT images after preprocessing of the projections (Feldkamp et al., 1984).

Although CBCT substantially improves anatomic visibility relative to 2D images, it also has several limitations:

1. Scatter effects. The amount of X-ray scatter is proportional to the amount of volume being irradiated during imaging. Compared with fan-beam CT, CBCT irradiates a much larger image volume, which substantially increases the amount of scatter detected by the detector. Scatter has the doubly negative effect of increasing quantum noise while decreasing image contrast in the projection data. Thus, the contrast-to-noise ratio (CNR) in CBCT images is lower than the CNR in traditional fan-beam CT images. Scattered radiation also creates streak and cupping artifacts, which impairs the Hounsfield unit (HU) accuracies in the reconstructed CBCT images.
2. High imaging dose. The imaging dose of CBCT can be as high as approximately 3 c Gy due to the large number of X-ray projections acquired (400 to 600 projections). (Hyer et al., 2010) If CBCT is taken daily, the accumulated imaging dose over a 30-fraction treatment course will be close to 1 Gy, which is not negligible, especially considering this imaging dose is delivered through a large volume of the body. This high imaging dose causes concerns for secondary cancer induction, especially for young patients.
3. Long scanning time. A typical CBCT scan takes around 30 to 60 s depending on the scanning mode. The long scanning time makes CBCT prone to artifacts caused by respiratory motions of the patient.

CBCT images could also be generated using MV beams with the MV detector. Currently, the imaging beam energies range from 2.5 MV to 6 MV. MV images have advantages for

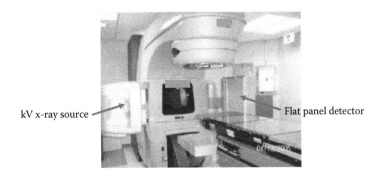

FIGURE 7.1   kV onboard imaging system.

imaging larger patient sizes and patients with metal implants. Potentially, both kV and MV beams could be used for CBCT image acquisition and could be reconstructed as dual-energy CBCT (Li et al., 2012, 2013).

7.4.2 Onboard MRI Imaging

Introduced by Lagendijk et al., magnetic resonance imaging (MRI) integrated with a radio-therapy unit has been introduced in recent years for target localization (Dempsey et al., 2006; Kirkby et al., 2008; Lagendijk et al., 2008; Crijns and Raaymakers, 2014). Compared to X-ray imaging techniques, MRI has much better soft-tissue contrast, which is critical for localizing tumors in the soft-tissue environment, for example, for prostate cancer (Noel et al., 2015). MRI also enjoys the benefits of no ionizing radiation dose to patients, which makes it an attractive modality for daily imaging for both inter- and intrafraction verification. Onboard MRI systems have been developed by integrating the MRI scanner with a linear accelerator or a ^{60}Co treatment machine (Dempsey et al., 2006; Fallone et al., 2009; Kron et al., 2006; Lagendijk et al., 2008; Raaymakers et al., 2009). Most of these MRI-radiotherapy machines uses low field MRI (0.2–0.35T) to reduce interference with radiation therapy delivery and decrease geometric distortion. High field (1.5T) MR-linac machines have also been introduced into clinics recently for initial clinical studies.

The cost of a radiotherapy machine with onboard MRI scanner is much higher than a radiotherapy machine with onboard kV/MV imagers, which may limit the availability of the system in clinics. Besides, MRI is not applicable for imaging patients with pacemakers, metal implants, or claustrophobia, and it is susceptible to MRI artifacts (Erasmus et al., 2004). The clinical impact of using MRI imaging for target localization, real-time tumor tracking, or adaptive radiotherapy needs to be investigated further (Figure 7.2).

7.4.3 Electromagnetic Monitoring System

The Calypso® System (Varian Medical Systems, Inc., Palo Alto, California) is a commercial device that has been developed and used mainly for target localization for prostate cancer treatments at present (Mantz, 2014). Three electromagnetic transponders, each with a different resonance frequency, are implanted near the base and apex of the patient's prostate before CT simulation. The locations of the three transponders are identified in the planning CT images, and imported into the Calypso 4D tracking station together with the treatment isocenter location. For onboard imaging, the Calypso detector antenna array contains source coils and receiver coils. The source coils generate an oscillating radio frequency (RF) field, inducing resonances in the transponders. When the field is switched off, the transponder emits a signal during relaxation that is detected by the receiver coils to establish the location of these transponders. The centroid of the three transponder locations is calculated and used to represent the location of the prostate.

Calypso can provide real-time monitoring of the prostate location before or during the radiotherapy treatment with no ionizing radiation dose to the patient. The signal from Calypso can be used for gated treatment or dynamic target tracking when intrafractional motion is significant, such as respiratory motion for lung cancer (Shah et al., 2013). However, using the system requires beacon implantation, which is an invasive procedure.

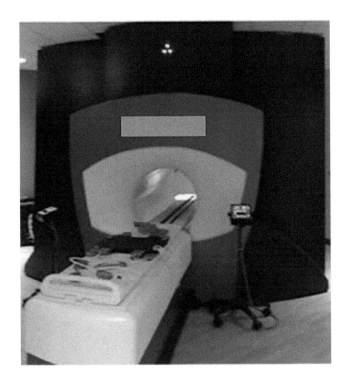

FIGURE 7.2 MRI-guided radiation therapy system combining MR scanner with a Co-60 radiotherapy unit.

The localization is based solely on the centroid of the three beacons locations, which may not be representative of the prostate location when there is deformation of the prostate or beacon migration. The tracking depth of Calypso is limited to between 16 and 20 cm. The calypso beacons also create artifacts in MRI images that affect the usage of MRI for target delineation or post-treatment assessment (Zhu et al., 2009c). The accuracy of this device depends on the calibration between the array and machine geometry (Figure 7.3).

Another electromagnetic tracking system being developed is the Raypilot system from the Micropos Medical. The system communicates with an implanted transmitter that is located in the region to be treated in the patient. The transmitter sends signals to a sensor plate 30 times per second to localize the position of the tumor being treated with submillimeter accuracy. A dosimeter can also be placed inside the transmitter to provide in situ dose in real time at a given position and accumulated dose over the treatment course.

7.4.4 Optical Imaging

Optical imaging tracks the patient surface movement using either infrared markers such as the Exactrac system (Brianlab, Inc), or optical surface imaging such as AlignRT (VisionRT, Inc), Catalyst (C-RAD), and Identify (humediQ). In the infrared marker–based system, several infrared markers are placed at noncoplanar locations on the patient surface. A set of infrared cameras mounted in the room tracks the locations of the markers to determine the patient motion and location. The optical-based surface imaging system uses three

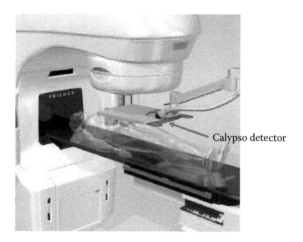

Calypso detector

FIGURE 7.3 Calypso tracking system. (Courtesy of Varian Medical System.)

FIGURE 7.4 AlignRT system to monitor patient surface motion. (Courtesy of VisionRT.)

ceiling-mounted cameras and a patterned light in order to reconstruct 3D surface data of the patient's body. During the treatment planning session, the system acquires a reference image when the patient is in the optimal position. When the patient is treated, a new image is taken and matched to the original reference image to determine the patient displacement in real time. Limitations of optical imaging techniques are that they require open skin surface, and they can only track the patient surface motion, which may not be fully correlated with the internal target motion (Figure 7.4).

7.4.5 Ultrasound Imaging

Ultrasound imaging has been used extensively for guidance of prostate brachytherapy (Davis et al., 2012). A transrectal ultrasound (TRUS) is used to guide the placement of the radioactive seeds at planned positions in the prostate. TRUS requires insertion of an ultrasound probe into the rectum of the patient. The probe sends and receives sound waves through the wall of the rectum to and from the prostate gland, which is situated right above the rectum. The ultrasound images of the prostate are produced in real time for implant guidance. Besides brachytherapy, ultrasound imaging has also been developed for external

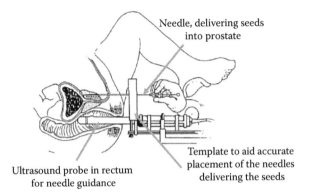

Needle, delivering seeds
into prostate

Template to aid accurate
placement of the needles
delivering the seeds

Ultrasound probe in rectum
for needle guidance

FIGURE 7.5   Ultrasound guided brachytherapy seed implant.

beam radiation therapy guidance for prostate cancer treatments, such as the Clarity system (Elekta, Inc.) (Lattanzi et al., 2000). The Clarity system places the ultrasound probe against the patient's perineum region to provide real-time monitoring of the prostate location during treatment for intra-fraction 4D verification, which can become important for hypofractionated treatments. In summary, ultrasound imaging has the advantages of providing real-time monitoring with no ionizing radiation dose to the patient. However, it has limited imaging depth due to the limited penetration depth of the high-frequency sound waves (Figure 7.5).

7.5 QUALITY ASSURANCE FOR DEVICES AND SYSTEMS

The accuracy of target localization relies on the accuracy of the imaging system used. Therefore, the performance of imaging devices and systems as part of the integrated radiation therapy system should be carefully monitored through quality assurance (QA) tests. The goal of QA is to ensure that there are no significant deviations of their performance characteristics of the imaging devices and systems, defined by physical parameters established during commissioning of the equipment. Typically, each localization device is required to go through a number of steps prior to clinical patient application, including acceptance testing after installation and major part replacement, commissioning for clinical utility and development of optimal application protocol, calibration of imaging system to correlate the coordinates between the localization system and treatment system, and performance of routine QA of the device. A QA program could be developed based on guidelines from the American Association of Physicists in Medicine (AAPM) Task Group reports (Kutcher et al., 1994; Klein et al., 2009; Yin et al., 2009; Benedict et al., 2010; Dieterich et al., 2011) to provide procedures that monitor the safety of patients and operators, physical parameters of the equipment, positioning and correction accuracy, and image quality. Quality levels obtained during the commissioning procedure should be used as the basis for specific tolerances. Safety tests should monitor the proper function of audio and visual alarms, collision detection and interlock, and so on. Various commercially available phantoms and tools could be used to check the physical parameters, accuracy, and image quality. A procedure to monitor those tests should be reproducible and consistent over an extended period. Once the QA program is developed, all procedures should be performed periodically to accumulate enough experience and confidence to ensure stability and accuracy of the system. The frequency of each test should be determined based on recommendations drawn from the literature (Low et al., 1996; Menon and Sloboda, 2004; Sharpe et al., 2006; Klein et al., 2009; Thomadsen et al., 2013) as well as a physicist's judgment and understanding of the system as a positioning verification device. It is important that all procedures are performed regularly to establish a guideline. Overall, QA should be established to check the system quality and performance routinely both for hardware and software used for control and imaging processing. End-to-end testing is highly recommended for each device.

7.6 CHALLENGES AND FUTURE DEVELOPMENT

Hypofractionation such as SRS and SBRT treatments is emerging as an effective treatment paradigm in radiation therapy, with promising clinical outcome (Sahgal, 2012; Rahimi et al., 2014; Chang et al., 2015; Tao et al., 2015) Compared to traditional fractionated radiotherapy, SRS/SBRT delivers much higher radiation dose per fraction, typically in no more than five fractions. This high radiation dose delivery requires high precision target localization. Several areas are under investigation to develop imaging techniques that can generate high-quality images with high efficiency, accuracy, and minimal imaging dose. The details of each area are explained in the following subsections.

7.6.1 Limited-Angle Imaging

One major limitation of CBCT is its high imaging dose and long scanning time due to the large scanning angle and number of projections acquired in a CBCT scan. Different imaging techniques are being developed to reconstruct images using only a limited angle scan, which can substantially reduce the imaging dose and scanning time.

7.6.1.1 Digital Tomosynthesis

Digital tomosynthesis (DTS) is a method for reconstructing 3D slices from 2D cone-beam X-ray projection data acquired with limited source angulation (e.g., 40°) (Pang and Rowlands, 2005; Godfrey et al., 2006a) By resolving overlying anatomy into slices, DTS greatly improves the visibility of both soft tissue and bone compared with kV or MV radiographic imaging. DTS requires less radiation exposure and unobstructed gantry rotation clearance, and it can be acquired with a much shorter scan time than CBCT (Dobbins and Godfrey, 2003; Godfrey et al., 2006b).

The DTS localization process includes the creation of reference DTS (RDTS) images from planning CT data, as well as the acquisition of onboard verification DTS images, acquired in the treatment room. Comparison of the two DTS image sets (reference and verification) allows for the determination of patient setup error. RDTS images are reconstructed from simulated limited angle cone-beam projections through a planning CT image volume. Onboard DTS slices are acquired in the same fashion as full CBCT, but with limited gantry rotation. The Feldkamp–Davis–Kress (FDK) back projection algorithm

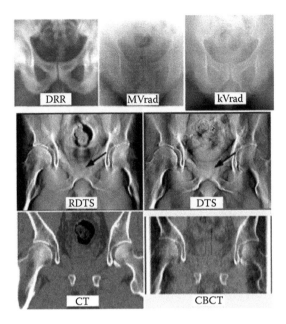

FIGURE 7.6   Setup fields from a prostate subject. Top row: DRR, MV radiograph, and kV radiograph. Middle row: Reference and onboard DTS coronal slice. An arrow points to the prostate. Bottom row: CT and CBCT coronal slice.

(Feldkamp et al., 1984) is typically used to reconstruct reference and onboard DTS images. Constraining the scan to 40° generally yields high-quality DTS slices with good soft-tissue visibility, while enabling the scan to be completed in less than 10 s with around one-ninth of the full rotation CBCT imaging dose. Individual DTS slices exhibit high resolution in the viewing plane, but the resolution in the third (plane-to-plane) dimension is limited by the narrow scan angle. As a result, CT HU values are not correctly rendered in DTS, and image contrast in DTS slices is inferior to that in full CBCT images. Figure 7.6 shows radiographic, DTS, and CT/CBCT coronal setup fields for a prostate subject. Although the effective slice profile of the DTS images is thicker than that of full CT/CBCT, soft-tissue visibility is reasonably good in the limited-angle DTS data and is markedly better than that provided by the traditional 2D radiographic fields with better localization accuracy, at least for spine, liver, breast, prostate, and lung cancer treatments (Godfrey et al., 2007; Yan et al., 2007; Ren et al., 2008a; Yoo et al., 2009; Zhang et al., 2009; Wu et al., 2011; Zhang et al., 2013a).

7.6.1.2 Limited-Angle Intrafraction Verification System
Although DTS imaging improved anatomical visibility compared to 2D X-ray imaging, it doesn't provide full volumetric information of the patient due to the limited scan angle used in acquisition. The lack of full 3D information may impair the localization accuracy of DTS when there is soft-tissue deformation of the patient. Novel image reconstruction techniques have been developed in recent years to reconstruct full volumetric images using limited number of projections (Ren et al., 2008b; Li et al., 2010, 2011; Ren et al., 2012a;

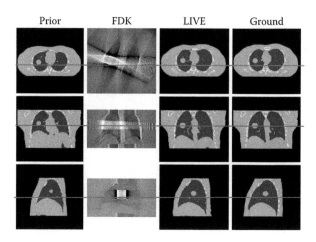

FIGURE 7.7 Comparison between images reconstructed by different methods using orthogonal 30° kV and BEV MV projections in the XCAT study.

Zhang et al., 2013b; Zhang et al., 2015). This method considers the onboard patient images as a deformation of prior images. So instead of solving the pixel values directly in reconstruction, it solves the deformation field that deforms the prior images to obtain the onboard images. The deformation field is usually solved iteratively using data fidelity constraint and motion modeling. Based on this method, a limited-angle intrafraction verification (LIVE) system was recently proposed to reconstruct patient intrafractional volumetric images based on limited angle kV and MV projections acquired during the treatment (Ren et al., 2014). Figure 7.7 shows the images reconstructed by the FDK back projection method and the LIVE system using data from the 4D digital extended-cardiac-torso (XCAT) phantom. The LIVE system was able to reconstruct high-quality volumetric images using only orthogonal 30° scan angles. The LIVE system is currently being evaluated under clinical trials to determine its efficiency and accuracy for target localization in lung SBRT.

7.6.2 Scatter Reduction and Correction Techniques for CBCT

As discussed in Section 3.1.2, scatter has been a major problem with CBCT imaging due to its large imaging field size. Many techniques have been proposed to address scattering. These techniques can be generally divided into two major types: (1) direct scatter reduction during image acquisition using antiscatter grids (Sorenson et al., 1980; Siewerdsen et al., 2004; Chang et al., 2010; Stankovic et al., 2014); and (2) postscan scatter correction, in which scatter distributions are measured and/or modeled for each patient and subtracted after the scan (Ning et al., 2004; Siewerdsen et al., 2006; Zhu et al. 2006; Zhu et al., 2009b). However, the efficiency of direct scatter reduction methods is frequently limited, and they often require increased dose to the patient. Postscan scatter correction methods do not address the scatter-related X-ray quantum noise and can potentially degrade the CNR (Siewerdsen et al., 2006; Zhu et al., 2009a). In addition, the traditional scatter correction methods using beam blocks typically require two scans. The first scan is acquired with the beam block attached so that the scatter signal can be measured under the blocked

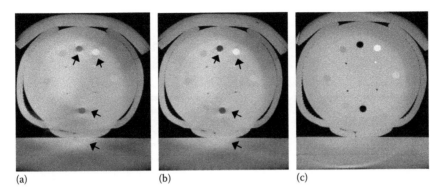

(a)   (b)   (c)

FIGURE 7.8 Effects of scatter reduction and correction in CBCT: (a) original CBCT, (b) CBCT after scatter reduction, and (c) CBCT after scatter reduction and correction.

region of the projections. The second scan is acquired with the beam block removed to acquire complete patient projections. The scatter signal measured in the first scan is then used to correct for scatter in the second scan. The two-scan acquisition scheme increases patient dose, and the accuracy of scatter correction is limited because scatter is measured under a condition different from the image acquisition condition. Recently, a moving grid system has been developed to provide both scatter reduction and scatter correction to overcome the issues mentioned above (Jin et al., 2010; Zhao et al., 2010; Ren et al., 2012b; Ren et al., 2015; Zhang et al., 2016). In this scheme, both scatter signal and patient projection data are acquired simultaneously in a single scan with the grid attached to the X-ray source, which provides physical scatter reduction during the acquisition. The scatter signal is measured under the blocked region, while the patient image is acquired under the unblocked region in the projection image. The scatter signal is then subtracted from the patient image for postscan scatter correction. The grid moves back and forth during the acquisition to acquire complementary projections to reconstruct complete CBCT images. This system achieves both scatter reduction and scatter correction using a single scan without increasing the imaging dose. Figure 7.8 shows an example of the scatter reduction and correction effects on CBCT, which demonstrates that this system can potentially reduce the scatter-related artifacts substantially while enhancing the CNR (Ren et al., 2012b).

7.6.3 4D Imaging

Localization of moving targets such as lung and liver tumors requires development of 4D imaging techniques to capture the target motion information. Four-dimensional CBCT techniques have been developed to reconstruct 3D CBCT images for each respiratory phase of the patient's breathing cycle. Specifically, 4D CBCT uses a slow gantry rotation to acquire cone-beam projections and then retrospectively sort all projections into different respiratory phases according to internal or external respiratory signal. The 4D CBCT is then obtained by reconstructing CBCT for each phase based on the projections sorted for the phase (Lu et al., 2007). Different reconstruction algorithms have been developed to use a deformation

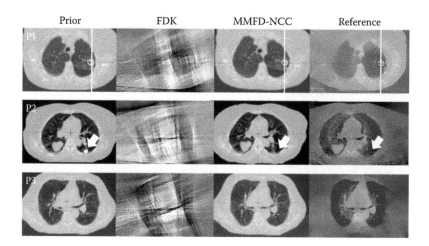

FIGURE 7.9   Slice cuts from the prior images, FDK images, MMFD NCC estimated images, and reference images of different patients (P1: Patient 1; P2: Patient 2; P3: Patient 3) at the end-expiration (50%) phase. Both FDK reconstructions and MMFD NCC estimations used orthogonal-view 30 projections around posterior-anterior and right-lateral directions. The onboard reference images were reconstructed by the FDK method using fully sampled 200 projections.

model or prior knowledge to improve the image quality of 4D CBCT (Wang and Gu, 2013; Zhang et al., 2010; Zhang et al., 2013b, 2015). Figure 7.9 shows a comparison between 4D CBCT reconstructed using both FDK and the prior knowledge-based motion modeling and free form deformation with normalized cross correlation (MMFD NCC) estimation method using only 30 orthogonal-view projections (Zhang et al., 2015). The prior-knowledge-based method achieved substantially higher-quality images even when compared to the reference images reconstructed by FDK using fully sampled 200 projections.

Four-dimensional MRI has also been developed through either prospective or retrospective approaches. Due to the limitations of hardware and software, prospective 4D MRI suffers from poor temporal resolution (approximately 1 s) and poor spatial resolution (4–5 mm) (Dinkel et al., 2009; Cai et al. 2011; Hu et al., 2013). Retrospective 4D MRI has better temporal and in-plane resolution, but it suffers from long acquisition time (5–30 min), and poor plane-to-plane resolution (3–5 mm slice thickness) (von Siebenthal et al., 2007; Remmert et al., 2007; Cai et al., 2011; Tryggestad et al., 2013). A volumetric cine (VC) MRI imaging technique is currently under development to use patient prior knowledge and motion modeling to generate 3D MRI images in real time (Harris et al., 2016). Figure 7.10 shows a comparison between 2D cine MR and estimated VC MRI images. VC MRI can potentially be used for real-time 3D target verification or tracking during radiation therapy delivery to minimize the treatment errors and improve the treatment outcome.

7.6.4 Onboard Functional Imaging

Functional imaging modalities such as positron emission tomography (PET) and single photon emission computed tomography (SPECT) rely on radiotracers to identify

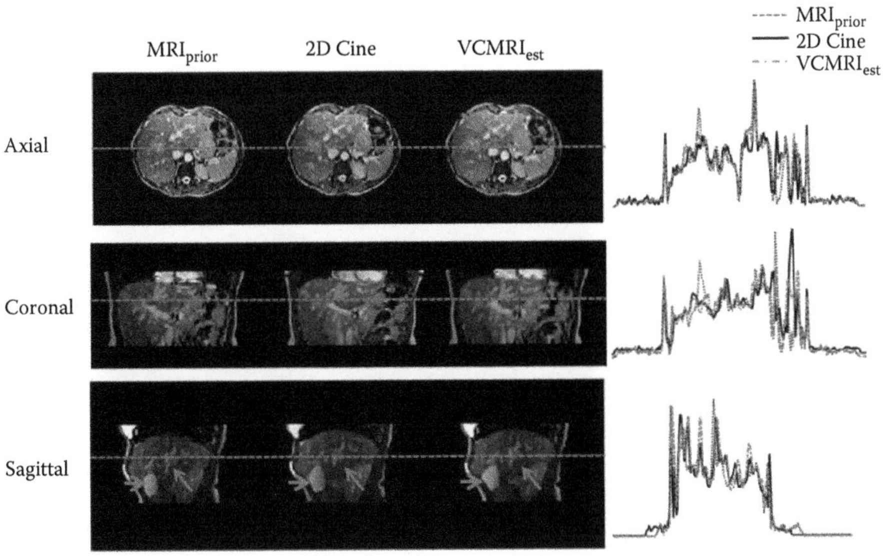

FIGURE 7.10   Prior MRI (MRI_{prior}), 2D cine MRI, and estimated VC-MRI for a liver cancer patient. The horizontal red dotted line corresponds to the location of the profile curves shown to the right of the images.

physiological processes—such as angiogenesis, apoptosis, hormone receptor status, hypoxia, and proliferation—that have important implications in cancer management. In the context of radiation therapy, such functional information can be used to define biological target volumes that are not visible in CT or MRI images. Onboard SPECT has been developed in recent years for localization of biological target volumes, which are potentially more accurate than anatomical target volumes (Roper et al., 2009) Developing an onboard SPECT system for target localization has several challenges, such as scan time, image quality, mechanical constraints, and so on. While these constraints have a negative impact on implementation of functional imaging capability in the treatment room, it is important to consider that the task for onboard SPECT would be different than with most diagnostic SPECT scans. Because target size and approximate location are known, it may be possible to localize functional targets accurately. Furthermore, SPECT detector trajectories can be optimized for imaging the treatment region, a factor that may enhance SPECT images in the region of interest (ROI) compared to diagnostic scans, which typically have to survey a larger volume. Different studies are being carried out to develop and optimize the onboard SPECT system in terms of accuracy, efficiency, and mechanical clearance (Roper et al., 2009; Bowsher et al., 2014; Yan et al., 2014). Figure 7.11 shows an example of a reconstructed onboard SPECT images for different tumor sizes and scan times in a computer-simulated NCAT phantom.

7.6.5 Real-Time Tumor Tracking

The ideal method for minimizing radiation dose to normal tissues is to track tumor motion in real time so that the PTV volume required to compensate for motion effects

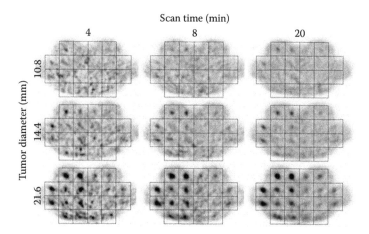

FIGURE 7.11 Sample reconstructed onboard SPECT images, smoothed with 14-mm-FWHM Gaussian, for different tumor sizes and scan times in a computer-simulated NCAT phantom.

could potentially be reduced. A number of techniques have been developed for tumor tracking in recent years, including 2D kV/MV fluoroscopic imaging, 2D MR cine imaging, or electromagnetic tracking with implanted beacons (Kitamura et al., 2002; Willoughby et al., 2006; Ng et al., 2012). Two radiation delivery techniques have been developed using these tracking techniques: (1) gated treatment in which the radiation beam is fixed and delivered only when the target moves into the beam aperture, and (2) delivery tracking in which the radiation beam is repositioned dynamically by adjusting the multileaf collimator (MLC) position to follow the motion of the target (Keall et al., 2006). Algorithms to predict the target motion have been developed for delivery tracking to minimize lagging between target motion and MLC repositioning (Yan et al., 2006). Overall, both delivery techniques allow us to reduce the PTV volume and thus reduce the radiation dose to surrounding healthy tissues, which in turn allows potential dose escalation to enhance the tumor control probability. However, comprehensive QA process is required to ensure the accuracy and robustness of the tumor tracking (Keall et al., 2006).

7.7 CONCLUSION

The quality of patient setup and target localization is vital to the success of radiation therapy. The selection of the techniques for setup and target localization need to be carefully optimized by evaluating different aspects of the techniques, including patient comfort, setup repeatability, efficiency, imaging dose, localization accuracy, and so on. This optimization needs to be considered within the context of the overall clinical flow for each individual patient because setup and target localization are highly correlated to other clinical parameters in radiation therapy, such as dose fractionation schemes, PTV margin, and treatment delivery techniques. Staff training and QA need to be addressed so that the selected techniques are implemented effectively to maximize their benefits to the patient. Future advancements in localization techniques can lead to lower dose, more efficient, and more accurate 3D/4D imaging techniques.

REFERENCES

Benedict, S. H., K. M. Yenice, D. Followill, J. M. Galvin, W. Hinson, B. Kavanagh, P. Keall et al. 2010. Stereotactic body radiation therapy: The report of AAPM Task Group 101. *Med Phys* 37 (8):4078–4101.

Bowsher, J., S. Yan, J. Roper, W. Giles, and F. F. Yin. 2014. Onboard functional and molecular imaging: A design investigation for robotic multipinhole SPECT. *Med Phys* 41 (1):010701. doi:10.1118/1.4845195.

Cai, J., Z. Chang, Z. Wang, W. Paul Segars, and F. F. Yin. 2011. Four-dimensional magnetic resonance imaging (4D-MRI) using image-based respiratory surrogate: A feasibility study. *Med Phys* 38 (12):6384–6394. doi:10.1118/1.3658737.

Chang, J., S. Kim, D. Y. Jang, and T. S. Suh. 2010. A static multi-slit collimator system for scatter reduction in cone-beam CT. *J Appl Clin Med Phys* 11 (4):3269.

Chang, J. Y., S. Senan, M. A. Paul, R. J. Mehran, A. V. Louie, P. Balter, H. J. Groen et al. 2015. Stereotactic ablative radiotherapy versus lobectomy for operable stage I non-small-cell lung cancer: A pooled analysis of two randomised trials. *Lancet Oncol* 16 (6):630–637. doi:10.1016/S1470-2045(15)70168-3.

Crijns, S., and B. Raaymakers. 2014. From static to dynamic 1.5T MRI-linac prototype: impact of gantry position related magnetic field variation on image fidelity. *Phys Med Biol* 59 (13):3241–3247. doi:10.1088/0031-9155/59/13/3241.

Davis, B. J., E. M. Horwitz, W. R. Lee, J. M. Crook, R. G. Stock, G. S. Merrick, W. M. Butler et al. 2012. American Brachytherapy Society consensus guidelines for transrectal ultrasound-guided permanent prostate brachytherapy. *Brachytherapy* 11 (1):6–19. doi:10.1016/j.brachy.2011.07.005.

Dempsey, J., B. Dionne, J. Fitzsimmons, A. Haghigat, J. Li, D. Low, S. Mutic, J. Palta, H. Romeijn, and G. Sjoden. 2006. A real-time MRI guided external beam radiotherapy delivery system. *Med Phys* 33 (6):2254.

Dieterich, S., C. Cavedon, C. F. Chuang, A. B. Cohen, J. A. Garrett, C. L. Lee, J. R. Lowenstein et al. 2011. Report of AAPM TG 135: Quality assurance for robotic radiosurgery. *Med Phys* 38 (6):2914–2936.

Dinkel, J., C. Hintze, R. Tetzlaff, P. E. Huber, K. Herfarth, J. Debus, H. U. Kauczor, and C. Thieke. 2009. 4D-MRI analysis of lung tumor motion in patients with hemidiaphragmatic paralysis. *Radiother Oncol* 91 (3):449–454. doi:10.1016/j.radonc.2009.03.021.

Dobbins, J. T., and D. J. Godfrey. 2003. Digital x-ray tomosynthesis: Current state of the art and clinical potential. *Phys Med Biol* 48 (19):R65–R106.

Eccles, C. L., L. A. Dawson, J. L. Moseley, and K. K. Brock. 2011. Interfraction liver shape variability and impact on GTV position during liver stereotactic radiotherapy using abdominal compression. *Int J Radiat Oncol Biol Phys* 80 (3):938–946. doi:10.1016/j.ijrobp.2010.08.003.

Erasmus, L. J., D. Hurter, M. Naude, H. G. Kritzinger, and S. Acho. 2004. A short overview of MRI artefacts. *SA Journal of Radiology* 8 (2):13–17.

Fallone, B. G., B. Murray, S. Rathee, T. Stanescu, S. Steciw, S. Vidakovic, E. Blosser, and D. Tymofichuk. 2009. First MR images obtained during megavoltage photon irradiation from a prototype integrated linac-MR system. *Med Phys* 36 (6):2084–2088.

Feldkamp, L. A., L. C. Davis, and J. W. Kress. 1984. Practical cone-beam algorithm. *Journal of the Optical Society of America a-Optics Image Science and Vision* 1 (6):612–619. doi:10.1364/Josaa.1.000612.

Fuss, M., B. J. Salter, P. Rassiah, D. Cheek, S. X. Cavanaugh, and T. S. Herman. 2004. Repositioning accuracy of a commercially available double-vacuum whole body immobilization system for stereotactic body radiation therapy. *Technol Cancer Res Treat* 3 (1):59–67.

Godfrey, D. J., H. P. McAdams, and J. T. Dobbins. 2006. Optimization of the matrix inversion tomosynthesis (MITS) impulse response and modulation transfer function characteristics for chest imaging. *Med Phys* 33 (3):655–667.

Godfrey, D. J., L. Ren, H. Yan, Q. Wu, S. Yoo, M. Oldham, and F. F. Yin. 2007. Evaluation of three types of reference image data for external beam radiotherapy target localization using digital tomosynthesis (DTS). *Med Phys* 34 (8):3374–3384.

Godfrey, D. J., F. F. Yin, M. Oldham, S. Yoo, and C. Willett. 2006. Digital tomosynthesis with an on-board kilovoltage imaging device. *Int J Radiat Oncol Biol Phys* 65 (1):8–15. doi:10.1016/j.ijrobp.2006.01.025.

Han, K., P. Cheung, P. S. Basran, I. Poon, L. Yeung, and F. Lochray. 2010. A comparison of two immobilization systems for stereotactic body radiation therapy of lung tumors. *Radiother Oncol* 95 (1):103–108. doi:10.1016/j.radonc.2010.01.025.

Harris, W., L. Ren, J. Cai, Y. Zhang, Z. Chang, and F. F. Yin. 2016. A technique for generating volumetric cine-magnetic resonance imaging. *International Journal of Radiation Oncology Biology Physics* 95 (2):844–853.

Hu, Y., S. D. Caruthers, D. A. Low, P. J. Parikh, and S. Mutic. 2013. Respiratory amplitude guided 4-dimensional magnetic resonance imaging. *Int J Radiat Oncol Biol Phys* 86 (1):198–204. doi:10.1016/j.ijrobp.2012.12.014.

Hyer, D. E., C. F. Serago, S. Kim, J. G. Li, and D. E. Hintenlang. 2010. An organ and effective dose study of XVI and OBI cone-beam CT systems. *J Appl Clin Med Phys* 11 (2):3183.

Jaffray, D. A. 2005. Emergent technologies for 3-dimensional image-guided radiation delivery. *Semin Radiat Oncol* 15 (3):208–216.

Jaffray, D. A., and J. H. Siewerdsen. 2000. Cone-beam computed tomography with a flat-panel imager: Initial performance characterization. *Med Phys* 27 (6):1311–1323.

Jin, J. Y., L. Ren, Q. Liu, J. Kim, N. Wen, H. Guan, B. Movsas, and I. J. Chetty. 2010. Combining scatter reduction and correction to improve image quality in cone-beam computed tomography (CBCT). *Med Phys* 37 (11):5634–5644.

Keall, P. J., H. Cattell, D. Pokhrel, S. Dieterich, K. H. Wong, M. J. Murphy, S. S. Vedam, K. Wijesooriya, and R. Mohan. 2006. Geometric accuracy of a real-time target tracking system with dynamic multileaf collimator tracking system. *Int J Radiat Oncol Biol Phys* 65 (5):1579–1584. doi:10.1016/j.ijrobp.2006.04.038.

Kirkby, C., T. Stanescu, S. Rathee, M. Carlone, B. Murray, and B. G. Fallone. 2008. Patient dosimetry for hybrid MRI-radiotherapy systems. *Med Phys* 35 (3):1019–1027.

Kitamura, K., H. Shirato, S. Shimizu, N. Shinohara, T. Harabayashi, T. Shimizu, Y. Kodama et al. 2002. Registration accuracy and possible migration of internal fiducial gold marker implanted in prostate and liver treated with real-time tumor-tracking radiation therapy (RTRT). *Radiother Oncol* 62 (3):275–281.

Klein, E. E., J. Hanley, J. Bayouth, F. F. Yin, W. Simon, S. Dresser, C. Serago et al. 2009. Task Group 142 report: Quality assurance of medical accelerators. *Med Phys* 36 (9):4197–4212.

Kron, T., D. Eyles, S. L. John, and J. Battista. 2006. Magnetic resonance imaging for adaptive cobalt tomotherapy: A proposal. *J Med Phys* 31 (4):242–254. doi:10.4103/0971-6203.29194.

Kutcher, G. J., L. Coia, M. Gillin, W. F. Hanson, S. Leibel, R. J. Morton, J. R. Palta et al. 1994. Comprehensive QA for radiation oncology: Report of AAPM Radiation Therapy Committee Task Group 40. *Med Phys* 21 (4):581–618.

Lagendijk, J. J., B. W. Raaymakers, A. J. Raaijmakers, J. Overweg, K. J. Brown, E. M. Kerkhof, R. W. van der Put, B. Hardemark, M. van Vulpen, and U. A. van der Heide. 2008. MRI/linac integration. *Radiother Oncol* 86 (1):25–29. doi:10.1016/j.radonc.2007.10.034.

Lattanzi, J., S. McNeeley, A. Hanlon, T. E. Schultheiss, and G. E. Hanks. 2000. Ultrasound-based stereotactic guidance of precision conformal external beam radiation therapy in clinically localized prostate cancer. *Urology* 55 (1):73–78.

Letourneau, D., J. W. Wong, M. Oldham, M. Gulam, L. Watt, D. A. Jaffray, J. H. Siewerdsen, and A. A. Martinez. 2005. Cone-beam-CT guided radiation therapy: Technical implementation. *Radiother Oncol* 75 (3):279–286. doi:10.1016/j.radonc.2005.03.001.

Li, H., W. Giles, L. Ren, J. Bowsher, and F. F. Yin. 2012. Implementation of dual-energy technique for virtual monochromatic and linearly mixed CBCTs. *Med Phys* 39 (10):6056–6064. doi:10.1118/1.4752212.

Li, R., X. Jia, J. H. Lewis, X. Gu, M. Folkerts, C. Men, and S. B. Jiang. 2010. Real-time volumetric image reconstruction and 3D tumor localization based on a single x-ray projection image for lung cancer radiotherapy. *Med Phys* 37 (6):2822–2826.

Li, R., J. H. Lewis, X. Jia, T. Zhao, W. Liu, S. Wuenschel, J. Lamb, D. Yang, D. A. Low, and S. B. Jiang. 2011. On a PCA-based lung motion model. *Phys Med Biol* 56 (18):6009–6030. doi:10.1088/0031-9155/56/18/015.

Li, H., B. Liu, and F. F. Yin. 2013. Generation of virtual monochromatic CBCT from dual kVMV beam projections. *Med Phys* 40 (12):121910. doi:10.1118/1.4824324.

Low, D. A., E. E. Klein, D. K. Maag, W. E. Umfleet, and J. A. Purdy. 1996. Commissioning and periodic quality assurance of a clinical electronic portal imaging device. *Int J Radiat Oncol Biol Phys* 34 (1):117–123.

Lu, J., T. M. Guerrero, P. Munro, A. Jeung, P. C. Chi, P. Balter, X. R. Zhu, R. Mohan, and T. Pan. 2007. Four-dimensional cone beam CT with adaptive gantry rotation and adaptive data sampling. *Med Phys* 34 (9):3520–3529.

Lymberis, S. C., J. K. deWyngaert, P. Parhar, A. M. Chhabra, M. Fenton-Kerimian, J. Chang, T. Hochman et al. 2012. Prospective assessment of optimal individual position (prone versus supine) for breast radiotherapy: Volumetric and dosimetric correlations in 100 patients. *Int J Radiat Oncol Biol Phys* 84 (4):902–909. doi:10.1016/j.ijrobp.2012.01.040.

Mantz, C. 2014. A Phase II trial of stereotactic ablative body radiotherapy for low-risk prostate cancer using a non-robotic linear accelerator and real-time target tracking: Report of toxicity, quality of life, and disease control outcomes with 5-year minimum follow-up. *Front Oncol* 4:279. doi:10.3389/fonc.2014.00279.

Menon, G. V., and R. S. Sloboda. 2004. Quality assurance measurements of a-Si EPID performance. *Med Dosim* 29 (1):11–17. doi:10.1016/j.meddos.2003.09.002.

Nelson, C., G. Starkschall, P. Balter, M. J. Fitzpatrick, J. A. Antolak, N. Tolani, and K. Prado. 2005. Respiration-correlated treatment delivery using feedback-guided breath hold: A technical study. *Med Phys* 32 (1):175–181. doi:10.1118/1.1836332.

Ng, J. A., J. T. Booth, P. R. Poulsen, W. Fledelius, E. S. Worm, T. Eade, F. Hegi, A. Kneebone, Z. Kuncic, and P. J. Keall. 2012. Kilovoltage intrafraction monitoring for prostate intensity modulated arc therapy: First clinical results. *Int J Radiat Oncol Biol Phys* 84 (5):e655–e661. doi:10.1016/j.ijrobp.2012.07.2367.

Ning, R., X. Tang, and D. Conover. 2004. X-ray scatter correction algorithm for cone beam CT imaging. *Med Phys* 31 (5):1195–202.

Noel, C. E., P. J. Parikh, C. R. Spencer, O. L. Green, Y. Hu, S. Mutic, and J. R. Olsen. 2015. Comparison of onboard low-field magnetic resonance imaging versus onboard computed tomography for anatomy visualization in radiotherapy. *Acta Oncol* 54 (9):1474–1482. doi:10.3109/0284186X.2015.1062541.

Oldham, M., D. Letourneau, L. Watt, G. Hugo, D. Yan, D. Lockman, L. H. Kim, P. Y. Chen, A. Martinez, and J. W. Wong. 2005. Cone-beam-CT guided radiation therapy: A model for on-line application. *Radiother Oncol* 75 (3):271–278. doi:10.1016/j.radonc.2005.03.026.

Pang, G., and J. A. Rowlands. 2005. Just-in-time tomography (JiTT): A new concept for image-guided radiation therapy. *Phys Med Biol* 50 (21):N323–N330. doi:10.1088/0031-9155/50/21/N05.

Park, Y. K., S. Kim, H. Kim, I. H. Kim, K. Lee, and S. J. Ye. 2011. Quasi-breath-hold technique using personalized audio-visual biofeedback for respiratory motion management in radiotherapy. *Med Phys* 38 (6):3114–3124. doi:10.1118/1.3592648.

Raaymakers, B. W., J. J. Lagendijk, J. Overweg, J. G. Kok, A. J. Raaijmakers, E. M. Kerkhof, R. W. van der Put et al. 2009. Integrating a 1.5 T MRI scanner with a 6 MV accelerator: Proof of concept. *Phys Med Biol* 54 (12):N229–N237. doi:10.1088/0031-9155/54/12/N01.

Rahimi, A. S., A. Spangler, D. Garwood, C. Ding, S. Stevenson, R. Rao, M. Leitch et al. 2014. Phase 1 dose escalation trial using stereotactic body radiation therapy (SBRT) for partial breast irradiation (PBI). *Int J Radiat Oncol Biol Phys* 90 (1):S253.

Remmert, G., J. Biederer, F. Lohberger, M. Fabel, and G. H. Hartmann. 2007. Four-dimensional magnetic resonance imaging for the determination of tumour movement and its evaluation using a dynamic porcine lung phantom. *Phys Med Biol* 52 (18):N401–N415. doi:10.1088/0031-9155/52/18/N02.

Ren, L., Y. Chen, Y. Zhang, W. Giles, J. Jin, and F. F. Yin. 2015. Scatter reduction and correction for dual-source cone-beam CT using prepatient grids. *Technol Cancer Res Treat.* doi:10.1177/1533034615587615.

Ren, L., I. J. Chetty, J. Zhang, J. Y. Jin, Q. J. Wu, H. Yan, D. M. Brizel, W. R. Lee, B. Movsas, and F. F. Yin. 2012. Development and clinical evaluation of a three-dimensional cone-beam computed tomography estimation method using a deformation field map. *Int J Radiat Oncol Biol Phys* 82 (5):1584–1593. doi:10.1016/j.ijrobp.2011.02.002.

Ren, L., D. J. Godfrey, H. Yan, Q. J. Wu, and F. F. Yin. 2008. Automatic registration between reference and on-board digital tomosynthesis images for positioning verification. *Med Phys* 35 (2):664–672.

Ren, L., F. F. Yin, I. J. Chetty, D. A. Jaffray, and J. Y. Jin. 2012. Feasibility study of a synchronized-moving-grid (SMOG) system to improve image quality in cone-beam computed tomography (CBCT). *Med Phys* 39 (8):5099–5110. doi:10.1118/1.4736826.

Ren, L., Y. Zhang, and F. F. Yin. 2014. A limited-angle intrafraction verification (LIVE) system for radiation therapy. *Med Phys* 41 (2):020701. doi:10.1118/1.4861820.

Ren, L., J. Zhang, D. Thongphiew, D. J. Godfrey, Q. J. Wu, S. M. Zhou, and F. F. Yin. 2008. A novel digital tomosynthesis (DTS) reconstruction method using a deformation field map. *Med Phys* 35 (7):3110–3115.

Roper, J., J. Bowsher, and F. F. Yin. 2009. On-board SPECT for localizing functional targets: A simulation study. *Med Phys* 36 (5):1727–1735.

Sahgal, A. 2012. Technological advances in brain and spine radiosurgery. *Technol Cancer Res Treat* 11 (1):1–2.

Shah, A. P., P. A. Kupelian, B. J. Waghorn, T. R. Willoughby, J. M. Rineer, R. R. Manon, M. A. Vollenweider, and S. L. Meeks. 2013. Real-time tumor tracking in the lung using an electromagnetic tracking system. *Int J Radiat Oncol Biol Phys* 86 (3):477–483. doi:10.1016/j.ijrobp.2012.12.030.

Sharpe, M. B., D. J. Moseley, T. G. Purdie, M. Islam, J. H. Siewerdsen, and D. A. Jaffray. 2006. The stability of mechanical calibration for a kV cone beam computed tomography system integrated with linear accelerator. *Med Phys* 33 (1):136–144.

Siewerdsen, J. H., M. J. Daly, B. Bakhtiar, D. J. Moseley, S. Richard, H. Keller, and D. A. Jaffray. 2006. A simple, direct method for x-ray scatter estimation and correction in digital radiography and cone-beam CT. *Med Phys* 33 (1):187–197.

Siewerdsen, J. H., D. J. Moseley, B. Bakhtiar, S. Richard, and D. A. Jaffray. 2004. The influence of antiscatter grids on soft-tissue detectability in cone-beam computed tomography with flat-panel detectors. *Med Phys* 31 (12):3506–3520.

Soike, M., J. M. Kilburn, J. T. Lucas, D. Ayala-Peacock, A. W. Blackstock, W. T. Kearns, W. H. Hinson, A. T. Miller, W. J. Petty, and J. J. Urbanic. 2013. Image guided radiation therapy results in improved local control in lung cancer patients treated with fractionated radiation therapy for stage IIB-IIIB disease. *Int J Radiat Oncol Biol Phys* 87 (2):S547–S5548.

Sorenson, J. A., L. T. Niklason, and D. F. Knutti. 1980. Performance characteristics of improved antiscatter grids. *Med Phys* 7 (5):525–528.

Stankovic, U., M. van Herk, L. S. Ploeger, and J. J. Sonke. 2014. Improved image quality of cone beam CT scans for radiotherapy image guidance using fiber-interspaced antiscatter grid. *Med Phys* 41 (6):061910. doi:10.1118/1.4875978.

Tao, R., S. Krishnan, P. R. Bhosale, M. M. Javle, T. A. Aloia, R. T. Shroff, A. O. Kaseb et al. 2015. Ablative radiotherapy doses lead to a substantial prolongation of survival in patients with inoperable intrahepatic cholangiocarcinoma: A retrospective dose response analysis. *J Clin Oncol.* doi:10.1200/JCO.2015.61.3778.

Thomadsen, B. R., P. Dunscombe, E. Ford, S. Huq, T. Pawlicki, and S. Sutlief. 2013. Quality and safety in radiotherapy: Learning new approaches in task group 100 and beyond. *AAPM Monograph (AAPM, Madison, WI, 2013)* 36:95–112.

Tryggestad, E., A. Flammang, S. Han-Oh, R. Hales, J. Herman, T. McNutt, T. Roland, S. M. Shea, and J. Wong. 2013. Respiration-based sorting of dynamic MRI to derive representative 4D-MRI for radiotherapy planning. *Med Phys* 40 (5):051909. doi:10.1118/1.4800808.

von Siebenthal, M., G. Szekely, U. Gamper, P. Boesiger, A. Lomax, and P. Cattin. 2007. 4D MR imaging of respiratory organ motion and its variability. *Phys Med Biol* 52 (6):1547–1564. doi:10.1088/0031-9155/52/6/001.

Wang, J., and X. Gu. 2013. Simultaneous motion estimation and image reconstruction (SMEIR) for 4D cone-beam CT. *Med Phys* 40 (10):101912. doi:10.1118/1.4821099.

Willoughby, T. R., P. A. Kupelian, J. Pouliot, K. Shinohara, M. Aubin, M. Roach, L. L. Skrumeda et al. 2006. Target localization and real-time tracking using the Calypso 4D localization system in patients with localized prostate cancer. *Int J Radiat Oncol Biol Phys* 65 (2):528–534. doi:10.1016/j.ijrobp.2006.01.050.

Wu, Q. J., J. Meyer, J. Fuller, D. Godfrey, Z. Wang, J. Zhang, and F. F. Yin. 2011. Digital tomosynthesis for respiratory gated liver treatment: Clinical feasibility for daily image guidance. *Int J Radiat Oncol Biol Phys* 79 (1):289–296. doi:10.1016/j.ijrobp.2010.01.047.

Yan, H., L. Ren, D. J. Godfrey, and F. F. Yin. 2007. Accelerating reconstruction of reference digital tomosynthesis using graphics hardware. *Med Phys* 34 (10):3768–3776. doi:10.1118/1.2779945.

Yan, S., J. Bowsher, M. Tough, L. Cheng, and F. F. Yin. 2014. A hardware investigation of robotic SPECT for functional and molecular imaging onboard radiation therapy systems. *Med Phys* 41 (11):112504. doi:10.1118/1.4898121.

Yan, H., F. F. Yin, G. P. Zhu, M. Ajlouni, and J. H. Kim. 2006. Adaptive prediction of internal target motion using external marker motion: A technical study. *Phys Med Biol* 51 (1):31–44. doi:10.1088/0031-9155/51/1/003.

Yin, F. F., J. Wong, J. Balter, S. Benedict, J. Craig, L. Dong, D. A. Jaffray et al. 2009. The role of in-room kV X-ray imaging for patient setup and target localization. *Report of AAPM Task Group 104.*

Yoo, S., Q. J. Wu, D. Godfrey, H. Yan, L. Ren, S. Das, W. R. Lee, and F. F. Yin. 2009. Clinical evaluation of positioning verification using digital tomosynthesis and bony anatomy and soft tissues for prostate image-guided radiotherapy. *Int J Radiat Oncol Biol Phys* 73 (1):296–305. doi:10.1016/j.ijrobp.2008.09.006.

Zelefsky, M. J., M. Kollmeier, B. Cox, A. Fidaleo, D. Sperling, X. Pei, B. Carver, J. Coleman, M. Lovelock, and M. Hunt. 2012. Improved clinical outcomes with high-dose image guided radiotherapy compared with non-IGRT for the treatment of clinically localized prostate cancer. *Int J Radiat Oncol Biol Phys* 84 (1):125–129. doi:10.1016/j.ijrobp.2011.11.047.

Zhang, Q., Y. C. Hu, F. Liu, K. Goodman, K. E. Rosenzweig, and G. S. Mageras. 2010. Correction of motion artifacts in cone-beam CT using a patient-specific respiratory motion model. *Med Phys* 37 (6):2901–2909.

Zhang, H., L. Ren, V. Kong, W. Giles, Y. Zhang, and J. Y. Jin. 2016. An interprojection sensor fusion approach to estimate blocked projection signal in synchronized moving grid-based CBCT system. *Med Phys* 43 (1):268. doi:10.1118/1.4937934.

Zhang, J., Q. J. Wu, D. J. Godfrey, T. Fatunase, L. B. Marks, and F. F. Yin. 2009. Comparing digital tomosynthesis to cone-beam CT for position verification in patients undergoing partial breast irradiation. *Int J Radiat Oncol Biol Phys* 73 (3):952–957. doi:10.1016/j.ijrobp.2008.10.036.

Zhang, Y., L. Ren, C. C. Ling, and F. F. Yin. 2013. Respiration-phase-matched digital tomosynthesis imaging for moving target verification: A feasibility study. *Med Phys* 40 (7):071723. doi:10.1118/1.4810921.

Zhang, Y., F. F. Yin, T. Pan, I. Vergalasova, and L. Ren. 2015. Preliminary clinical evaluation of a 4D-CBCT estimation technique using prior information and limited-angle projections. *Radiother Oncol* 115 (1):22–29. doi:10.1016/j.radonc.2015.02.022.

Zhang, Y., F. F. Yin, and L. Ren. 2015. Dosimetric verification of lung cancer treatment using the CBCTs estimated from limited-angle on-board projections. *Med Phys* 42 (8):4783–4795. doi:10.1118/1.4926559.

Zhang, Y., F. F. Yin, W. P. Segars, and L. Ren. 2013. A technique for estimating 4D-CBCT using prior knowledge and limited-angle projections. *Med Phys* 40 (12):121701. doi:10.1118/1.4825097.

Zhao, L., W. Ji, L. Zhang, G. Ou, Q. Feng, Z. Zhou, M. Lei, W. Yang, and L. Wang. 2010. Changes of circulating transforming growth factor-beta1 level during radiation therapy are correlated with the prognosis of locally advanced non-small cell lung cancer. *J Thorac Oncol* 5 (4):521–525. doi:10.1097/JTO.0b013e3181cbf761.

Zhu, L., N. R. Bennett, and R. Fahrig. 2006. Scatter correction method for X-ray CT using primary modulation: Theory and preliminary results. *IEEE Trans Med Imaging* 25 (12):1573–1587.

Zhu, L., J. Wang, and L. Xing. 2009. Noise suppression in scatter correction for cone-beam CT. *Med Phys* 36 (3):741–752.

Zhu, L., Y. Xie, J. Wang, and L. Xing. 2009. Scatter correction for cone-beam CT in radiation therapy. *Med Phys* 36 (6):2258–2268.

Zhu, X., J. D. Bourland, Y. Yuan, T. Zhuang, J. O'Daniel, D. Thongphiew, Q. J. Wu, S. K. Das, S. Yoo, and F. F. Yin. 2009. Tradeoffs of integrating real-time tracking into IGRT for prostate cancer treatment. *Phys Med Biol* 54 (17):N393–N401. doi:10.1088/0031-9155/54/17/N03.

CHAPTER 8

Progress in Magnetic Resonance-guided Radiotherapy (MR-gRT) Unit Developments

Eric Paulson

CONTENTS

8.1 INTRODUCTION

Magnetic resonance imaging (MRI) produces images of exquisite soft-tissue contrast, which have been shown to improve delineation of target extent relative to proximal organs at risk (OARs) during radiation treatment planning (RTP). For this reason, along with the additional capability to acquire biological/functional images (van der Heide et al., 2012), Magnetic Resonance (MR) simulation is increasingly being used as an adjunct to computed tomography (CT) simulation (Devic, 2012; Paulson et al., 2015). Figure 8.1 demonstrates the soft-tissue contrast advantages of MRI for delineation of target (arrow in bottom center image) and OAR (arrow in top right image) in a locally advanced pancreatic adenocarcinoma patient.

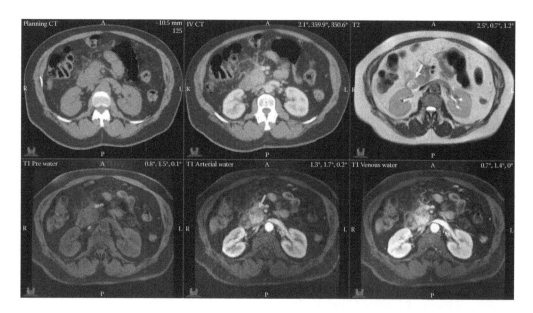

FIGURE 8.1 Offline MR-aided radiation treatment planning for a locally advanced pancreatic adenocarcinoma patient. Delineation of duodenal wall and target is challenged by the poor soft-tissue contrast of CT (top left and top middle). Delineation of these structure boundaries is facilitated using the T2-weighted and late arterial phase, fat-suppressed, postcontrast T1-weighted MR images (arrows in top right and bottom center images, respectively).

TABLE 8.1   Existing and Emerging Roles of MRI in Radiotherapy

Role of MRI	Features
Offline MR-aided radiation treatment planning	Utilizes a CT + MRI workflow (CT for dose calculation and MRI as contrast adjunct for delineation) during radiation treatment planning.
Offline MR-based radiation treatment planning	Utilizes an MRI-only workflow in which MRI is used for both delineation and dose calculation during radiation treatment planning. Compatible with existing linacs with CT-based IGRT systems.
Online MR-guided radiotherapy	Extends the soft tissue contrast, biological/functional imaging, and dynamic imaging advantages of MRI into the treatment room.
Online MR-guided brachytherapy	Real-time, high soft-tissue contrast imaging during brachytherapy needle and applicator insertion or positioning.
Response assessment	Utilizes morphological and biological/function changes from serial MRI scans to assess treatment response.
Risk-adaptive therapy assessment	Utilizes biological/functional imaging during an RT course to personalize treatment.

Over the past several years, the role of MRI in radiation oncology has evolved rapidly and is continuing to progress at a rapid pace. Table 8.1 provides a summary of the existing and emerging roles of MRI in radiotherapy and defines terminology in regard to how MRI is used in radiotherapy (RT). These roles include offline MR-aided RTP and imaging for response assessment, which are performed routinely in most clinics. Online MR-guided radiotherapy (MR-gRT) and online MR-guided brachytherapy (MR-gBT) have been used clinically at a few institutions at the time of this writing and are rapidly evolving. However, areas such as offline MR-based RTP and risk-adaptive therapy assessment are still being investigated at major research institutions. The term MR, rather than MRI, is used in these definitions to encompass both imaging and nonimaging (e.g., spectrosopy) MR technologies.

8.2 MAGNETIC RESONANCE SIMULATION

The prerequisite for modern RTP is high-fidelity, high-contrast resolution, three-dimensional (3D) images from which true disease extent and proximity relative to adjacent OAR can be accurately defined. MRI can fulfill these requirements. However, the use of MR for simulation, in either offline MR-aided RTP or offline MR-based RTP approaches, challenges conventional approaches of acquiring and utilizing images for delineation.

Compared to conventional imaging for diagnostic purposes, radiation therapy presents a new set of challenges and places additional constraints on MRI that, if not properly addressed, can undermine the advantages MRI offers for RTP. Major differences are introduced by the need to image patients in RT treatment position and control sources of geometric distortions. Use of alternative radio-frequency (RF) coil configurations to accommodate immobilized patients and higher readout bandwidths to mitigate patient-induced sources of distortion can reduce image quality, requiring additional optimization of sequence parameters. In addition, post-processing of MR images, including 3D corrections for gradient nonlinearity-induced distortions, is essential and must be performed prior to using MR images for delineation. Table 8.2 provides a summary of the general differences in MRI scanning protocols for diagnostic versus RTP purposes.

TABLE 8.2   Differences between MRI Scanning Protocols Used for Diagnostic versus Treatment Planning Purposes

	Diagnostic Radiology	Radiation Treatment Planning
Purpose	Detection, characterization, and staging of disease	Determination of true 3D disease extent and position relative to adjacent organs at risk
Field of view (FOV)	Can acquire with reduced FOV	Full cross section required on at least one scan for body contour in MR-based radiation therapy
Readout bandwidth (RBW)	Typically set as trade-off between fat/water shift and SNR	Intentionally set high to minimize chemical shift and susceptibility-induced spatial distortions while minimizing eddy currents
Slice thickness and interslice spacing	Typically 4–5 mm; may have interslice gaps 0–2 mm	Contiguous slices; thinner slices improve DRR image quality in MR-based treatment planning
Slice coverage	Prescribed overvolume of interest	Increased coverage required for target and OAR delineation (dose volume histograms), landmarks for registration, IGRT, and so on
Geometric distortion	Tolerated as long as diagnostic capability not affected	Required to be < 2 mm in all planes over the volume of interest
Image intensity non-uniformity	Tolerated as long as diagnostic capability not affected	Increased uniformity required for intensity-based image registration and image segmentation accuracy
Breath-holds	End of inspiration to maximize ability of patient to hold breath and minimize motion artifacts	Tailored to match gating windows used in gated radiotherapy delivery (e.g., end expiration for 50% phase)

In terms of peripheral equipment for simulation, immobilization devices optimized for radiotherapy may not necessarily be optimal for MRI (carbon fiber is an obvious example), and commercial MR-compatible immobilization devices are becoming more available. MR-compatible external laser systems are also commercially available. These systems are required for setup reference point definition in online MR-based RTP and may be useful for aligning and straightening patients in offline MR-aided RTP but may not be essential (Paulson et al., 2016). Some MRI vendors permit site-specific configuration of table offsets, facilitating table movement from the external laser isocenter to scanner isocenter. For other systems, these offsets can be configured within the imaging protocols.

Compared to the more conventional single-study-set delineation paradigm in RTP, MRI results in a shift to a multi-study-set (i.e., multicontrast), multiplanar, multidimensional, multi-time-point regime. This transition places additional demands on delineation software packages to handle large volumes of multiparametric MR images efficiently and safely. These demands will continue to increase with increasing utilization of MR-gRT.

8.3   FROM MAGNETIC RESONANCE SIMULATION TO MAGNETIC RESONANCE-GUIDED RADIOTHERAPY

The introduction of implanted fiducial markers along with in-room CT imaging was pivotal in the establishment of image-guided radiation therapy (IGRT). However, visualization can still be challenging for several tumor sites, including the esophagus, rectum, kidney, liver, and pancreas.

Recent improvements in MRI technology, including acceleration and triggering/gating capabilities, permit acquisition of high-quality body MR images in the presence of respiratory, cardiac, and peristaltic motions. Consequently, there has been increasing interest in extending the advantages of MRI into the RT treatment room, facilitating online MR-gRT for (1) daily position verification in regions that are challenging for existing CT-based IGRT systems (discussed above), (2) real-time imaging for exception gating (Crijns et al., 2012a), (3) dynamic target tracking (Crijns et al., 2012b), (4) "online" treatment adaptation in response to changes in tumor and normal tissue configuration, and (5) integrated assessment of therapy response. These capabilities present opportunities for local dose escalation, reduction of normal tissue toxicities, and risk-adaptive therapies.

8.4 MAGNETIC RESONANCE-GUIDED RADIOTHERAPY TECHNOLOGY OVERVIEW

The emerging online MR-gRT treatment cycle (Bol, 2015) is illustrated in Figure 8.2. The process involves acquisition of high-contrast MR images for target and OAR delineation followed by online adaptive treatment replanning and delivery, all in a continuous cycle. The rate at which this treatment cycle can occur depends on several factors, including MR-gRT implementation strategy (discussed below), required extent of human interaction, and high-speed online IMRT reoptimization (Kupelian and Sonke, 2012; Bol, 2015). Nonintegrated or "next-door" MR-gRT devices (Karlsson et al., 2009; Jaffray et al., 2014) support the treatment cycle as an interfraction process; the anatomy of the day is imaged and used to generate a plan of the day, which is subsequently delivered. Integrated or hybrid MR-gRT devices (Lagendijk et al., 2014; Mutic and Dempsey, 2014; Fallone, 2014; Keall et al., 2014) facilitate continuous, intrafraction imaging of translations, rotations, and deformations of targets and OARs. This information can then be utilized by online treatment planning systems (Kontaxis et al., 2015) to adapt a treatment plan dynamically, followed by direct delivery to the patient (see Figure 8.3). Online MR-gRT offers the potential for a dramatic paradigm shift from the current, sequential approach to radiotherapy (i.e., treatment preparation, treatment planning, treatment delivery), to the continuous feedback regime shown in Figure 8.2.

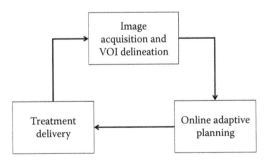

FIGURE 8.2 Emerging online MR-gRT treatment cycle. Integrated MR-gRT devices support continuous cycle updating.

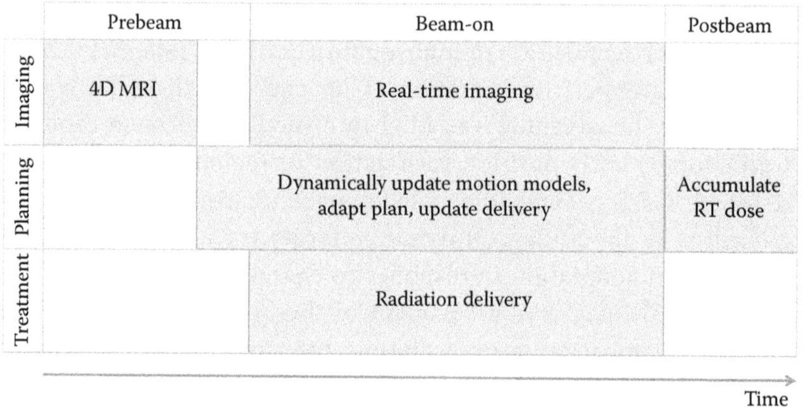

	Prebeam	Beam-on	Postbeam
Imaging	4D MRI	Real-time imaging	
Planning		Dynamically update motion models, adapt plan, update delivery	Accumulate RT dose
Treatment		Radiation delivery	

Time

FIGURE 8.3 Potential online MR-gRT treatment strategy. MR-gRT treatments are divided into prebeam, beam-on, and postbeam phases. During each phase, imaging, planning, and treatment subsystems can be executed independently or in parallel. In this example, the treatment plan is updated with motion models obtain from a daily prebeam 4D MRI scan, followed by real-time imaging and plan adaptation during 3D CRT, IMRT, or VMAT treatment delivery. After treatment, the delivered dose is compiled and accumulated on dose delivered in prior treatment fractions.

8.4.1 Design Criteria

Beyond the well-established challenges of integrating MRI in radiation oncology (e.g., minimization of patient- and system-induced geometric distortions (Reinsberg et al., 2005; Baldwin et al., 2009) and optimization of RT-specific scanning protocols (Paulson et al., 2015) discussed above, the design criteria for MR-gRT systems involves four major technical challenges: (1) decoupling the magnetic field from RT components (i.e., minimizing magnetic interference), (2) decoupling the RT components from MRI acquisition (i.e., minimizing broadband RF interference), and (3) beam transmission through the MRI (Lagendijk et al., 2014), (4) the Lorentz force on secondary electrons (Lagendijk et al, 2014). A summary of the basic design considerations for integrated MR-gRT devices is provided in Table 8.3. The magnetic field can affect an unshielded electron gun and waveguide, resulting in output degradation through current loss and detuning (Aubin et al., 2010; Constantin et al., 2011). The linac RF pulse modulator and multileaf collimator (MLC) motor brushes generate broadband RF noise detectable with the MRI-RF receive coils, resulting in spikes in k-space and degradation of the signal-to-noise ratio (SNR) (Lamey et al., 2012). Beam attenuation and Compton scatter to the patient can arise with transmission through the cryostat (Lagendijk et al., 2014; Keall et al., 2014).

8.4.2 Implementations

At the time of this writing, six MR-gRT implementations have been pioneered (Jaffray et al., 2014; Lagendijk et al., 2014; Mutic and Dempsey, 2014; Fallone, 2014; Keall et al., 2014). Each implementation takes a unique approach to addressing the technical challenges facing MR-gRT. Of the six implementations, five utilize an integrated MR-gRT architecture. The following sections discuss these implementations in detail. A comparison of the implementations is provided in Table 8.4.

TABLE 8.3   Basic Design Considerations for Integrated MR-gRT Devices

MRI	RT

Magnetic field strength:
 • *Lower*:
    • Reduced magnetic susceptibility distortions
    • Lower specific absorption rate
 • *Higher*:
    • Improved contrast-to-noise ratio
    • Permits higher acceleration factors and/or higher spatial resolution
 	
 Magnet geometry:
 • *Solenoidal*:
    • Potential for higher imaging performance
 • *Open (bi-planar)*:
    • May not require irradiation through cryostat
	Modality:
	 • *Radionuclide*:
	    • No electron gun or waveguide interactions with magnetic field (Constantin et al., 2011)
	    • No pulsatile irradiation induced currents in RF receive coils (Burke et al., 2012)
	 • *Linear accelerator*:
	    • Increased output and penetration depth
	    • Smaller penumbra
	    • Dose rate does not decrease with time
	    • No NRC compliance requirements
	    • Orientation of RT with Respect to Magnetic Field:
	 • *Inline (parallel)*:
	    • Less effect of magnetic field on electron gun and waveguide operations
	    • No electron return effect
	    • Lower exit dose
	 • *Perpendicular (transverse)*:
	    • No need to rotate magnet or patient
	    • Magnetic field sweeps away contaminating electrons, lowering skin dose

8.4.2.1 Implementation A

Implementation A addresses the MR-gRT design challenges using a next-door approach (Karlsson et al., 2009), which combines a conventional linac with a conventional MRI scanner suspended from the ceiling on rails (Jaffray et al., 2014). Prior to daily imaging, with the linac in standby, large shielding doors are opened and the MR scanner is moved into the treatment room. The linac couch is rotated and the MR is advanced over the patient for imaging. The MR-gRT treatment cycle in Figure 8.2 executes in single-loop mode prior to RT delivery, to manage interfraction variability. Implementation A permits sharing of the MRI scanner with adjacent imaging and/or therapy suites (e.g., MR simulation, additional linacs for MR-gRT, or MR-guided brachytherapy), acquisition of diagnostic-quality MR images for positioning verification, acquisition of images for response assessment using high-performance gradients, and use of the portal-imaging capabilities of the linac. A key limitation of implementation A is the lack of real-time MR imaging of moving anatomy during RT. However, real-time, linac-based kV imaging or third-party, motion-monitoring systems (e.g., VisionRT, Calypso, etc.) can be used.

8.4.2.2 Implementation B

Implementation B addresses the MR-gRT design challenges using an integrated combination of a low-field (0.35 T) magnet and three planetary radionuclide (Co-60) sources with independent MLCs arranged on a ring 120° apart (Mutic and Dempsey, 2014). The MLCs utilize a double-focused design, which aids in reducing the penumbra of the Co-60 beams. The radionuclide sources eliminate RF interference with MR imaging during RT delivery.

TABLE 8.4 Comparison of Current MR-gRT Implementations

Implementation	A	B	C	D	E	F
Configuration	Nonintegrated	Integrated	Integrated	Integrated	Integrated	Integrated
Magnetic field strength	1.5 T	0.35 T	0.35 T	1.5 T	0.5 T	1.0 T
Magnet geometry	Solenoid	Split solenoid	Split solenoid	Split solenoid	Bi-planar	Bi-planar
Magnetic field orientation	Horizontal	Horizontal	Horizontal	Horizontal	Dynamic	Dynamic
RT photon source	Conventional linac	(3) planetary Co-60 sources	Ring-mounted 6 MV Linac	Ring-mounted 7 MV Linac	6 MV Linac	6 MV Linac
RT field—magnetic field orientation	Independent	Orthogonal	Orthogonal	Orthogonal	Inline or orthogonal	Inline or orthogonal
Contained in one treatment room	No	Yes	Yes	Yes	Yes	Yes
Able to be installed in existing RT treatment rooms with modification	No (shielding doors req'd)	No (He vent req'd)	No (He vent req'd)	No (req's gantry clearance, He vent)	Yes	No (He vent req'd)
Supports MRI-based daily positioning verification	Yes	Yes	Yes	Yes	Yes	Yes
Supports offline adaptive re-planning	Yes	Yes	Yes	Yes	Yes	Yes
Supports online adaptive re-planning	Yes	Yes	Yes	Yes	Yes	Yes
Supports MRI-based real-time motion monitoring during treatment delivery	No	Yes	Yes	Yes	Yes	Yes
Supports functional/biological imaging	Yes	No	No	Yes	Yes	Yes
Supports response assessment imaging	Yes	No (limited sequence options)	No (limited sequence options)	Yes	No	No
Supports IMRT delivery	Yes	Yes	Yes	Yes	Yes	Yes

(Continued)

TABLE 8.4 (*Continued*) Comparison of Current MR-gRT Implementations

Implementation	A	B	C	D	E	F
Supports VMAT delivery	Yes	Yes	Yes	Yes	Yes	Yes
Supports use of noncoplanar beams	Yes	No	No	No	Dependent on magnet position	Dependent on magnet position
Susceptible to beam attenuation through RF coils	No	Yes	Yes	Yes	Yes	Yes
Susceptible to SNR reductions from radiation induced currents in RF receive coils	No	No	Yes	Yes	Yes	Yes
Susceptible to electron return effect	No	Yes	Yes	Yes	Depends on linac orientation	Depends on linac orientation
MRI-compatible patient and machine QA equipment, calibration factors, procedures required	No	Yes	Yes	Yes	Yes	Yes
NRC compliance required	No	Yes (special ruling)	No	No	No	No
U.S. Food and Drug Administration (FDA) approved at time of this report	Yes	Yes	510 k	No	No	No

In addition, with no linac, implementation B avoids interactions of the magnetic field with the electron gun and waveguide as well as SNR reductions induced by pulsatile irradiations. The magnet, gradient coil, and RF coils utilize a split core design, which provides a window for the radiation beam to pass through and moves any radiosensitive equipment outside the primary beam. Implementation B permits real-time MR imaging during delivery for intrafraction motion monitoring, providing the capability to execute the online MR-gRT treatment cycle in a continuous loop. Key limitations of implementation B include limited pulse sequence options, the properties of Co-60, and lower field strength, which can result in lower SNR images for the same imaging time and spatial resolution compared to higher field strength systems.

8.4.2.3 Implementation C

Implementation C largely mirrors implementation B, with the exception of the photon source; a ring-mounted 6 MV linac replaces the tri-Co-60 sources in implementation B. Specific details are scarce at the time of this writing, but the linac is housed within layers of carbon fiber and copper shielding to reduce RF interference, and additional magnetic shielding is used to reduce magnetic interference with RT hardware components. A key advantage of this implementation over implementation B is the elimination of regulatory requirements, which is a result of the removal of the Co-60 sources. However, similar to implementation B, lower SNR and limited pulse sequence options are key limitations of implementation C.

8.4.2.4 Implementation D

Implementation D addresses the MR-gRT design challenges using: (1) active shielding to minimize the fringe field over the space in which the RT hardware operates; (2) an RF shield integrated into the cryostat to ameliorate broadband RF interference; and (3) specially designed "windows" in the cryostat, gradient coils, and RF coils to minimize beam attenuation, scatter, and radiation-induced damage (Lagendijk et al., 2014). In addition, implementation D contains an integrated electronic portal imaging device that does not interfere with MR imaging (Raaymakers et al., 2011). Advantages of implementation D include real-time MR imaging of moving targets, which facilitates continuous and real-time looping of the MR-gRT treatment cycle (similar to implementations B and C). In comparison to implementations B and C, however, implementation D permits acquisition of diagnostic-quality morphological or functional images for response assessment or risk-adaptive therapy. These images can be acquired simultaneously during RT delivery for those cancer sites in which intrafraction motion monitoring may not be required (see Figure 8.4). The higher magnetic field strength used in implementation D can increase the severity of the electron return effect (ERE) (see Section 8.5), increases the magnitude of susceptibility induced distortions, and can result in higher specific absorption rates (SARs) during real-time imaging. However, each of these issues is manageable.

8.4.2.5 Implementation E

Implementation E addresses the MR-gRT design challenges using a whole-body 0.5 T biplanar (open) magnet with a linac that can be oriented either parallel or perpendicular

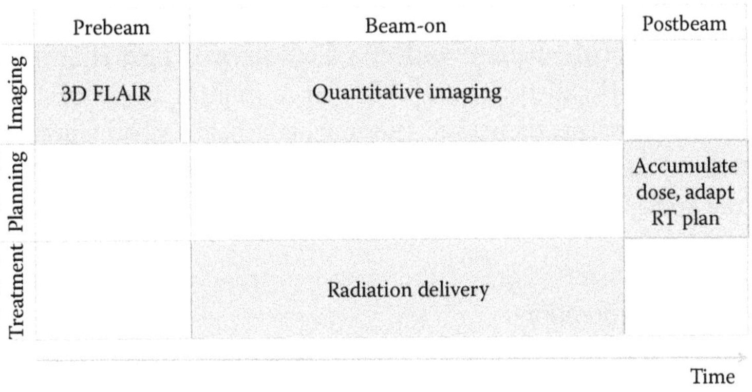

FIGURE 8.4 Potential MR-gRT treatment strategy for more stationary sites (e.g., brain). Contrary to the example shown in Figure 8.3, real-time MR imaging may not need to be performed in these cases, which permits concurrent acquisition of quantitative imaging during 3D CRT, IMRT, or VMAT treatment. The quantitative imaging can be combined with accumulated dose for offline response assessment or risk-adaptive therapy.

to the magnetic field (Fallone, 2014). In the perpendicular configuration, the linac is placed between the magnet planes. In the parallel, or "inline," configuration, the linac is directed through a central opening on one of the magnet planes. Both linac configurations eliminate transmission of the photon beam through a cryostat. To achieve different beam angles, the entire biplanar magnet/linac assembly rotates along the superior–inferior (S–I) axis of the patient. Shielding of the MLC motor assembly and cables and the addition of RF filters diminish RF interference (Lamey et al., 2012). The advantages of the inline orientation of the linac are a reduction of magnetic field effects on the electron gun and waveguide and avoidance of high exit dose and dosimetric hot spots resulting from the ERE (see Section 8.5). In addition, similar to implementations B and C, the low magnetic field strength reduces SAR, susceptibility-induced geometric distortions, and the magnitude of hot spots at lung/tissue interfaces in the perpendicular linac configuration. Because the magnet and RT assembly rotates about the patient together, variations in eddy currents arising from changes in RT hardware positions are avoided. Limitations of implementation E include magnetic field homogeneity variations arising from rotation around the patient, low contrast-to-noise ratio (CNR) due to the low magnetic field strength, and increased skin dose with the inline configuration.

8.4.2.6 Implementation F

Implementation F utilizes a similar approach as implementation E in addressing the MR-gRT design challenges, but with a few unique differences (Keall et al., 2014). First, the biplanar magnets have a field strength of 1 T. Second, each magnet plane has an 82 cm bore, which permits the patient to be oriented between the magnet planes (as in implementation E) or through the magnet planes (i.e., the magnetic field may run perpendicular or parallel to the S-I axis of the patient). Similar to implementation E, the biplanar magnet assembly may be rotated around the patient to achieve desired beam angles for treatment. Alternatively, rotation of the patient within a fixed magnet field is also being explored in this implementation (Keall et al., 2014). Similar to implementation E, the linac can be configured parallel or perpendicular to the magnetic field. Advantages and limitations of implementation F largely mirror those of implementation E. However, one additional advantage is the ability to investigate the biplanar magnetic field oriented parallel to the S–I axis of the patient.

8.4.3 Additional Considerations

Several, but not all, MR-gRT implementations support installation within existing treatment vaults. Large system components and the requirement of a helium exhaust vent may require room renovations in existing vaults or design of new vaults. Implementation E utilizes a cryocooler rather than cryogenic liquids to eliminate the requirement of an exhaust vent. Furthermore, it may be necessary to raise the existing vault floor in order to maintain patient access to the MR-gRT device while providing clearance for gantry rotation.

In theory, implementations B to F could fulfill additional use cases as MR simulators as well as MR-gRT devices. Simulation studies have demonstrated negligible effects of a static MLC on magnetic field uniformity as well as moving MLCs on magnetic field homogeneity (Kolling 2012). Although it may be possible to use lower field strength systems for MR simulation, field strengths below 1.5 T may not be optimal (Paulson et al., 2016). Acquiring high CNR images for delineation requires longer imaging times compared to higher field strength systems. In addition, advanced imaging techniques (e.g., perfusion and diffusion) useful for delineation during treatment planning may not be available across all MR-gRT platforms.

8.5 MAGNETIC RESONANCE-GUIDED RADIOTHERAPY DOSIMETRY

Although nonintegrated MR-gRT systems (i.e., implementation A) do not possess the capability to acquire dynamic MR images during treatment delivery, they avoid interactions of the magnetic field with secondary electrons generated in the patient. The following sections discuss specific dosimetry issues relevant to MR-gRT.

8.5.1 Titled Dose Kernel

The photon beam is not affected by the magnetic field; however secondary electrons experience the Lorentz force. Orthogonal magnetic and irradiation fields (e.g., implementations B, C, D, and potentially E and F) can result in (1) altered buildup distances, (2) asymmetric beam profiles, and (3) smeared penumbra through tilting of the dose kernel (Raaijmakers et al., 2008; Figure 8.5).

8.5.2 Electron Return Effect

At tissue–air interfaces, large impacts on the dose distribution can arise due to reentrance of secondary electrons with magnetic field-induced helical trajectories, a phenomenon known as the ERE. The ERE can result in local dose increases at air–tissue interfaces (e.g., lung–chest wall interface, rectum–prostate interface, etc.) as well as increased exit dose (Figure 8.6).

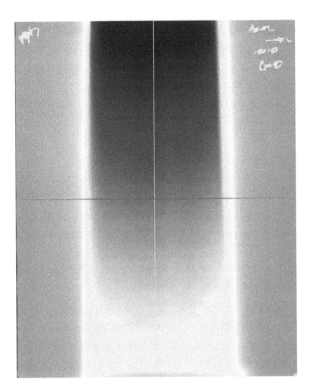

FIGURE 8.5 Cross-plane beam profile obtained with radiochromic film on a high-field MR-gRT device with orthogonal magnetic and irradiation fields. The asymmetric beam profile, smeared penumbra, and electron return effect at the exit are discernible.

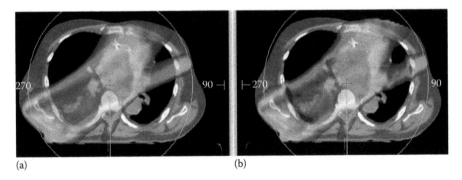

(a)   (b)

FIGURE 8.6 Demonstration of the electron return effect. Original treatment plan calculated without magnetic field (a). Recalculated treatment plan calculated with orthogonal 1.5 T magnetic field (b). The electron return effect results in local dose increases at air-tissue interfaces.

The ERE depends on magnetic field strength and direction, orientation of the irradiation field with respect to the magnetic field, irradiation field size, surface orientation, and structure motion (Raaijmakers et al., 2005, 2008). Dose inhomogeneities around cylindrical air cavities are minimal if the radius of the cavity is small relative to the in-air radius of the secondary electron trajectories. For larger cavities with radius greater than 1 cm, dose inhomogeneities exist across magnetic field strengths ranging from 0.2 to 3 T (Raaijmakers et al. 2008). The ERE can be compensated to varying extents by orienting the irradiation field parallel to the magnetic field (the inline configuration) (Kirkby et al., 2010), using opposed treatment fields and IMRT optimization for static air cavities (Bol et al., 2015), and using exception gating and plan reoptimization for nonstationary cavities (Bol, 2015; Figure 8.6).

8.5.3 Surface Dose

Contaminating electrons from the cryostat and RT delivery hardware can contribute to increased surface dose in MR-gRT. Increased surface dose may also arise from electrons generated in RF receive coils placed on the patient for imaging. In implementations B, C, and D, and depending on the linac orientation in implementations E and F, the transverse magnetic field can sweep away contaminating electrons, thus lowering surface dose (Mutic and Dempsey, 2014).

8.5.4 Beam Attenuation from Radio-Frequency Coils

In addition to increased surface dose, a 2.2% underestimation of the dose due to beam attenuation through a flexible RF coil (Hoogcarspel et al., 2013) has been reported. For these reasons, it is necessary to (1) move all RF coil digitization hardware (e.g., preamps) out of the direct radiation beam path, (2) model the RF coils along with the patient in MRI-gRT treatment planning systems, or (3) perform both.

8.5.5 Dose Calculation Algorithms

Optimization and dose calculation for online MR-gRT treatment planning incorporating magnetic field strength and direction have been performed using Monte Carlo (Bol et al., 2012) as well as deterministic solutions to the first-order Boltzmann transport equations (St Aubin et al., 2015). Current calculation times range from 15 s to several minutes. For the online MR-gRT treatment cycle (see Figure 8.2) to execute in a real-time continuous loop, significant improvements in computational speed and/or algorithm development are needed.

8.6 STRATEGIES FOR ADAPTIVE MAGNETIC RESONANCE-GUIDED RADIOTHERAPY

It is well known that there may be translation, rotation, and/or deformation of targets and OAR daily or, potentially, even in real time. In addition, target volumes may change due to treatment response. All of these conditions warrant adaptive radiotherapy. Common concerns regarding adaptive radiotherapy, including recontouring time, reoptimization time, how to properly quality assurance (QA) adapted treatment plans, also exist for MR-gRT. However, additional issues arise in online adaptive MR-gRT due to lack of electron density information and magnetic field effects. This section touches briefly on strategies to facilitate handling of inter- and intrafraction motions.

8.6.1 Virtual Couch Shift

The restriction of couch movement along only the superior-inferior direction in many MR-gRT implementations precludes lateral and vertical shifting of the patient for daily positioning. One fast method of addressing translations and rotations is the virtual couch shift (VCS) (Bol et al., 2013), in which the field apertures of the pretreatment dose distribution are translated rather than physically moving the patient. The VCS result depends on MLC orientation and is more appropriate for small shifts. Large shifts in the presence of flattening filter free (FFF) beams may result in large dosimetric changes and thus upper limits on the shifts that can be handled using VCS. In these cases, or in cases of major deformations, alternative strategies may be required.

8.6.2 Preshifted Plan Library and Segment Aperture Morphing

Reoptimization may be required for large translations in the presence of FFF beams. However, the time required to reoptimize a treatment plan fully in the presence of a magnetic field with the patient on the table may not be clinically acceptable. Warm-start optimization (WSO) using the dose from the original treatment plan may be one approach to speed up the process. In this process, a preshifted plan (PSP) library is produced by precalculating dose distributions at different isocenter positions on the anatomy obtained during simulation. These PSP library dose distributions then guide WSO. This process could be combined with online segment aperture morphing (SAM) to account fully for translations, FFF beams, and deformations in an online adaptive MR-gRT strategy (Ahunbay et al., 2016).

8.6.3 Continuous Adaptation

While the previous process can control for interfraction variability, integrated MR-gRT devices possess the inherent capability to control for intrafraction variabilities by continuously adapting a treatment plan with each new anatomy update from the MRI. One possible strategy to achieve continuous plan adaptation utilizes an adaptive sequencer, in which segment-by-segment optimization is performed based on an ideal fluence with direct and dynamic updating of MLC positions (Kontaxis, et al., 2015).

8.6.4 Additional Considerations

Limitations in the maximum velocity of MLCs may challenge the accuracy of MR-gRT systems to track targets in the presence of abrupt, irregular, intrafraction motions. In addition, depending on the magnitude of motions, the MLCs may move to high-gradient regions of FFF beams during tracking. In these cases, accurate dose accumulation following treatment delivery is essential to compensate for intrafractional variations (Kontaxis et al., 2015).

8.7 QUALITY ASSURANCE

Integrated MR-gRT systems may require alternative magnet, gradient coil, and RF coil designs in order to meet the design challenges listed in Table 8.3. In addition, the presence of the magnetic field may affect the response of equipment used for reference, relative, and in vivo dosimetry. Considerations for these issues are discussed in the following subsections.

8.7.1 Imaging Subsystem

The nature of integrated MR-gRT systems requires additional and more rigorous quality-assurance (QA) activities compared to MR simulator QA. Table 8.5 lists several imaging QA considerations for acceptance testing, commissioning, and monitoring of integrated MR-gRT systems. Additional QA activities (e.g., image quality checks, characterizations of gradient nonlinearity induced geometric distortions, etc.) common to MR simulators must also be performed. The majority of these activities arise from the presence of RT hardware and the delivering beam. For example, rotation of the RT components within the gantry may result in magnetic field changes (see Figure 8.7) and eddy current variations that can have an impact on image quality. In addition, pulsed irradiations can induce currents in RF coils that can degrade image SNR (Burke et al., 2012). In general, the activities in Table 8.5 can be carried out by comparison of images acquired with and without gantry rotation and with and without beam delivery (see Figure 8.8). It may also be possible to make some comparisons against conventional MR simulation systems. At the time of this writing, it is unclear how frequently monitoring activities should be performed to evaluate radiation-induced image quality degradation or equipment failures.

8.7.2 Radiotherapy Subsystem

Due to the Lorentz force many of the traditional tests for linac QA need to be redesigned for integrated MR-gRT devices. This can be accomplished by incorporating correction factors into existing QA procedures or developing new procedures specifically for integrated MR-gRT systems. Electron-dense materials (e.g., copper) have a shorter electron path length and therefore demonstrate a reduced magnetic field effect. Van Zijp showed that isocenter size and beam profiles can be determined accurately in the presence of a magnetic field using copper and film (van Zijp et al., 2016).

TABLE 8.5 Imaging Quality Assurance Considerations for Integrated MRI-gRT Devices

Acceptance/Commissioning	Monitoring
• Effect of gantry angle on magnetic field homogeneity (B0)	• Effect of radiation-induced damage to RF coils on SNR
• Effect of gantry rotation on imaging (variable eddy currents)	• Effect of radiation-induced damage to gradient coils on performance
• B0 drifts due to gradient and cryostat heating of shim coils	• System stability
• Effect of pulsed radiation induced currents in RF coils on SNR	
• MV-MR isocenter alignment	
• Leakage of integrated Faraday cage or RF cloaking mechanism	
• Accuracy and latency of motion tracking capabilities	
• Sequence-specific performance issues	
• End-to-end test of complete MR-gRT chain	

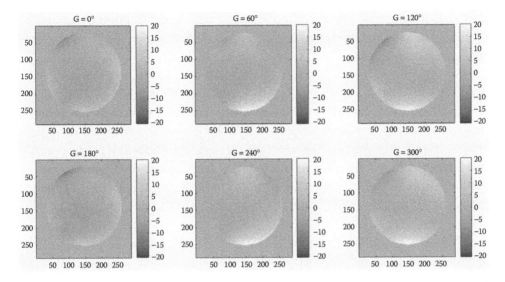

FIGURE 8.7 Magnetic field variation with gantry rotation on a high-field, integrated MR-gRT device with orthogonal magnetic and irradiation fields. Magnetic field maps were measured over a 30-cm spherical flood field phantom at six gantry angles. The color bars denote field inhomogeneity in the range of ±20 Hz.

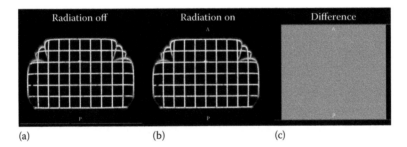

FIGURE 8.8 Effect of concurrent irradiation on MR image quality. A large geometric phantom was imaged on a high-field, integrated MR-gRT device with orthogonal magnetic and irradiation fields before (a) and during (b) irradiation. No discernible effects are apparent in the difference image (c), indicating control over magnetic and radiotherapy component interference has been achieved.

8.7.3 Reference and Relative Dosimetry

In terms of absolute dosimetry, the response of ion chambers depends on the chamber geometry, orientation within the magnetic field, and the strength of the magnetic field (Meisjsing et al., 2009). This results in the need for additional correction factors to be used in protocols employed to calibrate radiation beams for MR-gRT. However, linearity, reproducibility, and moving the chambers during irradiation are not affected by the magnetic field (Smit et al., 2014). In addition, the presence of the cylindrical bores in some MR-gRT implementations challenges water phantom sizes for profile measurements used in relative dosimetry. Consequently, horizontally measured percentage depth-dose curves using ultrasonic stepper motors through a thin water phantom window have been proposed for MR-gRT (Smit et al., 2014).

8.7.4 In Vivo Dosimetry

Solid-state devices are often used for in vivo dosimetry, and several groups have begun investigating the response of these devices in MR-gRT. Goddu reported no significant differences in the response of TLDs and OSLs at low field strength (Goddu et al., 2012). Knutson reported a 5% increase in metal oxide semiconductor field effect transistor (MOSFET) response during cine imaging at 0.35 T and recommended using handheld readers to avoid imaging interference (Knutson et al., 2014).

8.7.5 Patient-Specific Quality Assurance

In addition to imaging and dosimetry QA considerations, there are also considerations for the patient under treatment. The RF coils used for imaging are modeled in the treatment planning system and must be positioned in the same manner each day of treatment. Commercial MRI-compatible devices are now available that permit accurate patient-specific QA of treatment plans in the presence of magnetic fields (Li et al., 2015; Houweling et al., 2016). In addition, the feasibility of using gel dosimetry for 3D dose verification has also been reported (Zhang et al., 2014).

8.8 REMAINING CHALLENGES

Several remaining challenges must be overcome before the full clinical benefits of MR-gRT can be realized. One of these challenges includes accurate accumulation of delivered dose in the presence of moving OARs and targets, for use in adaptive replanning of subsequent treatment fractions. Overcoming this challenge requires advancements in the acceleration of volumetric MRI acquisitions. Another challenge involves limitations in computational speed. Fast optimization and dose calculation in the presence of the magnetic field are required to operate the MR-gRT treatment cycle in a continuous loop. The optimization would also benefit from advancements in biological optimization based on multiparametric MR images acquired throughout a treatment course. Finally, MR-gRT results in an increase in the overall dimensionality of imaging used for RT. New infrastructures are required to handle and utilize efficiently large volumes of multiparametric MR images acquired at multiple times during RT.

8.9 CONCLUSION

The success of radiotherapy largely depends on the precise alignment of the radiation beam with the target. MR-gRT extends the soft-tissue contrast and nonionizing advantages of MRI into the treatment room, facilitating management of inter- and intrafraction variabilities. Online MR-gRT offers the potential for a dramatic paradigm shift from the current, sequential approach to radiotherapy to a continuous feedback regime in which motion and biology changes can be incorporated throughout a treatment course.

These capabilities present new opportunities for local dose escalation, reduction of normal tissue toxicities, and risk-adaptive therapies. Different MR-gRT implementations have been proposed to address the design challenges facing integration of an MRI scanner with a radiotherapy treatment device. With future improvements in computational speed, fast acquisition of volumetric MR imaging, and a more complete understanding of how to adapt treatment plans to changes in tumor biology, MR-gRT is destined to be a direct hit.

REFERENCES

Ahunbay E, Ates O, Li XA. 2016. An online replanning method using warm start optimization and aperture morphing for flattening-filter-free beams. *Med Phys*; 43:4575–4584.

Aubin J, Steciw S, Fallone BG. 2010. Waveguide detuning caused by transverse magnetic fields on a simulated in-line 6 MV linac. *Med Phys*; 37:4751–4754.

Baldwin LN, Wachowicz K, Fallone BG. 2009. A two-step scheme for distortion rectification of magnetic resonance images. *Med Phys*; 36:3917–3926.

Bol GH, Hissoiny S, Lagendijk JJW, Raaymakers BW. 2012. Fast online monte carlo-based IMRT planning for the MRI linear accelerator. *Phys Med Biol* 2012; 57:1375–1385.

Bol GH, Lagendijk JJ, Raaymakers BW. 2013. Virtual couch shift (VCS): Accounting for patient translation and rotation by online IMRT re-optimization. *Phys Med Biol*; 58:2989–3000.

Bol GH, Lagenkijk JJW, Raaymakers BW. 2015. Compensating for the impact of non-stationary spherical air cavities on IMRT dose delivery in transverse magnetic fields. *Phys Med Biol*; 60:755–768.

Bol GS. 2015. Doctoral thesis. UMC Utrecht, the Netherlands.

Burke B, Wachowicz K, Fallone BG, Rathee S. 2012. Effect of radiation induced current on the quality of MR images in an integrated linac-MR system. *Med Phys*; 39:6139–6147.

Constantin D, Fahrig R, Keall PJ. 2011. A study of the effect of in-line and perpendicular magnetic fields on beam characteristics of electron guns in medical linear accelerators. *Med Phys*; 38:4174–4185.

Crijins SPM, Kok JGM, Lagendijk JJW, Raaymakers BW. 2012a. Towards MRI-guided linear accelerator control: Gating on an MRI accelerator. *Phys Med Biol*; 56:4815–4825.

Crijins SPM, Raaymakers BW, Lagendijk JJW. 2012b. Proof of concept of MRI-guided tracked radiation delivery: tracking one-dimensional motion. *Phys Med Biol*; 57:7863–7872.

Devic S. 2012. MRI Simulation for radiation treatment planning. *Med Phys*; 39:6701–6711.

Fallone BG. 2014. The rotating biplanar linacmagnetic resonance imaging system. *Semin Radiat Oncol*; 24:200–202.

Goddu S, Green OP, Mutic S. 2012. TG-51 calibration of first commercial MRI-guided IMRT system in the presence of 0.35 T magnetic field. *Med Phys*; 39:3968.

Hoogcarspel SJ, Crijins SPM, Lagendijk JJW, van Vulpen M, Raaymakers BW. 2013. The feasibility of using a conventional flexible RF coil for an online MR-guided radiotherapy treatment. *Phys Med Biol*; 58:1925–1932.

Houweling AC, de Vries JH, Wolthaus J, Woodings S, Kok JG, van Asselen B, Smit K, Bel A, Lagendijk JJ, Raaymakers BW. 2016. Performance of a cylindrical diode array for use in a 1.5 T MR-Linac. *Phys Med Biol*; 61:N80–N89.

Jaffray DA, Carlone MC, Milosevic MF, Breen SL, Stanescu T, Rink A, Alasti H, Simeonov A, Sweitzer MC, Winter JD. 2014. A facility for magnetic resonance-guided radiation therapy. *Semin Radiat Oncol*; 24:193–105.

Karlsson M, Karlsson MG, Nyholm T, Amies C, Zackrisson B. 2009. Dedicated magnetic resonance imaging in the radiotherapy clinic. *Int J Radiat Oncol Biol Phys*; 74:644–651.

Keall PJ, Barton M, Crozier S. 2014. The Austrailian magnetic resonance imaging-linac program. *Semin Radiat Oncol*; 24:203–206.

Kirkby C, Murray B, Rathee S, Fallone BG. 2010. Lung dosimetry in a linac-MRI radiotherapy unit with a longitudinal magnetic field. *Med Phys*; 37:4722–4732.

Knutson N, Li H, Rodriguez V, Hu Y, Kashani R, Wooten H, Tanderup K, Mutic S, Green O. 2014 A MOSFET-Based In-Vivo Dosimetry System for MR Image-Guided Radiation Therapy (MR-IGRT).*Med Phys*; 41(6):340.

Kolling S. 2012. Quantifying the impact of the multileaf collimator on the imaging capabilities of a prototype MRI-linear accelerator. Masters degree thesis. University of Heidelberg, Heidelberg, Germany.

Kontaxis C, Bol GH, Lagendijk JJ, Raaymakers BW. 2015. Towards adaptive IMRT sequencing for the MR-linac. *Phys Med Biol*; 60:2493–2509.

Kupelian P, Sonke JJ. 2012. Magnetic resonance-guided adaptive radiotherapy: A solution to the future. *Semin Radiat Oncol*; 24:227–232.

Lagendijk JJW, Raaymakers BW, van Vulpen M. 2014. The magnetic resonance imaging-linac system. *Semin Radiat Oncol*; 24:207–209.

Lamey M, Burke B, Blosser E, Rathee S, De Zanche N, Fallone BG. 2012. Radiofrequency shielding for a linac-MRI system. *Phys Med Biol*; 55:995–1006.

Li HH, Rodriguez VL, Green OL, Hu Y, Kashani R, Wooten O, Yang D, Mutic S. 2015. Patient-specific quality assurance for the delivery of Co-60 intensity modulated radiation therapy subject to a 0.35 T lateral magnetic field. *Int J Radiat Oncol Biol Phys*; 91:65–72.

Meisjsing I, Raaymakers BW, Raaijmakers AJE, Kok JGM, Hogeweg L, Liu B, Lagendijk JJW. 2009. Dosimetry for the MRI accelerator: The impact of a magnetic field on the response of a Famrer NE2571 ionization chamber. *Phys Med Biol*; 54:2993–3002.

Mutic S, Dempsey JF. 2014. The ViewRay system: Magnetic resonance-guided and controlled radiotherapy. *Semin Radiat Oncol*; 24:196–199.

Paulson ES, Crijns SPM, Keller BM, Wang J, Schmidt MA, Coutts G, van der Heide UA. 2016. Consensus opinion on MRI simulation for external beam radiation treatment planning. *Radiother Oncol*; 121:187–192.

Paulson ES, Erickson B, Schultz C, Li XA. 2015. Comprehensive MRI simulation methodology using a dedicated MRI scanner in radiation oncology for external beam radiation treatment planning. *Med Phys*; 42:28–39.

Raaijmakers AJE, Hardemark B, Raaymakers BW. 2007. Dose optimization for the MRI-accelerator: IMRT in the presence of a magnetic field. *Phys Med Biol*; 52:7045–7054.

Raaijmakers AJE, Raaymakers BW, Lagendijk JJW. 2005. Integrating a MRI scanner with a 6 MV radiotherapy accelerator: Dose increase at tissue-air interfaces in a lateral magnetic field due to returning electrons. *Phys Med Biol*; 50:1363–1376.

Raaijmakers AJE, Raaymakers BW, Lagendijk JJW. 2008. Magnetic field induced dose effects in MR-guided radiotherapy systems: Dependence on the magnetic field strength. *Phys Med Biol*; 53:909–923.

Raaymakers BW, de Boer JCJ, Knox C, Crijins CPM, Smit K, Stam ML, van den Bosch MR, Kok JGM, Lagendijk JJW. Integrated megavoltage portal imaging with a 1.5 T MRI Linac. *Phys Med Biol* 2011; 56:N207–N214.

Reinsberg SA, Doran SJ, Charles-Edwards EM, Leach MO. 2005. A complete distortion correction for MR images: II. Rectification of static-field inhomogeneities by similarity-based profile mapping. *Phys Med Biol*; 50:2651–2661.

Smit K, Sjoholm J, Kok JGM, Lagendijk JJW, Raaymakers BW. 2014. Relative dosimetry in a 1.5 T magnetic field: An MR-Linac compatible prototype scanning water phantom. *Phys Med Biol*; 59:4099–4109.

St Aubin J, Keyvanloo A, Vassiliev O, Fallone BG. 2015. A deterministic solution of the first order linear Boltzmann transport equation in the presence of external magnetic fields. *Med Phys*; 42:780–793.

Van der Heide UA, Houweling AC, Groenendaal G, Beets-Tan RG, Lambin P. 2012. Functional MRI for radiotherapy dose painting. *Magn Reson Imaging*; 9:1216–1223.

Van Zijp HM, van Asselen B, Wolthaus JWH, Kok JMG, de Vries JHW, Ishakoglu K, Beld E, Lagendijk JJW, Raaymakers BW. 2016. Minimizing the magnetic field effect in MR-linac specific QA-tests: The use of electron dense materials. *Phys Med Biol*; 61:N50–N59.

Zhang L, Du D, Green O, Rodriguez V, Wooten H, Xiao Z, Yand D, Hu Y, Li H. 2014. 3D Gel Dosimetry Using ViewRay On-Board MR Scanner: A Feasibility Study. *Med Phys*; 41(6):455.

CHAPTER 9

Advances in Charged Particle Therapy Machines

Jao Jang Su

CONTENTS

9.1 INTRODUCTION

A charged particle beam therapy system consists of a number of fully integrated subsystems to deliver a precise dose to a patient promptly. Most centers that are newly built or under construction have pencil beam scanning capability to perform intensity-modulated proton therapy (IMPT). To deliver three-dimensional (3D) spot scanning with high precision and speed, a hadron therapy system has to be capable of delivering beams with varying energy and from different angles. Intensity optimization of individual beams is also desired. At present, hospital-based hadron therapy centers use either a cyclotron or a synchrotron as a particle source for delivering high-energy beams to a single treatment room or multiple treatment rooms. Both the cyclotron and the synchrotron are mature technologies for delivering maximum proton energy at

235 MeV or higher. Synchrotron, however, is the only system available for carbon ion therapy in clinical settings.

A number of advanced accelerator technologies are under development, aiming to provide a more cost-effective tool with greater capability. Novel acceleration schemes have achieved great successes in the past decade on accelerating electrons and positrons. For examples, a dielectric wakefield accelerator (DWA) demonstrated 5 GeV/m acceleration field before breakdown and a plasma wakefield accelerator field delivered 90 GeV/m acceleration for electrons (Blumenfeld et al., 2007). However, accelerating ions is different from accelerating electrons or positrons because of the mass differences. Whereas the velocity of a 3 MeV electron reaches to about 99% of the speed of light, a 250 MeV proton travels just at about 60% of the speed of light. Thus, matching the phase velocity of acceleration field with the ions' velocity becomes a challenge in designing a high-acceleration field machine.

To accommodate slow-traveling ions, the DWA was proposed to match the ion velocity and the acceleration field phase using layers of Blumlein-type high-voltage pulse generators (Caporaso et al., 2007). To obtain a higher acceleration field for heavy ions in plasma-based ion acceleration, impulsive acceleration becomes a straightforward solution instead of traveling wave-based acceleration that is used for lightweight particles like leptons. The first impulsive ion acceleration scheme—Target Normal Sheath Acceleration (TNSA)—was proposed by Wilks et al. (2001), and it was followed by the breakout afterburner (BOA) (Yin et al., 2007), radiation pressure acceleration (RPA) (Esirkepov et al., 2004), and shock acceleration schemes (Haberberger et al., 2012). A research group in Germany has achieved 10^9 protons of energy above 1 MeV using a petawatt high-energy laser for heavy ion experiments (PHELIX, GSI, Germany) laser (Wagner et al., 2016). The experiment demonstrates realization of laser-based proton acceleration in a relatively large number of particles, which is a critical step to reach before clinical application of laser-driven ion acceleration.

Dielectric wall and laser driven accelerators aim to offer solutions for a compact hadron therapy unit by applying high-acceleration fields, whereas fixed field-alternating gradient accelerator (FFAG) and rapid-cycle synchrotron promise the capability of rapid energy change. While all these novel technologies have not yet been tested outside laboratories, superconducting cyclotrons have gradually outnumbered normal conducting cyclotrons in new installations. Ironless superconducting cyclotrons are being developed that may further reduce the weight of a cyclotron and the beam transport elements significantly when the technology becomes mature. National Institute of Radiological Sciences (NIRS, Japan) and Pronova have developed superconducting magnets for gantries that largely reduce the overall gantry size and weight. Lawrence Berkeley National Laboratory (LBNL, United States), Paul Scherrer Institute (PSI, Switzerland), and Varian are also developing superconducting magnets for beam transport system.

In this chapter, the historical development of radio frequency (RF) based accelerators, the recent progress toward compact lightweight designs, and the novel accelerator concepts in future hadron therapy systems are reviewed. This chapter uses several abbreviations, which are listed in the table for readers' convenience.

ABBREVIATIONS

ADAM	Applications of detectors and accelerators to medicine
AVF	Azimuthally varying field
BINP	Budker Institute of Nuclear Physics, Novosibirsk, Russia
BNL	Brookhaven National Laboratory, United States
BOA	Breakout afterburner
CERN	European Organization for Nuclear Research, Switzerland
CONFORM	COnstruction of a Non-scaling FFAG for Oncology, Research and Medicine
DOE	Department of Energy, United States
DWA	Dielectric wall accelerator
EBIS	Electron beam ion source
ECR	Electron cyclotron resonance
EMMA	Electron machine for many applications
ESS	Energy degrader and selection system
FAAET	Facility for advanced accelerator experimental tests
FDF	Focusing-defocusing-focusing
FFAG	Fixed field-alternating gradient accelerator
FWHM	Full width at half maximum
HCL	Harvard Cyclotron Laboratory, Cambridge, MA, United States
HIMAC	Heavy Ion Medical Accelerator, Chiba, Japan
HIT	Heidelberg Ion Beam Therapy Center, Heidelberg, Germany
IMPT	Intensity-modulated proton therapy
iRCMS	Ion Rapid Cycling Medical Synchrotron
J-PARC	Japan Proton Accelerator Research Complex, Kyoto, Japan
KEK	High Energy Accelerator Research Organization, Japan
KURRI	Kyoto University Research Reactor Institute
LBNL	Lawrence Berkeley National Laboratory, Berkeley, CA, United States
LHC	Large Hadron Collider
LIGHT	Linac for Image Guided Hadron Therapy
LRL	Lawrence Radiation Laboratory
MRI	Magnetic resonance imaging
NIRS	National Institute of Radiological Sciences, Japan
NORMA	Normal-conducting Racetrack Medical Accelerator
PAMELA	Particle accelerator for medical application
PET	Positron emission tomography
PHELIX	Petawatt high-energy laser for heavy ion experiments, GSI, Germany
PIG	Penning or Philips ionization gauge
PSI	Paul Scherrer Institute, Switzerland
RAL	Rutherford Appleton Laboratory, United Kingdom
RIKEN	Rikagaku Kenkyūsho, Japan
RPA	Radiation pressure acceleration
RF	Radio frequency
RFQ	Radio frequency quadrupole
SLAC	Stanford Linear Accelerator Center
TNSA	Target normal sheath acceleration
TRIUMF	Canada's National Laboratory for Particle and Nuclear Physics and Accelerator-Based Science, British Columbia, Canada

9.2 CYCLOTRON

About 60% of charged particle therapy facilities that already built or under construction use a cyclotron as their particle source. Superconducting cyclotrons outnumber normal conducting cyclotrons in new installations. Cyclotrons are a fixed energy machine and require an energy degrader with an energy selection system to reduce the beam energy from the maximum energy to lower energies to comply with the range requirements. Sufficient beam intensity must be provided from the cyclotron to meet the clinical requirement for low-energy beams because the beam loss is large from the energy degrader and selection system (ESS). The transmission efficiency for a cyclotron can be as low as less than 1% when degrading a proton beam from 235 MeV down to 70 MeV. The interaction of the beam with the energy degrader produces secondary neutrons and gamma rays that must be shielded. The time for adjusting the energy step typically takes about 1 second. PSI has demonstrated adjusting beam energy within 80 to 100 ms per step, with each step equivalent to 5 mm range in water. The response time of energy step adjustment for cyclotrons has become comparable to the timing of multi-energy extraction synchrotrons.

9.2.1 Isochronous Cyclotron and Synchrocyclotron

Cyclotrons have been commonly used in isotope production and charged particle beam therapy. The classical Lawrence cyclotron alternates Dee voltage progressively to accelerate the charge particles to a desired energy in a static dipolar magnetic field before extraction. The revolution frequency (named cyclotron frequency) f is proportional to the charge q and the magnetic field B, and it is inversely proportional to the particle relativistic mass $m = \gamma m_o$. That is, $f = qB/2\pi\gamma m_o$, where γ is the Lorentz factor, $\gamma = 1/\sqrt{1 - v^2/c^2}$, v is the particle velocity, and m_o is the particle rest mass.

While almost constant for particles in very low energy, the cyclotron frequency is the inverse of the kinetic energy for high-energy particles. For example, the Lorentz factors of a 20, 200 MeV, and 1 GeV proton are $\gamma = 1.02, 1.21$, and 2.07, respectively; that is, the cyclotron frequency of a 1 GeV proton is about 50% of that of a 20 MeV proton. To overcome the relativistic factor correction, one can either alternate the RF acceleration field to match the decreasing frequency at higher energies (the synchrocyclotron) or split the magnet into radial or spiral sectors to create extra focusing forces, and vary the field strength radially to maintain isochronism (the isochronous cyclotron). Table 9.1 shows the type of synchrocyclotrons and isochronous cyclotrons for proton therapy.

9.2.2 Synchrocyclotron

Synchrocyclotron let the RF frequency ω decrease as the energy increases ($\omega = \omega_o/\gamma$ to match the effective mass $m = \gamma m_o$). But this condition can only be satisfied for an ion at a particular "phase angle" θ_s, where its energy gain per gap exactly matches the drop in frequency. The spread of a beam in revolution frequency with energy would quickly exploit the energy spread in the beam and disperse the particles away from the RF acceleration phase. This problem was resolved by the principle of phase stability by Veksler and McMillan working independently in 1945.

TABLE 9.1 Overview of the Early Cyclotron Compared to the Latest Commercial Synchrocyclotrons and Isochronous Cyclotrons for Particle Therapy

	184″ Berkeley cyclotron	Synchrocyclotron			Isochronous Cyclotrons	
		Harvard cyclotron	Mevion S250	IBA S2C2	IBA/SHI C230	Varian probeam
Energy (MeV)	p: 730 d: 460 He: 910 ^{3}He: 1140	160	250	230	230	250
Repetition rate (Hz)	64	300	500	1,000	Cont.	Cont.
Weight (ton)	> 4000	> 700	Approximately 20	Approximately 50	Approximately 220	Approximately 90
Dimension (m)	Main components approximately $15 \times 6 \times 8.4$	Main component approximately $7 \times 4.5 \times 3$	Approximately 1.8 diameter × 1 H	Approximately 2.5 diameter × 1.5 H	4.3 diameter × 2.1 H	3.1 diameter × 1.6 H
Coil	Resistive	Resistive	SC Nb$_3$Sn	SC NbTi	Resistive	SC NbTi
Maximum beam current (nA)			Approximately 30	400	Approximately 400	Approximately 800
Peak magnetic field (Tesla)	2.34	1.6	9	5.7	2.9	3.8

The operation mode of frequency modulation requires a single beam to be accelerated between the maximum and minimum phase of acceleration field, θ_{max} and θ_{min}, called the "RF bucket." The effect is that a bunch of particles with an energy spread can be kept bunched throughout the acceleration cycle by simply injecting them at a suitable phase of the RF cycle. This longitudinal focusing effect is strong enough that the frequency modulation in the synchrocyclotron does not have to be specially tailored and is simply sinusoidal. The 184-inch synchrocyclotron in Lawrence Radiation Laboratory (LRL) was designed to accelerate protons to the energy of 730 MeV, with pulse length at about 500 μsec, and at the rate of 64 pulses per second. The RF acceleration field modulation sequence, and the magnetic field radial profile of it, are shown in Figure 9.1.

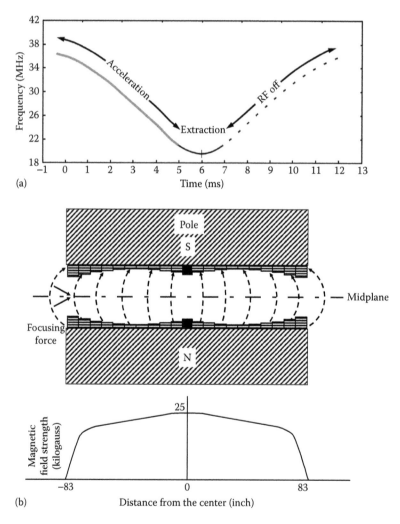

FIGURE 9.1 The magnetic field profile and RF frequency of the Lawrence Radiation Laboratory 184-inch synchrocyclotron. (a) RF cycle for accelerating protons, 64 cps, red line—acceleration phase, black line—extraction, broken line—beam off. The RF frequency swings from 36 to 18 MHz for proton beams and 18–13.5 MHz for Alpha particles. (b) (Bottom) magnetic-field strength versus radius, (top) magnetic field lines (broken arrows) and focusing forces (solid arrows).

The LRL synchrocyclotron accelerator gave researchers the opportunity to study ion beam therapy as originally proposed by Robert Wilson in 1946. There were 2,054 patients treated at Berkeley by using helium beams, 30 patients treated by proton beams, and 433 patients treated by other ions. A similar facility—the synchrocyclotron at Harvard Cyclotron Laboratory (HCL)—was commissioned in 1949 and later turned into a dedicated facility for ion beam radiotherapy. A total of 9,116 patients were treated at HCL from the 1960s until it was decommissioned in 2002 (Jermann, 2015).

The 184-inch synchrocyclotron contained 3,700 tons of steel in its magnet yoke and pole pieces, and 300 tons of copper in its exciting coils, providing 2.34 and 2.23 T magnetic field strengths at the center and radius of 82 in., respectively. The size and weight of modern synchrocyclotrons has been reduced significantly with the application of superconducting materials. For instance, a Mevion S250 unit has a central magnetic field of about 9 T using Nb_3Sn for the coil. IBA S2C2 uses NbTi coil producing 5.7 T central magnetic field. The weight of IBA S2C2 is about 50 tons and the size is about 2.5 m in diameter, a significant reduction in physical dimensions and mass compared to conventional cyclotrons. Table 9.1 is the overview of the main parameters of the early synchrocyclotrons and recent commercial medical synchrocyclotrons.

9.2.3 Isochronous Cyclotron

An isochronous cyclotron increases its average magnetic field as a function of radius to compensate for the relativistic mass increase. The radial varying symmetric magnetic field does not provide focusing force to maintain charge particles in the midplane. The first isochronous cyclotron with focusing force was demonstrated by Heyn and Tat in 1958 based on the model of a periodic orbit in a magnetic field varying with polar angle proposed by Thomas in 1938. It is known as azimuthally varying field (AVF) cyclotrons. Thomas proposed the magnetic field consisting of an azimuthal periodic structure of hills and valleys, as illustrated in Figure 9.2. It deforms the orbit to a noncircular shape and creates vertical forces that push particles back to the median plane. If the transition from a weak field (valley) to a strong field (hill) is not orthogonal to the trajectory, the beam can be focused to provide "edge focusing" or "defocusing" for particles traveling from a strong field to a weak field. The vertical focusing F_z can be obtained by radial velocity component V_R and the magnetic field azimuthal component B_θ at the edge of the varying field structure,

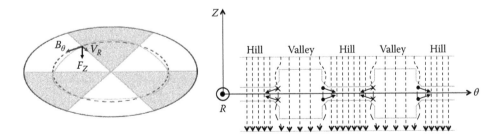

FIGURE 9.2 Azimuthally varying field (AVF)—focusing effect due to variation of the field with polar angle. Particle has a large trajectory curvature in the weak field valley regions and a smaller trajectory curvature in the strong field hill zones. The trajectory is not circular; orbit is not perpendicular to hill-valley edge either.

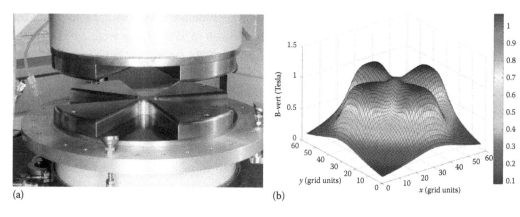

FIGURE 9.3 (a) The Rutgers 12-in. azimuthally varying field (AVF) cyclotron pole tips; (b) the measurement of the magnetic field. (Courtesy of Dr. Tim Koeth and Rutgers Cyclotron.)

as illustrated in Figure 9.2. Figure 9.3 shows the pole of an AVF cyclotron and the magnetic field measured in the midplane. The hills are the regions where a small gap and strong magnetic field exist, whereas the valleys contain a large gap and weak field. Separated sector cyclotrons are the extreme case of a design in which the magnetic field goes to zero in the valley. This concept was first proposed at Oak Ridge for a 900 MeV isochronous cyclotron in the 1950s, and the first separated sector cyclotron was built at PSI in the meson factory facility in the 1970s. Separated sector cyclotrons have individual yokes and coils; its valleys are magnetic field free for injection and extraction of ions, RF, and diagnostics. Compared to AVF cyclotrons, separated sector cyclotrons require a larger area to accommodate sector magnets, box resonators, and an external injection subsystem. The advantage of the separated sector cyclotron is its high extraction efficiency and power. Examples of separated sector cyclotrons are TRIUMF (520 MeV, proton), RIKEN Superconducting Ring Cyclotron (SRC) (440 MeV/nucleon, light ions; 350 MeV/nucleon, very heavy ions), and PSI Ring Cyclotron (592 MeV, proton). The RIKEN SRC, which delivers approximately 80 GeV heavy ions, provides the highest energy out of any type of continuous beam accelerators.

The idea that sectors do not need to be symmetric, which is the case with the fixed field-alternating gradient accelerator (FFAG), was proposed by (Kerst et al., 1954; Symon et al., 1956). The magnetic field with a tilted edge of angle θ provides more focusing for valley-to-hill transition and less defocusing for hill-to-valley transition. The stronger focusing and less defocusing result in the net effect of focusing for spiral sectors. A stronger vertical focusing is the result of the hill-to-valley boundaries being modified from the simple diametric lines to spiral poles, as shown in Figure 9.4. At a radius R, the boundaries between the hills and valleys are inclined at an angle $\theta(R)$ with respect to a diameter. Spiral-shaped poles lead to additional inclination of magnitude $\theta(R)$ between the particle equilibrium orbit and the boundary. The edge fields from the spiral inclination act to alternate focusing and defocusing of the particles. The focusing properties of a uniform field sector with different inclined boundaries are illustrated in Figure 9.4—the horizontal trajectories of initially parallel particle orbits on the main orbit (solid line) and the trajectories of displaced particles from the main orbit (dashed line). Focusing forces arise from the shape of the sector magnet

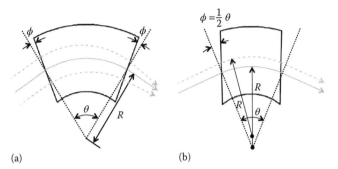

FIGURE 9.4 (a) Horizontal trajectories of initially parallel particles, solid line—main orbit; dashed lines—offset orbits displaced from the main orbit. Inclination angle $\phi < 1/2\theta$ gives a net radial focusing to particles. (b) No focusing for $\phi < 1/2\theta$.

boundaries. The orbit in the hill's magnetic region is a circular section of radius R that swipes across an angle (with inclination angle) at the boundary, assuming the inclination angles are the same at the entrance and exit. Figure 9.4 shows the radial focusing effects through the adjustment of inclination angles. If the boundary is parallel to the line of the midplane of the magnet, the gyrocenters of particle orbits all fall on this line. As a result, the orbits remain in parallel to each other throughout the magnet region, and there is no focusing for this configuration. The gyrocenter of an off-axis a particle with an inclination angle $\phi < 1/2 \theta$ will be moved to the upper left, or the gyrocenter will be moved to the lower right for $\phi > 1/2 \theta$ resulting in the particles emerge from the sector focused toward the axis.

All modern AVF isochronous cyclotrons have spiraled pole edges to maintain the vertical stability and radial focusing of the circulating proton beam for high energies. The ion's orbit has an inclination at the boundaries of high-field regions to provide vertical confinement by the edge focusing. The combined effects of edge focusing and defocusing lead to an additional vertical confinement force. A 270° spiral sweeping design was demonstrated at Rutgers Cyclotron to minimize phase slippage and to preserve axial stability (see Figure 9.5).

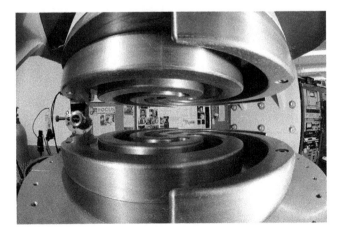

FIGURE 9.5 A four-sector Archimedean spiral sweeping 270° magnetic pole tip at Rutgers Cyclotron. (Courtesy of Dr. Tim Koeth and Rutgers Cyclotron.)

9.2.4 Superconducting Cyclotrons and Ironless Superconducting Cyclotrons

The first commercial superconducting medical cyclotron was developed by ACCEL (now a part of Varion) in 2001. The size and weight of the cyclotron is significantly reduced because of the higher magnetic field with the use of superconducting coils. The high magnetic fields are above the saturation level of the iron poles; therefore the magnetic dipoles in the poles are well aligned and the fields can be predicted more accurately. Low temperature superconductors such as NbTi and Nb_3Sn have been used in modern superconducting cyclotrons (see Table 9.1). All superconductors become superconductors only if their operational parameters—temperature, current density, and ambient magnetic field—are within certain bounds, as shown in Figure 9.6 (Bottura, 2000; Brüning and Collier, 2007). NbTi is used for the superconducting magnets of the Large Hadron Collider (LHC), where magnesium diboride (MgB_2) becomes superconducting at 39 K, the highest known transition temperatures (Tc) out of any superconductor. NbTi and Nb_3Sn are two types of superconducting materials that have been used in commercial medical cyclotrons. The advantage of Nb_3Sn is that the coil is capable of producing higher magnetic fields, thus reducing the overall size and lowering the weight of the accelerator. NbTi has been more commonly used in many applications, such as serving as MRI magnets. Nb_3Sn could be operated at a higher temperature than the temperature of NbTi, which would make the cost of the cryo-system and operation for Nb_3Sn less expensive. MgB_2 is a high-Tc superconducting material that makes magnets operational at a higher temperature without using liquid helium for cooling, working at approximately 4 T at 20 to 30 K. This enables the use of a single-stage cryocooler

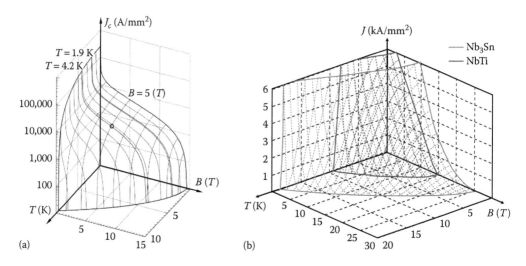

FIGURE 9.6 (a) The 1.9 K temperature of the NbTi for LHC allows 8.4 T magnetic field using a current density of 1.5 ~ 2 kA/mm² inside the superconducting cables. It uses superfluid helium as a coolant. (Courtesy of Dr. Luca Bottura, CERN). (b) The critical surface of Nb_3Sn (light gray), NbTi (dark gray) versus temperature T and magnetic field B are shown in light gray, dark gray and green, respectively. The superconducting material Nb_3Sn can generate a higher magnetic field than the field from NbTi operating at the same temperature. This illustrates the size and weight differences between Mevion S250 (Nb_3Sn) and IBA/Varian (NbTi) superconducting cyclotrons.

and simplifies cryostats. Currently, MgB_2 delivers only at moderate magnetic fields; it has better performance at higher fields, which has been demonstrated in thin film and in bulk material on the order of tens of Tesla. MgB2 wires and tapes are manufactured by means of the so-called powder-in-tube (PIT) method. Because of its relatively simple PIT deposition approach, MgB_2 may serve applications in the form of flexible flat ribbon cables for electronics and superconducting cavities for RF applications. Figure 9.6 illustrates the critical surface of a NbTi wire as a function of magnetic field, temperature, and current density. The critical surface expresses the maximum current that can be carried by the superconductor. For instance, the maximum current can be carried by a NbTi wire is about 3,000 A/mm^2 operating at temperature 4.2 K and 5 T magnetic field.

To keep the beam in the midplane as discussed above, the pole surface of magnets for cyclotrons or beamlines are shaped. Shims are added, and special grades of mild steel are selected to achieve high uniformities and fringe fields at the edge of dipoles. The iron core is the heaviest part of a cyclotron. Several groups have proposed using ironless superconducting magnets for cyclotrons. Some prototypes of low-energy ironless superconducting proton cyclotrons, such as a 10 μA current 12 MeV superconducting proton cyclotron, for PET isotope productions have been demonstrated (Alonso and Antaya 2012). The size and weight of an ironless cyclotron are significantly less than those of cyclotrons with an iron core. The Massachusetts Institute of Technology and the company ProNova Solutions jointly developed an ironless superconducting cyclotron for proton therapy under the support of the U. S. Department of Energy Accelerator Stewardship program. The ironless superconducting cyclotron can be either synchrocyclotron or isochronous cyclotron. There are primary coils, magnetic field shaping coils, and shielding coils needed to produce the desired magnetic fields. To make an isochronous cyclotron, spiral conduction coils are added for azimuthally fixed or varying fields—AVF type. While a Mevion S250 unit weighs 25 tons, the weight of a 250 MeV coreless superconducting cyclotron would be about 4 tons. The advantages of coreless cyclotron include reduced weight and fringe field; larger space for ion sources, RF, extraction, and cryostat systems; and less cryogenic load because of no internal iron. A technical challenge of an ironless cyclotron is that the cooling for the shield coils might require forced flow of helium because of the larger radius (size) of shield coils. Ironless superconducting cyclotrons can significantly change the designs of proton therapy facilities.

9.3 SYNCHROTRON

The first dedicated, hospital-based proton therapy center was built at Loma Linda University Medical Center using a 250 MeV synchrotron that was based on a Fermilab design. A synchrotron can provide rapid energy variation without maneuvering mechanical devices, unlike a cyclotron system that has to rely on a mechanical energy degrader and energy selection system to change the beam energy. Strong focusing synchrotrons consist of bending magnets, quadrupole magnets, quadrupole lens sets, and sextupole magnets to correct horizontal chromaticity and form a separatrix for beam extraction (see Figure 9.7a for a Hitachi compact synchrotron). The bending magnets of weak focusing synchrotrons have an FDF structure, where the focusing (F) and defocusing (D) sectors have bending

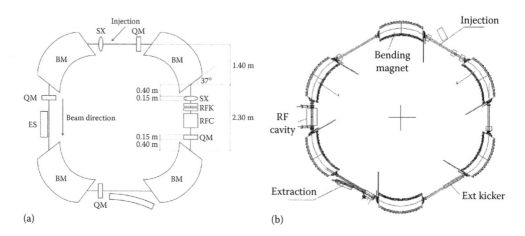

FIGURE 9.7 (a) Strong focusing synchrotrons consist of bending magnets, quadrupole magnets, quadrupole lens sets, and sextupole magnets to correct horizontal chromaticity and form a separatrix for beam extraction. (Courtesy of Hitachi). (b) The bending magnets of weak focusing synchrotrons have FDF structure where the focusing (F) and defocusing (D) sectors are with bending angles to replace quadrupole sets. Shown on the right. (Courtesy of BINP, Dr. E. Levichev.)

angles to replace quadrupole sets such as the Loma Linda, ProTom, and Budker Institute of Nuclear Physics (BINP) proton synchrotrons (see Figure 9.7b for a BINP booster). While both cyclotron and synchroton systems are commercially available for proton therapy facilities, only synchrotron systems are available for heavy ion therapy in the clinical setting.

A number of slow-cycle proton synchrotrons have been installed for clinical treatment. Heavy ion therapy synchrotrons follow the approach developed by NIRS and GSI. Both NIRS and GSI systems use an electron cyclotron resonance (ECR) ion source, a radiofrequency quadrupole (RFQ), and a drift tube linac as an injector followed by a synchrotron to accelerate ions (from proton to carbon) up to 400 MeV per nucleon. The synchrotron designed by BINP is different from an NIRS or GSI design. BINP uses an electrostatic tandem as an injector to deliver protons or heavy ions to a rapid-cycle booster synchrotron—which is capable of accelerating protons up to a maximum energy of 250 MeV and the carbon ions up to 30 MeV/u. The proton beam from the booster is extracted to the extraction channel and transported via beamline to treatment rooms. The carbon beam is injected into the main ring for its further acceleration. An electron cooling device is installed in the main synchrotron. The use of an electron cooling system provides extremely small transverse emittance and energy spread of the ion beam. The electron cooling device enables the control of the extracted beam energy by varying the energy of the electron beam simultaneously with the synchrotron magnetic field.

9.3.1 Ion Source and Injector

Several types of ion source are commonly used for accelerators: Penning or Philips ionization Gauge (PIG) ion source, ECR ion source, and electron beam ion source (EBIS). While investigating laser ion acceleration, GSI has installed a 10 MeV proton beamline using laser-driven protons as a test platform for new-generation ion sources (see Section 9.7).

The Penning ion source was once popular for cyclotrons. It consists of a cylinder anode with a magnetic field parallel to the cylinder axis. Electrons emitted from the cathode run into the gas and ionize it. The ions diffuse to a pinhole of the anode wall and are extracted by a ground electrode. The principle of ECR ion source is based on plasma heating by RF or microwave matching the electron cyclotron frequency in a magnetic field, $\omega = eB/m_e$. Microwaves are sent to a gas immersed in a magnetic field in an ECR chamber. The magnetic field has a typical mirror configuration where the field is at a minimum in the center and at a maximum at both ends. Charged particles are stored in the magnetic bottle and collide multiple times; highly charged ions are produced as a result. For example, Au^{24+}, Pb^{27+} are produced at GSI. PIG ion source is inadequate for highly charged heavier ions; thus, ECR has become a better ion source option for accelerators.

EBIS is an ion source based on the principle of electron impact ionization to produce highly charged ions. An intense electron beam is produced and enters a solenoidal magnetic field, decelerates, and is dumped in the electron collector. There is an ion trap region in an EBIS with a series of cylindrical electrodes in the solenoid. Electrostatic potential barriers are induced to constrain ions by applying positive voltage on both ends of the electrodes. Ions are also radially confined by the electron beam space charge. The ions are released to the beamline when the desired charge state is reached via reducing the exit end electrode voltage.

Two types of negative ion source production schemes—surface production and volume production—are generally implemented. The principle of surface negative ion source is based on the desired ion specie to be absorbed onto metals of low work function. The probability is very high that the desired ion specie captures an electron from the low work function metal when it leaves the surface. Volumatic negative ion source production is possible in a plasma.

9.3.2 Electron Cooling for Heavy Ions

A heavy ion therapy system is more expensive and physically much larger than a proton therapy system. An electron cooling mechanism was proposed by BINP to reduce the aperture of a heavy ion system, which in turn would reduce the system cost and provide better quality heavy ion beams, that is, very low emittance, at the same time. A cooled ion beam has much a smaller transverse size and energy spread. Electron cooling allows the increase in beam intensity, resulting in reduced operational cost.

The electron cooling technique developed at BINP provides a good control of the ion beam parameters, as illustrated in Figure 9.8. The basic mechanism of electron cooling is the heat exchange of an ion beam circulating in the storage ring, with an electron current through a common section, then the separation of heated electrons to a dump. Cooling hot ion beams with cold electron currents takes place through Coulomb collisions. Ion beam temperature gradually decreases to the effective temperature of electrons. The process is shown in Figure 9.8. The top figure shows a hot ion beam (red) entering the cooling section from the left. A cold electron current is sent from the electrode from the top and mixes

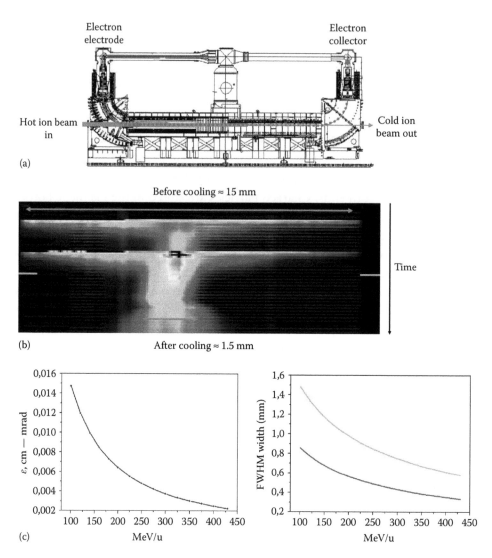

FIGURE 9.8 (a) Schematic diagram of electron cooling system. An electron beam is produced at the upper left and passes the accelerating tube, the toroidal section then enters the horizontal section traveling together with the ion beam. (b) The initial beam size gradually reduced from approximately 15 to approximately 1.5 mm. (c) The ion beam emittance and spot size after electron cooling system are shown in the bottom figures. Left: Transverse emittance of carbon ions of intensity $n = 10^{10}$ in the extraction energy range from 100 to 430 MeV/u. Right: The rms FWHM of a beam at the extraction septum in horizontal plane (dark gray line) and in vertical plane (light gray line). (Courtesy of BINP, Dr. E. Levichev.)

with the ion beam in the same beam pipe. The electron beam is extracted at the end of the heat exchanger section at the right back to the electron collector. The ion beam temperature is reduced each time it passes the cooling section. The picture in the middle shows that the ion beam radius gets reduced rapidly after the cooling process. The final beam size as a function of the beam energy is shown in the bottom of Figure 9.8. With electron beam cooling, the size of a 250 MeV proton beam is about 4.5 mm FWHM, which is about one-third that of a typical 250 MeV cyclotron proton beam (approximately 12 mm). In addition to potential clinical advantages, lower emittance and smaller spot size significantly reduce the aperture of the beam transport line, magnets, and other supporting infrastructures. Electron cooling systems have been installed to cool the hadrons for accumulation and preparation of bunches in several nuclear physics research laboratories such as Fermilab, CERN, and HIMAC, and so on.

9.3.3 Extraction—Multiple Energy Extraction from Single Spill

A particle beam is injected from the injector to a synchrotron and travels in the synchrotron ring guided by bending magnets and focusing lattices. The particle beam gains energy each time that it passes through an RF cavity in the ring. When the beam reaches the desired energy, it is extracted by either a fast extraction (single-turn extraction) scheme or a slow extraction (multi-turn extraction) scheme, as illustrated in Figure 9.9. Slow extraction is necessary for accurate dose delivery of therapeutic applications. In the fast extraction scheme, the magnetic kicker deflects the whole beam in a single turn, as shown in the upper frame of Figure 9.9. There are two types of extraction mechanism for slow extraction: nonresonant multi-turn extraction and resonant multi-turn extraction. In the nonresonant case, the beam is knocked closer to a septum, and part of the beam is chopped off turn by turn, as shown in the bottom frame of Figure 9.9. The resonant extraction reduces the regions of resonance and allows the particles orbiting at amplitudes outside the stable region to be extracted. Theoretically, a synchrotron beam can be accelerated or decelerated

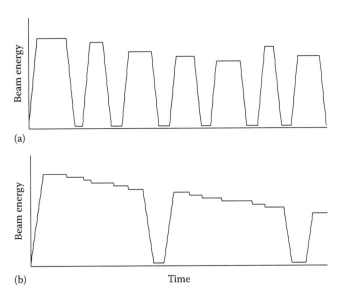

(a)

(b)

Time

FIGURE 9.9 (a) Pulse to pulse energy change and (b) multi-energy per spill extraction. Prompt energy change is important for a higher dose rate in 3D conformal therapy and IMPT. The beam energy can be changed with pulse-to-pulse operation, avoiding beam loss and minimizing secondary radiation from mechanical degraders. Multiple energy extraction from a spill further optimizes the usage of the beam.

each time it passes the RF cavity. It is possible to perform multi-energy extraction from a single spill to meet the requirements of the 3D pencil or raster scanning beam technique. Hitachi takes the advantage of implementing multi-energy extraction from a single spill. Implementing acceleration and deceleration during the extraction process enables multi-energy extraction from a spill with small discrete energy steps. This optimizes the use of available protons in the beam without deceleration and injection of the next spill.

9.4 RAPID-CYCLE SYNCHROTRONS, FFAG, DWA, AND LINAC

Typical circulation time of one cycle in a synchrotron is less than 1 μsec for a proton beam and 8 μsec for a carbon beam. The idea of rapid-cycle synchrotron is based on fast extraction—extracting the whole beam with a strong focusing and high repetition rate (Peggs et al., 2002). The rapid cycling synchrotron for proton or carbon-ion therapy, proposed by the Brookhaven National Laboratory (BNL) group, has a maximum cycling frequency of 30 Hz. It allows the use of simple magnets and power supplies, and possibly a pulse-by-pulse energy variation at the cycling frequency during the acceleration phase. A rapid-cycle synchrotron is a pulse machine. It is suitable for spot scanning irradiation, so controlling the number of protons (or heavy ions) both in each pulse and during beam transport must be very precise. The rapid-cycle synchrotron design has not yet been used for particle therapy. BNL—in partnership with Best Medical International, which is based in Virginia—is developing the "ion Rapid Cycling Medical Synchrotron" (iRCMS) for clinical use. Several rapid-cycle synchrotrons are installed in nuclear physics research laboratories such as the ISIS pulsed neutron and muon source at Rutherford Appleton Laboratory (RAL) in the United Kingdom, and the Japan Proton Accelerator Research Complex (J-PARC) and High Energy Accelerator Research Organization (KEK) in Japan.

The idea of an FFAG was suggested in the 1950s. FFAG machines can be seen as a generalized synchrotron or cyclotron that uses a constant magnetic field—hence the name "fixed field"—with the "alternating gradient" principle to achieve an increased focusing strength. FFAG has benefited from progresses in various areas, including beam dynamics simulations; magnet designs; manufacturing processes; and fast swept-frequency, broadband, and high-gradient RF technologies; thus, FFAG has become a candidate in many challenging applications, including particle radiotherapy.

Two types of FFAGs—scaling FFAG and nonscaling FFAG—are identified based on the transverse dynamics. The scaling FFAG is similar to isochronous cyclotrons, and the AVF cyclotron can be treated as a radial sector of FFAG. A near-zero chromaticity can be achieved by a combination of sectoral layout and constant field, resulting in geometrically similar orbits. The magnetic field of a scaling FFAG increases with radius; the field is given by $B_\perp = B_o \left(r/r_o \right)^k$, where k is the field index $k = -\rho / B_\perp \, \partial B_\perp / \partial B$, ρ is the bending radius, and r_o is the reference radius. Like the cyclotron, the orbit radius varies with the particle energy for FFAG and requires large aperture magnets to accommodate a desired energy range as a result.

A number of FFAG machines have been built in Japan and in Europe, such as the KURRI FFAG (Tanigaki et al., 2006) complex in Japan, the Electron Model with Many Applications (EMMA) (Barlow et al., 2010; Machidal et al., 2012) in the United Kingdom, and projects under investigation like the Particle Accelerator for Medical Application—PAMELA (Peach, 2014) and the normal-conducting racetrack medical accelerator NORMA projects (Garland et al., 2015). EMMA is an FFAG demonstration machine commissioned in 2010/2011 to investigate some of the key dynamics issues and technical issues involved in realizing the technology. It aims to demonstrate the viability of the nonscaling FFAG technology for high-power proton accelerators.

The nonscaling FFAG uses strong nonlinear magnets to cover a wide range of energy while maintaining the use of reasonably small aperture magnets. EMMA is a nonscaling FFAG electron machine designed, constructed, and tested at the Daresbury Laboratory in the United Kingdom to demonstrate the technology of a nonscaling FFAG accelerator. EMMA demonstrated no beam blow despite resonance crossing for electron beams. EMMA has shown stable rapid acceleration, from 12 to 18 MeV in six turns, in the nonscaling FFAG EMMA machine. The result of rapid acceleration implies a possible practical realization of nonscaling FFAG for wider applications; one of the ongoing projects, PAMELA or NORMA for charged particle therapy, is an example. Acceleration of protons requires more and slower turns than the acceleration of electrons in EMMA. Any small errors in the lattice contributes to kicks to the beam and potentially damages the beam. PAMELA was a design study of nonscaling FFAG for hadron therapy as part of the Construction of a nonscaling FFAG for Oncology, Research and Medicine (CONFORM) project. The aim of the PAMELA project is to design a highly efficient hadron therapy system using the features of nonscaling FFAG technology. The fixed field allows more rapid acceleration with variable energy extraction from 50 to 250 MeV (protons) or 70 to 450 MeV/u (carbon). NORMA is a design study following the PAMELA design using normal conducting magnets instead of superconducting magnets suggested by PAMELA.

Instead of employing metallic structures to produce the electric fields for accelerating charge particles, the DWA concept was proposed to use a novel insulation tube to deliver a longitudinal electric field for the acceleration process. The DWA high-voltage pulses are based on the concept developed by Alan Blumlein in 1941. Because of the fast switching characteristics of the duration, on the order of a few nanoseconds, the tube allows higher acceleration gradients on the order of 100 MeV/m. A DWA is constructed by stacking a series of rings of high-gradient insulator with conducting sheets in between and connected to a Blumlein-type high-voltage pulse generator. By properly arranging the triggering time of high-voltage pulses, a tuneable phase velocity traveling acceleration field can be generated to match the velocity of the charged particle beam. The technology can potentially deliver 200 MeV protons with a unit only several meters in length and mounted on a 2-m robotic arm. The DWA approach offers a significant reduction in size, but it has yet been demonstrated in a clinical setting.

Linac-based ion systems using a cyclotron or an RFQ and a drift tube section as an injector have been developed (Lennox, 1991; Degiovanni and Amalfi 2014). The advantage of using a high-frequency linac proton or heavy ion accelerator is that the beam energy can be varied continuously pulse by pulse. Three-dimensional spot scanning can be achieved by adjusting the RF power of a number of klystrons or the power of the final several klystrons. The capability of pulse-to-pulse energy variation eliminates the need for energy degrading and selection system for fast and precise control of dose delivery. The energy step adjustment of the linac systems can be varied in less than 5 ms, so that the system is capable of painting tumors in three dimensions at a very high rate (approximately 200 Hz). Ideally, this system is more flexible for overcoming issues related to organ motions. The technique had been taken up by a CERN spin-off company—Applications of Detectors and Accelerators to Medicine (ADAM). A UK company, Advanced Oncotherapy, later acquired ADAM to commercialize the linac-based proton therapy system—Linac for Image Guided Hadron Therapy (LIGHT).

9.5 LASER-DRIVEN ION BEAMS

The acceleration field of an RF accelerator is limited by the wall breakdown constraints subject to the RF frequency and cavity material. For example, the acceleration field of a resistive 106 MHz RF cavity is about 2 MV/m where an ingot-niobium-based superconducting RF cavity has demonstrated a record accelerating gradient of $E = 45.6$ MV/m (Kneisel et al., 2015). A high acceleration gradient is important for developing compact accelerators for various applications such as colliders for particle physics; compact accelerators for radiation oncology, nuclear medicine, and pharmacology; homeland security; transmutation of nuclear wastes; and so on.

A plasma is a quasi-neutral ionized gas consisting of approximately equal numbers of positive charged ions and negative charged electrons. There is no wall breakdown limit like there is with an RF cavity because a plasma is ionized. A disturbance in a plasma can create regions of charge separation where the uneven charge distribution builds up an electric field from positive regions to negative regions. Such a disturbance can be driven by an intense laser pulse or by a charged particle beam exciting a traveling plasma wakefield velocity equal to the velocity of the driver. The plasma accelerator concept was first proposed by Tajima and Dowson in 1979 for accelerating electrons by laser-driven plasma Langmuir waves. The acceleration field of a plasma wakefield is approximately $E \sim \sqrt{n}$ eV/cm (n is the plasma density—particle/cm^3; for example, $E = 1 \times 10^{11}$ eV/m for $n = 10^{18}$ particle/cm^3, which provides a peak acceleration field 3 orders of magnitude higher than a field from a superconducting RF cavity). Significant progress has been made on plasma accelerators for accelerating electron beams and positron beams over the past decades. Generation of 4.2 GeV beams with 6% energy spread, approximately 4×10^7 electrons per pulse, and 0.3 mrad rms divergence has been made from a 9-cm-long capillary with a plasma density of approximately 7×10^{17}/cm^3. A 27 GeV energy gain of a positron beam in a 20-cm-long high-ionization-potential gas (argon), boosting its initial 20.35 GeV energy by 130%, was demonstrated in a particle beam–driven experiment at the Stanford Linear Accelerator Center's (SLAC's) Facility for Advanced Accelerator Experimental Tests (FAAET) (Corde et al., 2015).

Following the success of accelerating leptons by plasma waves, ion acceleration driven by intense laser has attracted investigations as an alternative to RF accelerators for proton and heavy ion beam therapy systems (Bulanov et al., 2002). Experiments have demonstrated generation of up to tens of MeV proton and ion beams over a wide range of laser and target parameters (Snavely et al., 2000; Wilks et al., 2001; Busold et al., 2014).

While most works were demonstrated based on the original TNSA, several acceleration mechanisms have been proposed and demonstrated in theory, numerical simulation, and experiment. However, the protons or ions generated from intense laser target interaction come with a wide spread of logarithm power law energy spectrum, $I(E) \sim E^{-n}$, where n varies with the energy ranges and acceleration mechanisms. The mechanism of accelerating ions is very different from plasma wakefield acceleration for electrons and positrons.

Acceleration of electrons and positrons in a plasma wakefield can be treated as the analogy of acceleration in a traveling wave linac. For example, electron or position acceleration in an $n = 10^{18}$ particle/cm^3 plasma can be treated as in a 10 THz, $E_{peak} = 1 \times 10^2$ GV/m traveling wave linear accelerator. The acceleration mechanism of ions driven by a laser is impulsive, with a typical interaction duration on the order of tens of femtoseconds ~ picoseconds. As a result, the ion energy spectra are broadband, typically with an inverse power law profile up to cutoff because there is no "RF bucket" to confine ions in energy spread. Figure 9.10 illustrates an intense laser interacting with an overdense or a subcritical density plasma. Electrons are preferentially heated to a few MeV and accelerated forward toward the rear side of the target. Relativistic electrons departing from the rear side of the target

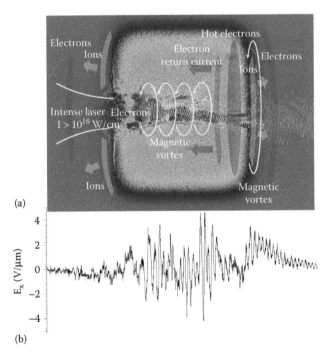

(a)

(b)

FIGURE 9.10 (a) Schematic illustration of an intense laser interacting with an overdense or a subcritical density plasma. Electrons are preferentially heated to a few MeV and accelerated forward toward the rear side of the target. Relativistic electrons departing from the rear side of the target results in an electrostatic field to pull ions out from the surface, known as target normal sheath acceleration (TNSA). (b) The electric field, E_x shows approximately 3 MeV/μm at the front side (to the right) of the target volume. About 10^{11} protons are accelerated to energy greater than 5 MeV, and some protons reach maximum energy greater than 15 MeV.

results in an electrostatic field to pull ions out from the surface. Subject to the parameters of the laser and plasma, electrons return current streams back to the entrance side of the target, then bounce back to the exit side (rear side) and oscillate back and forth until electrons in the region are thermalized. The return current forms a toroidal quasi-steady-state magnetic field of a typical magnitude, $|B| \sim m_e c \omega_o / e$. The magnetic vortex is about 100 MG for an $I = 10^{19}$ W/cm^2 Ti:Sapphire laser hitting a density 10^{21} cm^{-3} hydrogen target. The bottom figure shows the acceleration electric field $|E| \sim 3$ MeV/μm at the exit end of the target volume. About 10^{11} protons are accelerated to energy greater than 5 MeV, while some protons reach their maximum energy, which is greater than 15 MeV.

Several acceleration mechanisms are proposed for producing quasi-monoenergetic ions, RPA, collisionless shock acceleration, hole boring acceleration, and acceleration in relativistic transparency regimes. Proton beams with a narrow energy spread driven by collisionless shocks up to the energy of 20 MeV were demonstrated by a UCLA group using CO_2 laser hitting a hydrogen gas jet (Haberberger et al., 2012). RPA of thin foils with circularly polarized laser pulses was proposed for the generation of quasi-monoenergetic ions (Henig et al., 2009; Steinke et al., 2013). RPA acceleration scheme is a thin nontransparent media being accelerated by reflecting the impinging laser light. The nontransparent media can be a thin film, or a ultra-high density gas jet. The laser radiation pressure pushes the media's electrons forward and these departing electrons pull the ions along via the electric field between them. This is the basis of light sail regime of RPA, which would theoretically be capable of producing quasi-monoenergetic protons. Experiments with RPA show separated protons and carbon ions that accumulate to pronounced peaks at around 2 MeV/u that contain as much as 6.5% of the laser energy (Steinke et al., 2013). However, the RPA acceleration mechanism is still impulsive compared to the acceleration in an RF cavity or traveling wave. Thus, the ion energy spectra still remain with a very large energy spread, which will require energy selection devices to meet the clinical requirement.

The optimum goal of laser ion acceleration is to replace bulky and expensive conventional accelerators for energetic proton or heavier ion generation. A lightweight fiber laser may provide a flexible solution for laser beamline transport (Mourou et al., 2013). Considering the advantage of high optical-to-optical conversion efficiency and the flexibility of optical fibers, assembling arrays of fiber lasers in parallel can be used in the future for laser-driven accelerators. Besides the fiber laser option, the technology of high-repetition-rate laser has become more mature in the past several years. A Yb:YAG laser was demonstrated as being able to deliver 1 J, 500 Hz pulse continuously over 30 min (Baumgarten et al., 2016). With such a Joule level, the Yb:YAG laser can energize 10^9 protons to the energy 5 ~ 10 MeV.

Using ions generated directly from laser target interaction for therapeutic applications still needs further investigations. Conventional RF-based accelerators, especially for heavy ions, can benefit from a compact laser-driven system providing ultrashort low-emittance ion sources as an injector for linacs, synchrotrons, or synchrocyclotrons. Energetic ions driven by an intense laser can be a good ion source for conventional RF accelerators, as demonstrated by the GSI 10 MeV beamline (Steinke et al., 2013). The GSI new beamline demonstrates a good alternative ion source for future particle therapy systems. A laser operated at a 500 Hz repetition rate is potentially capable of delivering protons equivalent to 80 nA to an accelerator. The laser-driven ion source for synchrotrons can be a near-term practical stepping-stone prior to the final realization of laser ion acceleration for particle beam therapy.

9.6 SUPERCONDUCTOR FOR GANTRY BENDING MAGNET

The Loma Linda in Loma Linda, California, facility pioneered the clinical use of a rotating gantry to irradiate proton beams into a patient from different angles. It gives radiotherapy planners flexibility on treatment planning and minimizes the need for rotating patients. A gantry consists of a mechanical support structure, magnets, vacuum vessels, beam diagnostics, and other supporting devices. Most commercial gantries are isocentric. The heaviest part of a conventional gantry is the magnet system, which determines the construction of supporting structures. Typically, the total mass of a conventional isocentric proton gantry weights between 100 and 200 tons. For heavy ions (e.g., carbon) the gantry is significantly heavier and larger. The heavy ion gantry of the Heidelberg Ion Beam Therapy Center (HIT), with its normal conducting magnets, requires a gantry about 15 m in diameter, 19 m in length, and 600 tons in total mass. The magnets alone weigh 135 tons. As a comparison, the total weight of a Loma Linda gantry for proton therapy is about 65 tons.

Using superconducting materials for gantry magnets to reduce the size and weight of a particle therapy system has been investigated by several groups. Lawrence Berkeley National Laboratory, the Paul Scherrer Institute, and Varian are collaboratively developing a superconducting magnet system for a lightweight gantry that will reduce the size and weight by a factor of 10. This project is partially supported by DOE Accelerator Stewardship program. NIRS has installed the first superconducting gantry for heavy ion therapy (Iwata et al., 2013), and ProNova Solutions offers superconducting gantry solutions for its proton therapy systems. The design of modern superconducting magnets for accelerators (dipoles, quadrupoles, and higher multipoles) has reverted to the simple and elegant configurations: to have current distributions located around a cylindrical aperture.

The bending magnet dipole fields of NIRS superconducting magnet are 2.9 ~ 3.3 T, which reduces the radius of bending to about 2 m for 425 MeV/u carbons. As a result, the NIRS superconducting rotating gantry dimensions is about 13 m long, 4 ~ 5.5 m in radius, and 200 tons in weight, as shown in Figure 9.11. The NIRS superconducting gantry magnets are designed to match the beam energy within 200 ms for each of over 200 energy step adjustments for carbon ions between 430 and 56 MeV/u. A Pronova proton therapy system comes with superconducting gantries capable of ramping fields of 2 T for 70 MeV to 4 T for 250 MeV in 1 min.

The nonscaling FFAG gantry using superconducting materials for hadron therapy was proposed as an alternative approach (Trbojevic et al., 2007). The advantage of the FFAG gantry is that there is no need to change the field while the beam energy changes. The disadvantage is that the beam position changes as the beam energy varies. It requires further integration with patient treatment planning and nozzle control. The clinical energy range is larger than the FFAG lattice setting; that is, it requires a number of magnet field settings to cover the entire range and cooperate with scanning magnets for treatment.

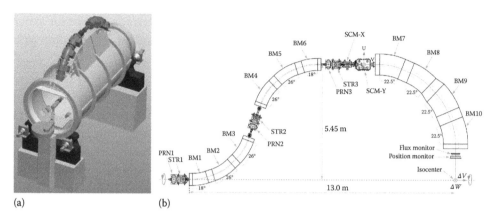

(a) (b)

FIGURE 9.11 (a) Three-dimensional image of the NIRS superconducting rotating gantry for heavy-ion therapy. (b) Layout of the superconducting rotating gantry. The gantry consists of 10 superconducting magnets (BM1-10), a pair of the scanning magnets (SCM-X and SCM-Y), and three pairs of beam profile-monitor and steering magnets (STR1-3 and PRN1-3). (Courtesy Dr. Yoshiyuki Iwata/NIRS.)

9.7 CONCLUSION

A number of technologies have been applied to hadron therapies, as discussed in this chapter. Superconducting cyclotrons gradually surpass normal conducting cyclotrons in new installations. Multi-energy extraction techniques allow a synchrotron to respond to the energy change in a fraction of a second without using any mechanical devices. A technique of energy step adjustment developed at PSI is capable of adjusting beam energy equivalent to the 5 mm range in water within 80 to 100 ms. Installing the PSI energy step adjustment design in a cyclotron system becomes a counterbalance to a synchrotron system on the energy step adjustment time issue.

The current variation between the maximum energy and the minimum energy of a synchrotron is smaller than the variation of a cyclotron system. Both the cyclotron system and the synchrotron system can deliver the clinically required dosage with accuracy within the same timeframe. Besides cyclotrons and synchrotrons, the linac system has become a commercially available solution for proton therapy. Selection of a proton therapy system for a new clinical proton therapy center is usually driven by economic considerations. However, the equipment cost is only one part of the total project expense of a proton therapy center. Reducing the footprint and weight of equipment and minimizing secondary radiation by reducing the shielding wall thickness can also lower the total project cost significantly.

Compact synchrotrons that use normal conductor materials have been in operation at many proton therapy sites. Although a compact synchrotron ring is about 4 to 5 m in diameter, the total weight (approximately 20 tons) and secondary radiation is much lower than those of a cyclotron system (90 ~ 200 tons). Superconducting technologies have shown dramatic reduction in size and weight for cyclotrons. The proposed coreless superconducting cyclotrons will further reduce weight. NIRS and ProNova have demonstrated the benefit of using superconducting materials for gantry magnets.

Novel acceleration schemes with compact lightweight solutions may make hadron therapy system more affordable and accessible. Quasi-monoenergetic ions can be generated from an intense laser interacting with a relatively thin target that is either a gas jet or a thin soil foil. Potentially, a parallelized fiber laser can produce a pulse with an energy of 10 J at approximately a 10 kHz repetition rate. High repetition rate laser and fiber laser technologies may provide the required laser configuration to match the plasma target parameters for delivering repeatable ion beams with specified energy and intensity.

Besides the promised technical performance, FFAG and rapid-cycle synchrotron still need to be competitive in price to be selected for hadron therapy. Superconducting technology raises the bar for non-RF-based accelerators entering the market. Cyclotron-, synchrotron-, and possibly linac-based systems with superconducting components will be the major workhorse of hadron therapy in the foreseeable future. Synchrotrons with superconducting gantries will likely to continue to dominate as heavy ion therapy systems.

REFERENCES

Alonso, J. and T. Antaya. 2012, Superconductivity in medicine, in *Reviews of Accelerator Science and Technology*, A. Chao and W. Chou (Eds.), pp. 227–263. Singapore: World Scientific Publishing Company.

Barlow, R, J. Berg, C. Beard, et al. 2010, EMMA—The world's first non-scaling FFAG, *Nucl. Instr. Meth. Phys. A* 624:1–19.

Baumgarten, C., B. Reagan, M. Pedicone et al. 2016, Demonstration of a compact 500 Hz repetition rate joule-level chirped pulse amplification laser, *Conference on Lasers and Electro-Optics*, San Jose, CA: OSA *Technical Digest*, STu3M.3.

Blumenfeld, I., C. E. Clayton, F. Decker, et al. 2007, Energy doubling of 42 GeV electrons in a metre-scale plasma wakefield accelerator, *Nature* 445:741–744.

Blumlein, A. D. 1941, U K Patent 589127.

Bottura, L. 2000, A practical fit for the critical surface of NbTi, *IEEE Trans. Appl. Supercon.* 10:1054–1057.

Brüning, O., and P. Collier. 2007, Building a behemoth, *Nature* 448:285–289.

Bulanov, S., T. Esirkepov, V. Khoroshkov, et al. 2002, Oncological hadrontherapy with laser ion accelerators, *Phys. Lett. A* 299:240–247

Busold, S., D. Schumacher, O. Deppert, et al. 2014, Commissioning of a compact laser-based proton beam line for high intensity bunches around 10 MeV, *Phys. Rev. STAB* 17:031302–031306.

Caporaso, G. J., S. Sampayan, Y. Chen, et al. 2007, Compact accelerator concept for proton therapy, *Nucl. Instr. Meth. Phys. B* 261:777–781.

Corde, S., E. Adli, J. M. Allen, et al. 2015, Multi-gigaelectronvolt acceleration of positrons in a self-loaded plasma wakefield. *Nature* 524:442–445.

Degiovanni, A., and U. Amalfi. 2014, Proton and carbon linacs for hadron therapy, *Proceedings of 27th International Linear Accelerator Conference (LINAC14)*, Geneva, Switzerland, August 31–September 5, 2014, pp. 1207–1212.

Esirkepov, T., M. Borghesi, S.V. Bulanov, G. Mourou, and T. Tajima. 2004, Highly efficient relativistic-ion generation in the laser-piston regime, *Phys. Rev. Lett.* 92:175003, doi:10.1103/PhysRevLett.92.175003.

Garland, J., R. B. Appleby, H. Owen, and S. Tygier. 2015, Normal-conducting scaling fixed field alternating gradient accelerator for proton therapy, *Phys. Rev. STAB* 18:094701.

Haberberger, D., S. Tochitskyl, F. Fiuza, et al. 2012, Collisionless shocks in laser-produced plasma generate monoenergetic high-energy proton beams, *Nat. Phys.* 8:95–99.

Henig, A., S. Steinke, M. Schnürer, et al. 2009, Radiation pressure acceleration of ion beams driven by circularly polarized laser pulses, *Phys. Rev. Lett* 103:245003.

Heyn, F. A., and K. K. Tat. 1958, Operation of a radial sector fixed-frequency proton cyclotron, *Rev. of Sci. Instruments* 29:662.

Iwata, Y., K. Noda, T. Murakami, et al. 2013, Development of a superconducting rotating-gantry for heavy-ion therapy, *Nucl. Instr. Meth. Phys. B* 317:793.

Jermann, M. 2015, Particle therapy statistics in 2014, *Int. J. Part. Ther.*, 2:50–54.

Kerst, D. W., K. M. Terwilliger, L. W. Jones, and K. R. Symon. 1954, A fixed field-alternating gradient accelerator with spirally ridged poles, MURA-DWK/KMT/LWH/KRS-3, Midwestern Universities Research Association, November 12, Madison, Wisconsin.

Kneisel, P., G. Ciovati, P. Dhakal, et al. 2015, Review of ingot niobium as a material for superconducting radiofrequency accelerating cavities, *Nucl. Instr. Meth. Phys. Res. A*, 774:133–150.

Lennox, A. 1991, Hospital-based proton linear accelerator for particle therapy and radioisotope production, *Nucl. Instr. Meth. Phys. B* 56–57:1197–1200.

Machida1, S., R. Barlow, J. S. Berg, et al. 2012, Acceleration in the linear non-scaling fixed-field alternating-gradient accelerator EMMA, *Nat. Phys.* 8:243–247.

McMillan, E. M. 1945, The synchrotron—A proposed high energy particle accelerator, *Phys. Rev.* 68:143–144.

Mourou, G., B. Brocklesby, T. Tajima, and J. Limpert. 2013, The future is fibre accelerators, *Nat. Photon.* 7:258–261.

Peach, K. 2014, Accelerators for charged particle therapy: PAMELA and related issues, *Int. J Mod. Phys. A* 29: 14020011–14020013.

Peggs, S., D. Barton, J. Beebe-Wang, et al. 2002, The rapid cycling medical synchrotron, RCMS, *Proceedings of 8th European Particle Accelerator Conference (EPAC02)*, Paris, France, pp. 2754–2756.

Snavely, R., M. Key, S. Hatchett, et al. 2000, Intense high-energy proton beams from petawatt-laser irradiation of solids, *Phys. Rev. Lett.* 85:2945–2948.

Steinke, S., P. Hilz, M. Schnürer, et al. 2013, Stable laser-ion acceleration in the light sail regime, *Phys. Rev. STAB* 16:011303.

Symon, K. R., D. W. Kerst, L. W. Jones, L. J. Laslett, and K. M. Terwilliger. 1956, Fixed-field alternating-gradient particle accelerators, *Phys. Rev.* 103:1837–1859.

Tajima T., and J. Dawson. 1979, Laser electron accelerator, *Phys. Rev. Lett.* 43:267–270.

Tanigaki, M., Y. Mori, M. Inoue, et al. 2006, Present status of FFAG accelerators in KURRI for ADS study, *Proceedings of EPAC*, Edinburgh, Scotland, pp. 2367–2369.

Thomas, L. H. 1938, The paths of ions in the cyclotron, *Phys. Rev.* 54:580–598.

Trbojevic, D., R. Gupta, B. Parker, et al. 2007, Superconducting non-scaling FFAG for carbon/proton cancer therapy, *2007 IEEE Particle Accelerator Conference (PAC)*, Albuquerque, New Mexico, NM,pp. 3199–3201.

Veksler, V. 1945, Concerning some new methods of acceleration of relativistic particles, *Phys. Rev.* 69:244–244.

Wagner, F., O. Deppert, C. Brabetz, et al. Maximum proton energy above 85 MeV from the relativistic interaction of laser pulses with micrometer thick CH2 targets, 2016, *Phys. Rev. Lett.* 116:205002.

Wilks, S. C., A. B. Langdon, T. E. Cowan, et al. 2001, Energetic proton generation in ultra-intense laser–solid interactions, *Phys. Plasmas* 8:542–549.

Wilson, R. 1946, Radiological use of fast protons, *Radiology* 47:487–491.

Yin, L., B. J. Albright, B. M. Hegelich, et al. 2007, Relativistic buneman instability in the laser breakout afterburner, Phys. Plasmas 14:056706.

CHAPTER 10

Nonconventional Therapy

Yang-Kyun Park, Taeho Kim, and Siyong Kim

CONTENTS

10.1 INTRODUCTION

Several different external beam cancer therapy approaches are not based on conventional radiation treatment but could be effective. This chapter describes three such nonconventional therapy methods: (1) microbeam radiation therapy (MRT), (2) MRI-guided focused ultrasound therapy, and (3) tumor treating fields (TTFs). The authors believe that these technologies can potentially be effective and are worthy of introduction to readers.

10.2 MICROBEAM RADIATION THERAPY

10.2.1 Background

MRT is a therapy technique using multiple narrow beams with much higher dose, and the regions between the beams have much lower doses. The typical width of the narrow beams is 25–75 μm, and the center-to-center (c-t-c) spacing is 100–400 μm. MRT was of significant clinical and research interest more than 5 decades ago, when it was noted that it achieved better normal tissue sparing while imparting more damage to malignant tissues at high doses compared to broad beam therapy (Curtis, 1967; Slatkin 1992). The approach has received renewed attention recently with the advent of new irradiation technologies. Normal tissue sparing during radiation therapy is especially important for pediatric patients where treatment-related toxicities can significantly degrade quality of life (Rutkowski et al., 2005). MRT remains a promising technique for pediatric cancer patients or highly radiation-resistant tumors (Miura et al., 2006).

Microbeam was first investigated in 1967 to simulate the effects of heavy cosmic-ray particles on astronauts during space travel (Curtis, 1967). A 25 μm wide deuteron microbeam was incident on a mouse skull that was surgically exposed. An unprecedented dose effect was found in this study, in which a microbeam dose up to 3000 Gy left persistent damage to the brain, while broad beam required only 150–300 Gy to do equivalent damage. The hypothesis for this effect was that the blood vessel cells were largely missed by the narrow mircrobeams, and even irradiated cells could be replaced by the division of the nearby undamaged cells. Because the 22 MeV deuteron microbeam could only penetrate 1.5 mm in tissue, the more penetrating X-ray microbeam had a potential advantage over the deuteron-based approach. Straile and Chase (1963) irradiated mouse skin with an absorbed dose of approximately 60 Gy using a 200 kVp and 150 μm-wide X-ray microbeam. They observed much less severe damage in the microbeam group compared to the control group that was irradiated with a seamless 5 mm beam.

Slatkin (1992) hypothesized that the dose ratio between the intensely irradiated regions ("peaks") and the adjacent low-dose regions ("valleys") is an important element for the therapeutic effect of MRT. This peak-to-valley dose ratio (PVDR) is illustrated in Figure 10.1. The higher PVDR was expected to provide smaller normal-tissue toxicity.

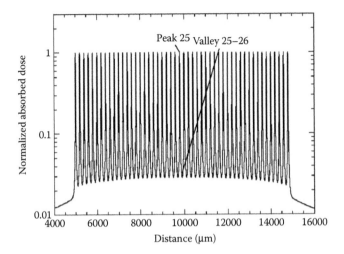

FIGURE 10.1   A Monte Carlo simulated microbeam array dose profile. A phantom was considered to be irradiated by an array of 50 parallel, 25-mm-wide X-ray microbeams. The inter-microbeam distance used was 200 μm. The short oblique line points to the peak absorbed dose for microbeam number 25. The long oblique line (labeled "Valley 25–26") points to the absorbed dose in a valley region. (From Siegbahn, E., et al., *Med. Phys.*, 36, 1128–1137, 2009.)

To keep the PVDR as high as possible, three desired characteristics of microbeam were proposed: (1) high fluence rate, (2) small divergence, and (3) short ranges of secondary electrons in tissue. The high fluence rate is related to short exposure time to avoid the blurring of irradiated zones due to body or organ movements. The small divergence and the short ranges of secondary electrons are also important to keep valley doses below a certain threshold level even when X-ray microbeams penetrate into a deep-seated target region.

A synchrotron-generated X-ray beam was hence considered promising (Slatkin, 1992). First, the synchrotron could generate X-ray beams with a very high fluence rate by using a beam wiggler system (Thomlinson et al., 1988); such a fluence rate would not be tolerated by a conventional X-ray system due to the limited heat capacity of the tube. The small divergence and the short range of secondary electrons could also be satisfied by using 50–600 keV synchrotron X-rays.

Starting from Slatkin's Monte Carlo (MC) study in 1992, many MRT studies have been conducted using synchrotron-generated microbeams (Stepanek et al., 2000; Bräuer-Krisch et al., 2003; De Felici et al., 2005; Spiga et al., 2007; Ohno et al., 2008; Martínez-Rovira and Prezado, 2011; Martinez-Rovira et al., 2012a, 2012b; Prezado et al., 2012). The useful X-ray source for MRT could be produced from a third-generation electron synchrotron such as the European Synchrotron Radiation Facility (ESRF) in Grenoble, France; the Brookhaven National Laboratory in Upton, New York; and Spring-8 in Himeji, Japan.

Slatkin (1992) presented three possible geometries for the microbeam: single cylindrical, multiple cylindrical, and multiple planar. In most studies so far, the multiple planar microbeam has been the most preferred option for practical reasons, such as ease of manufacturing and detection of microscopic tissue lesions after irradiations (Brauer-Krisch et al., 2010). To generate multiple planar microbeams with small divergence and steep dose gradients, a multi-slit collimator (MSC) using a sequence of tungsten slits obtained by electro-erosion has been widely used (Brauer-Krisch et al., 2009).

The use of MSC in conjunction with adequate beam energy and divergence was shown to preserve the sharp dose gradients between peaks and valleys even after 15 cm of penetration of the microbeams in the tissue (Siegbahn et al., 2006). Figure 10.2 illustrates the typical designs of the MSC for microbeam generation.

Many updates on the biological effects of MRT have been made with the use of synchrotron-generated X-rays. Even though fundamental research is still needed to understand the mechanisms of MRT, it seems obvious that MRT can spare microsegmental lesions in the vasculature of normal tissues while more damage remains in tissues with immature, fast-growing blood vessels (Dilmanian et al., 2002; Dilmanian et al., 2003; Serduc et al., 2006). Similarly, in malignant tumor tissues associated with immature and poorly differentiated neovasculature, the repair process after microbeam irradiation may not be as efficient as normal tissue and ischemic necrosis takes place (Blattmann et al., 2005). Therefore, the microbeam is now believed to have a "preferential tumoricidal effect" (Dilmanian et al., 2002), which can be a great advantage for radiation therapy. Regarding the biological findings, however, it should be noted that there have been no human or other primate research to date. Figure 10.3 demonstrates a histologic evidence of cell destruction along the microbeam paths.

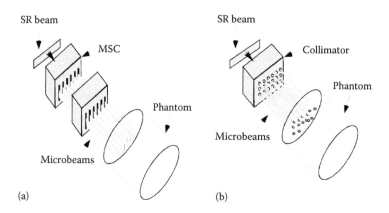

FIGURE 10.2 Illustrations for collimator design of microbeam generated from a synchrotron radiation (SR) beam. (a) A multi-slit collimator (MSC) producing the planar microbeams. (b) A collimator producing cylindrical microbeams. (From Siegbahn, E.A., et al., 2006, *Med. Phys.*, 33, 3248–3259, 2006.)

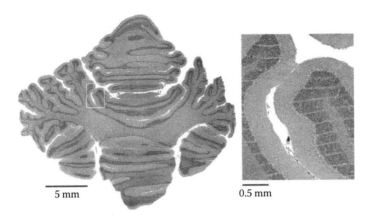

FIGURE 10.3   Piglet cerebellum about 15 months after irradiation (skin-entrance dose was 300 Gy). The 27-μm wide "stripes" of damaged tissue were observed, but no apparent damage to cells just outside the microbeam path was found. (From Laissue, J.A., et al., *Dev. Med. Child Neurol.*, 49, 577–581, 2007.)

10.2.2 Monte Carlo Study on MRT

Due to the need for a very high dose rate (1000 Gy/s at beam entrance), available MRT source is so far limited to synchrotron-wiggler-generated X-rays. An MRT experiment using such a heavy system is not only expensive but also inaccessible for general researchers. MC simulation has been a valuable tool for MRT studies investigating dosimetric characteristics, system design, and performance. Several MC codes are freely available for MRT studies, and they have been well validated through comparison between difference codes and measurements (Slatkin, 1992; Stepanek et al., 2000; De Felici et al., 2005; Siegbahn et al., 2006; Spiga et al., 2007). Therefore, one can conveniently choose one of the benchmarked MC codes, validate his or her customization by comparing it with previous study data with the same simulation parameters, and then perform his or her own MRT simulations. The following sections cover details of MC-based MRT simulation as well as reviews of previous studies.

10.2.2.1 Physics in MRT MC Simulation

Photon and electron interactions with matter are mostly of concern in MC simulations for MRT. Appropriate X-ray energy for MRT ranges from 50 to 200 keV, which can provide sharp fall-off around peak and valley regions (Slatkin, 1992; Stepanek et al., 2000). Therefore, it is crucial to check that an MC package is established for the energy range of the X-ray sources to be simulated. For photon simulations, the relevant physical processes for MRT are mainly photoelectric effects and Compton scattering (Spiga et al., 2007). Coherent scattering (Rayleigh effects) also contributes to the direction changes of photons with relatively low probability. The electron interactions to be taken into account in MRT MC simulation are the elastic scattering and the ionization (Siegbahn et al., 2006). The elastic scattering of low-energy electrons from Compton interactions and photoelectric absorption is of great importance in MRT MC simulation because this determines how far electrons are transported into the valley region (Siegbahn et al., 2006).

Spiga et al. (2007) demonstrated the effects of each physical process on lateral dose profiles of MRT simulations performed with 100 keV, 25 μm diameter beam (see Figure 10.4).

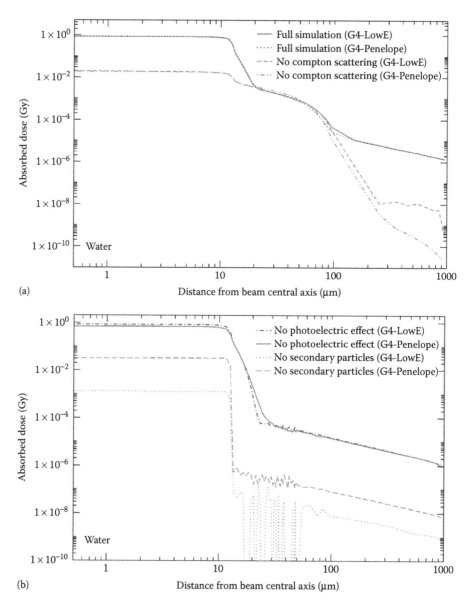

FIGURE 10.4 The curves demonstrating the impact of different types of interactions on the lateral-dose profile. The simulations were performed using both the GEANT4-*LowE* libraries and the GEANT4-PENELOPE model. (a) Lateral profiles from full simulations and from simulations with Compton scattering switched off. (b) Lateral profiles obtained with suppressing photoelectric effect and electron transport. (From Spiga, J. et al., *Med. Phys.*, 34, 4322–4330, 2007.)

In full simulations, two shoulders and a tail are observed along the lateral direction from the collimator edge. When switching off the Compton scattering, the first shoulder is weakened and the tail disappears. The second shoulder is eliminated with photoelectric effect off. Finally, only the rapid fall-off region and the tail remain if secondary electron transport is off.

The X-ray beam used in MRT is a polychromatic beam to yield high fluence rate. To remove the low-energy components of the spectrum, the beam is filtered with an aluminum block (De Felici et al., 2005). In contrast to the characteristic shown in Figure 10.1, where the monochromatic microbeam was simulated, the polychromatic beam generates a smeared dose profile, as shown in Figure 10.5b.

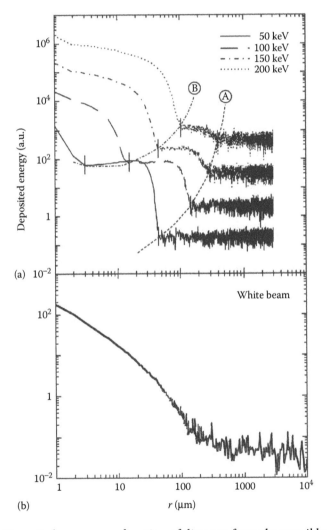

FIGURE 10.5 (a) Deposited energy as a function of distance from the pencil beam calculated for different energies of the incident beam. The (A) and (B) dashed lines indicate the maximum extension range of the photoelectric and Compton electron clouds, respectively. (b) Dose profile generated with the continuous spectrum. It is demonstrated that the energy dependent dosimetric features are all smeared out by the spectrum beam. (From De Felici, M. et al., *Med. Phys.*, 32, 2455–2463, 2005.)

The synchrotron X-ray beam exhibits a linear polarization in the plane of the electron storage ring. This effect is the largest in the first Compton scattering, and its magnitude gradually diminishes as the beam undergoes subsequent scattering processes (De Felici et al., 2005). De Felici et al. demonstrated that the polarization effect is negligible in the center of the microbeam array and causes up to 16% difference at the far edges of the array (De Felici et al., 2005). Even though the polarization effect may not have a significant impact on treatment planning for MRT, better accuracy would be expected if taking it into account in MC simulations. Several MC codes support simulation option for the polarization effect (Anderson et al., 2012; De Felici et al., 2005).

10.2.2.2 Benchmark MRT Studies Using Different MC Codes

In his pioneering article on MRT, Slatkin (1992) first calculated synchrotron X-ray-based MRT dose distribution in water using an early version of EGS4. Following this study, Stepanek et al. (2000) used PSI-version of GEANT3 code and compared their result to that of the previous study. Noticeable discrepancies between the results from the two studies were observed, and they were due to the "single-collision" model in the electron transport simulations employed by PSI-GEANT3 and the "condensed history" model used in the early version of EGS4. De Felici et al. (2007) used KEK version of EGS4, where the accuracy of low-energy electron transport is improved by reducing the size of the electron path step, and they found a good agreement with the previous PSI-GEANT3 results. Siegbahn et al. (2006) first employed PENELOPE for an MRT MC simulation and demonstrated good agreement with the previous results from PSI-GEANT3 and KEK-EGS4. Spiga et al. used GEANT4 and found good agreement with the previous results from PENELOPE, especially when the same library for low energy electron transport was used. Ohno et al. (2008) used PHITS and showed good agreement with their experimental results acquired with a phosphor detector and a charge-coupled device. From these previous studies, it could be concluded that most of the recently available MC codes comply well with MRT MC simulations. Table 10.1 summarizes the typical PVDR results calculated from some of the abovementioned studies.

10.2.2.3 MC Simulation Geometry for MRT

A full geometry of an MRT MC simulation includes a synchrotron storage ring, wiggler magnets, filters, collimators, and phantom/patient. Even though including all the components in a simulation sounds ideal, a lot of time is required to reduce the statistical uncertainty below a certain level (e.g., less than 2%). It was reported that, after traversing all the beam modifiers along 42 m of source to the target distance, only 0.04% of the primary photons generated in the wiggler reached the patient position (Martinez-Rovira et al., 2012a). Therefore, most previous studies considered the photons to be starting on top of the phantom (Slatkin, 1992; Stepanek et al., 2000; De Felici et al., 2005; Siegbahn et al., 2006; Spiga et al., 2007). In those studies, the X-ray microbeam (which is either monochromatic or polychromatic) impinges perpendicularly to the flat surface of the phantom.

Two microbeam geometries can be considered: (1) multiple cylindrical and (2) multiple planar (Slatkin, 1992). While the multiple planar microbeams are more relevant to practical applications, the multiple cylindrical microbeams are still worth considering for a benchmark

TABLE 10.1   Comparison of Published PVDRs from Various MC Codes*

| | | | PVDR | | | |
| | | | Energy (keV) | | | |
Study	MC Code	Depth of Slice (cm)	50	100	150	ESRF Spectrum
Spiga et al. (2007) et al.	GEANT4	0–1	343	743	614	568
		7–8	166	400	363	319
		15–16	161	416	375	347
Siegbahn et al. (2006) et al.	PENELOPE	0–1	286	657	542	504
		7–8	159	387	336	306
		15–16	168	390	347	312
Stepanek et al. (2000) et al.	PSI-GEANT3	0–1	318	763	630	
		7–8	144	386	369	
		15–16	147	368	376	
Slatkin (1992)	INHOM(EGS4)	0–1	513	834	826	
		7–8	173	391	468	
		15–16	111	337	396	

* The PVDRs were calculated at the center of a 1 × 1 cm² bundle of equally spaced 25 μm diameter cylindrical microbeams with the center-to-center distance of 200 μm.

study based on synchrotron-generated microbeams (Stepanek et al., 2000; De Felici et al., 2005; Spiga et al., 2007). The diameter of each cylindrical microbeam ranges from 25 to 75 μm (Spiga et al., 2007; Martínez-Rovira and Prezado, 2011). On the other hand, an array of parallel and rectangular fields is considered for the multiple planar microbeams. The width of the rectangles is comparable to the diameter of the cylindrical beam, and the height corresponds to the array size. C-t-c distance is another important factor determining PVDR. In synchrotron-based MRT, the practical value of c-t-c distance ranges from 200 to 400 μm (Martínez-Rovira and Prezado, 2011; Anderson et al., 2012). The array size of the microbeams may vary according to the size of the treatment region. Its typical value ranges from 0.4 to 3 cm (Anderson et al., 2012). Figure 10.6 illustrates the definitions of the geometry parameters of microbeams.

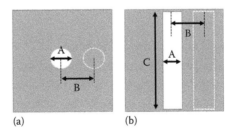

(a)   (b)

FIGURE 10.6   Geometry of microbeam in an MC MRT simulation. (a) Beam geometry for multiple cylindrical microbeams. (b) Beam geometry for multiple planar microbeams. Beam width/diameter, c-t-c distance and beam height are illustrated by A, B, and C, respectively. The microbeam geometry is defined at the top of the phantom where the parallel microbeams enter. For the sake of computational efficiency, a single microbeam dose distribution is superimposed to constitute a full-dose distribution of a microbeam array.

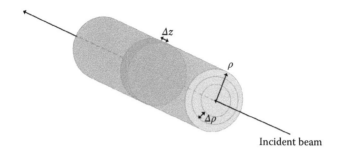

FIGURE 10.7 A dose scoring geometry in the 20-cm diameter, 20-cm long cylindrical phantom used in MC MRT simulations.

A cylindrical phantom with 16–20 cm diameter and 16–20 cm length was used in classical MC MRT simulations (Stepanek et al., 2000; Spiga et al., 2007). The X-ray microbeam impinges perpendicularly to the flat surface of the phantom, as seen in Figure 10.7. The phantom can be composed of either uniform materials such as water and polymethyl methacrylate (PMMA), or even heterogeneous media (Martinez-Rovira et al., 2012a).

In an MC MRT simulation, the dose scoring technique may vary according to the microbeam geometry. Cylindrical shells and parallelepipeds can be used for scoring doses from a cylindrical microbeam and a planar microbeam, respectively (Siegbahn et al., 2006). To score the lateral dose profiles that strongly relate to PVDR, the cumulated dose over a 1 cm depth bin is collected along the radial direction with decreasing resolution at the outer positions (Siegbahn et al., 2006). The relevant depths from the phantom surface would be a potential tumor depth (7–8 cm) and surface (0–1 cm). The percentage depth dose of 200 keV, 25 μm microbeam X-ray at 7–8 cm was found to be 38.4% (Stepanek et al., 2000).

If an MC MRT simulation is performed on a CT image set rather than on a synthetic phantom, a careful approach is needed for dose scoring. Some MC codes associate scoring bin to the mm-sized CT voxel size, so one would not be able to observe peaks and valleys in the radial dose profile. Therefore, decoupling between an image grid and a scoring grid may have to be performed in the transversal direction before the simulation (Martinez-Rovira et al., 2012b).

10.2.2.4 MC Simulations with Complicated Geometry

Even though the most of the previous studies employed the simplified geometry (i.e., beam simulation starts beyond a collimator), few studies have attempted to take more complicated geometries into account to improve the accuracy. Nettelbeck et al. (2009) investigated the effects of beam divergence, finite source size, and the MSC. It was shown that beam divergence has a profound impact on the microbeam profiles, increasing valley dose by 10%. The effects of source size and scattering at MSC on the valley dose were found to be relatively small (less than 5%). Martinez-Rovira et al. (2012a) also included a full geometry in their simulations. To improve the computation efficiency, however, they acquired phase-space state variables (i.e., energy, spatial position, and direction of flight) of the photons before the MSC structure and used it as an input to generate a photon beam model.

10.2.2.5 Dose Superposition and Computation Time

Calculating dose from the full number of microbeams in an array is time consuming and impractical. Therefore, previous studies used the superposition of the dose profile from a single microbeam to generate a full-array dose distribution (Slatkin, 1992; Stepanek et al., 2000; De Felici et al., 2005; Siegbahn et al., 2006; Spiga et al., 2007). This approach has been validated by the Nettelbeck et al. (2009) study, where dose distributions from superposition and a full-array simulation were compared.

In general, MC simulation results are presented with statistical uncertainties. Simulations are discontinued when the statistical uncertainty goes below a certain level. In an MRT MC simulation, the uncertainty varies dramatically between peak and valley regions due to the difference in collection efficiency (Nettelbeck et al., 2009). Therefore, one can design the uncertainty goal of a simulation based on the peak region, the valley region, or an average of both. Typical uncertainty levels of published papers are 0.5% for peak regions (Nettelbeck et al., 2009; Schreiber and Chang, 2012), which corresponds to 2σ, and 2.5% as an average of both regions (Martinez-Rovira et al., 2012a). To achieve the abovementioned uncertainties, 1.0×10^9 to 2.7×10^9 photon histories were simulated in previous studies (Bräuer-Krisch et al., 2003; Nettelbeck et al., 2009; Anderson et al. 2012), mainly determining the computation time. Cutoff energies, so-called ECUT and PCUT for electron and photon transport, respectively, are also crucial factors to determine trade-offs between computation time and accuracy. Because the region of interest in an MRT MC simulation is of µm order, appropriate cutoff energies should be set according to phantom materials (De Felici et al., 2005). Typical ECUT and PCUT values are 10 and 1 keV, respectively (Stepanek et al., 2000). ECUT of 1 keV corresponds to an electron range of approximately 0.1 µm. In some MC codes, a cutoff threshold in length can be set directly (e.g., 1 µm) instead of using the cutoff energies (Spiga et al., 2007).

10.2.3 Compact Microbeam Therapy System Using Carbon Nanotube-Based X-Ray Technology

In the past couple of decades, most MRT studies were conducted in a few synchrotron facilities capable of ultra-high fluence rate X-ray. To facilitate MRT-related research, smaller and more affordable devices that could produce similar beams to the synchrotron-generated X-rays were needed. In 1995, it was found that carbon nanotubes (CNTs) have field emission ability, making them a good electron source (De Heer et al., 1995). This breakthrough discovery prompted a new approach to generating X-rays: one with spatially distributed CNT cathodes that can dramatically increase X-ray fluence without heat-induced damage to the system (Yue et al., 2002). With the potential of the CNT-based X-ray in mind, a multidisciplinary research team from the University of North Carolina (UNC) pioneered a compact MRT irradiation system. Promising results have been shown in preliminary studies (Schreiber and Chang, 2012; Hadsell et al., 2013; Chtcheprov et al., 2014; Hadsell et al., 2014; Zhang et al., 2014), including a biological study (Yuan et al., 2015).

10.2.3.1 Rationale of CNT-Based X-Ray Technology

In 1991, Iijima (1991) first discovered CNTs. A CNT has closed tubular structures consisting of nested cylindrical graphite layers with fullerene-like cap structures and a hollow internal cavity. Several fascinating properties were found in CNT, including semiconduction, extreme stiffness, and suitability as a tip for scanning probe microscopy (Dai et al., 1996; Falvo et al., 1997; Salvetat et al., 1999). CNTs were also found to be good electron sources due to their field emission properties (De Heer et al., 1995). Unlike thermionic electrons produced from a heated metal filament, field emission electrons are generated by a quantum tunneling effect induced by the application of an external electric field (Fowler and Nordheim, 1928; Gomer, 1961). Figure 10.8 illustrates how field emission electrons are generated from a CNT.

The physics of field emission can be formulated in terms of the Fowler-Nordheim (FN) equation (Fowler and Nordheim, 1928), as follows:

$$J = \frac{a\beta^2 E^2}{\phi} \exp\left(\frac{-b\phi^{3/2}}{\beta E}\right)$$

where:
  J is the emission current
  E is the field voltage
  ϕ is the substance-dependent work function
  β is the geometric enhancement factor
  a and b are constants

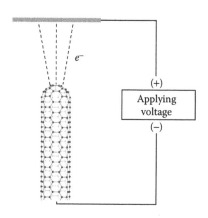

FIGURE 10.8 Illustration of field emission from a CNT. When a potential is applied between a carbon nanotube and a counter electrode, electrons are emitted from the surface of CNT due to a quantum tunneling phenomenon.

Because CNTs have atomic-scale sharp tips and large aspect ratios, the β value for CNTs is much larger than other materials, which results in a much lower emission threshold field even in a compact device (Schreiber and Chang, 2012).

The principle of generating X-ray using CNT technology is similar to that of conventional X-ray tubes except that arrays of CNTs are used as cathodes rather than a single electron gun. Due to its high efficiency in generating electrons, a CNT-based cathode can provide a high degree of flexibility in the design, such as dimension, number of sources, geometric configuration of an individual source, and operation mode (Zhang et al., 2014). CNT-based cathodes can be fabricated in various shapes, such as a segmented pixel array where each segment can be individually controlled (Wang et al., 2007; Bordelon et al., 2008). Because the CNT field emission takes place instantly when a voltage potential is applied, response time to switching on/off is much faster than with a conventional X-ray tube, which could be a great advantage for gated exposure. The abovementioned advantages (i.e., design flexibility, individual controllability, and instantaneous response) of the CNT-based X-ray system first drew attention for its potential use in imaging technology. Using the CNT-based X-ray systems, several studies demonstrated the feasibility of fast volumetric imaging without mechanical motion of sources (Zhang et al., 2005; Maltz et al., 2009; Qian et al., 2012; Gonzales et al., 2013).

10.2.3.2 Compact MRT System Using CNT-Based X-Ray Technology
Even though the initial applications of CNT-based X-rays were for imaging technology, its strong potential of generating high X-ray fluences also drew attention from the MRT research community. The high X-ray fluence from the CNT X-ray system is enabled by the improved design of both the cathode and anode: (1) A spatially distributed cathode array can produce a much higher number of electrons than a thermionic emission cathode can produce (Bordelon et al., 2008; Wang et al., 2011), and (2) with a long-stationary anode design, anode heat can be distributed across a long focal line, efficiently keeping the anode heat loading below a certain limit (Hadsell et al., 2013). Due to the relatively small size of the CNT-based X-ray source assembly, the concept of a compact MRT system was introduced as an alternative to the few large synchrotron-based MRT facilities (Hadsell et al., 2013).

In parallel to the effort to develop a prototype compact MRT system, Schreiber and Chang (2012) performed the first MC simulation study on a compact MRT irradiation system. In their study, a microbeam of 30–200 μm in width and 5–30 mm in length at an 18.6 cm source-to-axis distance was simulated with a copper collimator. Assuming that the highest tube voltage feasible in the prototype is 160 kVp and that the maximum electron current to a segment anode is 1 A, the maximum dose rate of approximately 12 Gy/s per a segment source was calculated to be achievable. Even though this dose rate is much higher than that of a conventional X-ray generator, it is still far below that of the synchrotron X-ray sources, which are more than 1,000 Gy/s. It was projected that additional work, perhaps with a stereotactic beam configuration and physiological gating, would be required to compensate for this disadvantage. For example, a 24-source ring configuration was proposed to deliver 280 Gy/s to an isocenter, as shown in Figure 10.9.

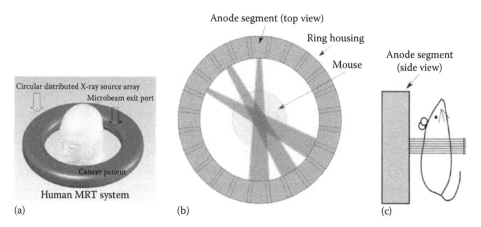

FIGURE 10.9   A multi-source ring configuration to increase the dose rate of CNT-based compact MRT system. (a) The basic concept of the ring-type compact MRT device. The linear circular X-ray microbeam source is located within the circular X-ray tube housing. (b) A top view of the segmented ring with equally spaced 24 linear segment sources. (c) A single linear segment anode producing multiple parallel diverging microbeams. (From Schreiber, E.C. and Chang, S.X., *Med. Phys.*, 39, 4669–4678, 2012.)

The PVDR ranging from 10 to 100 with c-t-c spacing varying from 150 to1000 μm could be achieved in the simulation, which is comparable to that of a realistic synchrotron-based MRT system, where PVDR of 25–65 at c-t-c spacings of 200–400 μm were achieved (Martínez-Rovira and Prezado, 2011).

Hadsell et al. (2014) conducted a pilot study using a CNT-based micro-CT scanner to demonstrate the feasibility of the compact MRT system. They successfully generated a single 220 μm × 25 mm planar microbeam using a steel collimator and obtained multiple peak and valley dose distributions by laterally translating a phantom. The film dosimetry result showed that a PVDR ranging from 5.1 to 9.6 was measured according to the varying c-t-c spacing of 600–1200 μm. Because a low-power and low-energy X-ray tube (0.5 mA and 50 kVp) dedicated to an imaging system was used in this feasibility study, however, the achieved dose rate was only 0.00154 Gy/s. Under the assumption that the tube current and anode voltage can be scaled up to 1.0 A/160 kVp as planned for the prototype MRT device, they calculated an achievable dose rate of 100 Gy/s.

In 2013, the first report on the prototype compact CNT-based MRT system for small animal studies was presented by the UNC research group (Hadsell et al., 2013). In this system, a linear array of CNT cathodes was fabricated to create a linear array of narrow focal line segments on a tungsten anode, as shown in Figure 10.10.

A single-slit collimator was made from two tungsten carbide parallels placed with a fixed gap of 175 μm. A specially designed collimator alignment system aligned the slit to the appropriate microplane of X-ray fields that emanates from the anode focal line. Only a single planar beam is available in this prototype system, so a motorized stage that precisely translates a sample according to the specified c-t-c spacing was employed to provide the equivalent dose distribution to that from multiple planar microbeams. The developed

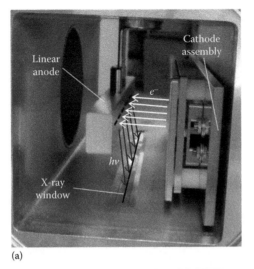

(a)

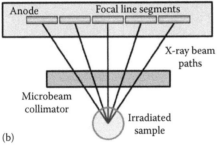

(b)

FIGURE 10.10 (a) Photograph of the inner structure of the compact microbeam irradiator. Indicators are shown for electron trajectories from the cathode assembly (shown as white area in the X-ray window), the location of the segmented focal line on the anode (dashed line in dark gray color), X-ray trajectories from the anode (black arrow marks), and the projected focal line on the window (dark black line). (b) Diagram showing how the multiple line segments with a microbeam collimator produce microbeam paths that irradiates a sample from different angles. (From Hadsell, M. et al., *Appl. Phys. Lett.*, 103, 183505, 2013.)

system was operated with a constant anode voltage of 160 kV and tube current of 70 mA for a single cathode, which yields an entrance dose rate of approximately 2 Gy/s with all 5 cathodes turned on simultaneously. Even though a much higher dose rate was expected from the previous simulation and pilot studies (Schreiber and Chang, 2012; Hadsell et al., 2014), the achieved dose rate was substantially limited, presumably due to the anode heat loading. The developed device was shown to provide microbeams with a width of 300 μm, a c-t-c spacing of 900 μm, and a PVDR of approximately 17. The research group is currently constructing a second-generation compact MRT irradiator that may provide several orders of magnitude increase in dose rate over the present system (Zhang et al., 2014).

The present compact MRT system, with a lower dose rate compared to that of synchrotron microbeams, would cause physiological motion-induced microbeam blurring. To minimize this effect, physiologically gated delivery was proposed by Chtcheprov et al. (2014). Their study demonstrated that the respiratory gated irradiation of a CNT-based

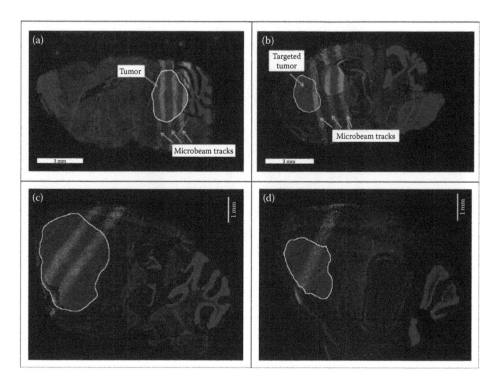

FIGURE 10.11   Sliced brain tissues stained with γ-H2AX after microbeam irradiation. The tissues were irradiated with three microbeams of 48 Gy/beam (a and b), two microbeams of 108 Gy/beam (c), and a single microbeam of 138 Gy (d). The tumor sites are delineated with yellow contours. (From Zhang, L. et al., *Phys. Med. Biol.*, 59, 1283–1303, 2014.)

microbeam is feasible using a respiration sensor that monitors the chest cavity motion of an animal. To localize an animal accurately with respect to the CNT-based microbeam, a rigorous image guidance strategy was proposed (Zhang et al., 2014). Using the proposed image guidance method, promising histological results were achieved in tumor-bearing mice irradiated with the prototype compact MRT system, for which per-beam dose varied from 48 to 138 Gy, as shown in Figure 10.11.

The first biological assessment study using the compact CNT-based MRT system was conducted by Yuan et al. (2015). Consistent with the results from synchrotron MRT studies, significant normal tissue sparing and preferential killing of tumor cells were observed in the mouse group irradiated with the compact MRT system compared to those irradiated with a broad beam.

In summary, the feasibility of a compact CNT-based MRT system with great potential to support broad MRT research has been demonstrated. For some dosimetric characteristics such as the PVDR, c-t-c spacing, and beam width, the CNT-based MRT system was shown to achieve comparable or slightly inferior results to that of synchrotron-based MRT studies. However, the dose rate is much lower than the synchrotron-based MRT (approximately 2 Gy/s versus approximately 1000 Gy/s), even though it is still significantly higher than a conventional X-ray irradiator. The UNC research group is currently focusing on improving the dose rate for their second-generation compact MRT system. After technical improvements are made, more biological studies will be conducted to support the effectiveness of the compact CNT-based MRT system.

10.2.4 Perspectives in the Clinical Application of MRT

Even though MRT has not yet been used for any human patient treatment, it is still drawing the attention of researchers and patients struggling with challenging cancers, such as recurrent tumors and glioblastoma multiform (GBM), where conventional radiotherapy has not been satisfactory. The main hurdles for clinical application of MRT can be summarized as follows:

1. The exact biological mechanism of MRT still remains unknown.

2. Adequate MRT beam sources with ultra-high fluence rate are few worldwide.

3. Technical infrastructures, such as hardware and software, need to be developed for MRT clinical trials.

4. Prior to MRT implementation for human patients, a careful and multidisciplinary evaluation needs to be performed to consider epidemiological, medical, logistical, and ethical aspects (Grotzer et al., 2015).

Currently, clinical MRT applications are actively under development mainly by Australian Synchrotron in Melbourne, Australia, and by ESRF in Grenoble, France (Grotzer et al., 2015).

10.3 MRI-GUIDED FOCUSED ULTRASOUND SURGERY

10.3.1 Background

MRI-guided focused ultrasound surgery (MRgFUS) has recently emerged as a noninvasive thermal ablation technique with MR image guidance (Cline et al., 1992; Hynynen et al., 1993; Jolesz and McDannold, 2008). MRgFUS is the integration of diagnostic MRI with therapeutically focused ultrasound to define and localize target volume, monitor *in vivo* temperature in real time, and confirm local control of the target volume after the FUS procedure (Figure 10.12). MRI utilizes the magnetic properties of the human body in high magnetic fields (1.5–3 T) to produce comprehensive anatomic images for disease diagnosis, treatment planning, and *in vivo* thermometry (Parker et al., 1983; Brown et al., 2014; Abdullah et al., 2010). FUS utilizes high-intensity focused ultrasound to heat and maintain the localized target volume to temperatures above 55°C for varying periods of time longer than 1 s, with a sharp thermal boundary, for healthy tissue sparing (Haar and Coussios, 2007). For precise target localization and thermal boundary management, FUS requires real-time image guidance and thermometry.

In clinical practice, MRgFUS combined with conventional treatment techniques has the potential to improve patient care and can even replace conventional invasive methods. For example, MRgFUS has been utilized as a noninvasive treatment technique in uterine

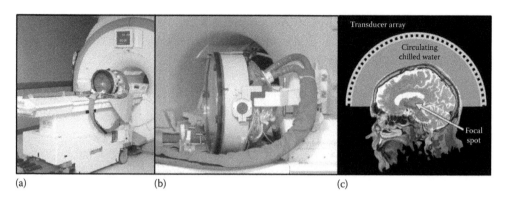

FIGURE 10.12 Examples of MRgFUS systems: (a) ExAblate® Neuro system from Insightec Ltd; (b) transducer with mechanical positioning unit (White paper: Transcranial Magnetic Resonance-guided Focused Ultrasound Surgery); and (c) ultrasound transducer and acoustically transparent fluid bath. (From Ghanouni, P. et al., *AJR Am. J. Roentgenol.*, 205, 150–159, 2015.)

fibroids and bone metastases; additional multiple clinical trials are ongoing (Abdullah et al., 2010; Gianfelice et al., 2008; Ticca et al., 2014). The effectiveness of MRgFUS for uterine fibroids has been reported with the clinical outcome of fibroid shrinkage at 6-month follow-up (Ticca et al., 2014; Lénárd et al., 2008). The nonperfused volume (NPV) of the fibroid immediately after the treatment was also used as a valid predictor of the clinical outcome using contrast-enhanced MR imaging techniques. In addition to tumor ablation, MRgFUS can be used to reduce the pain from bone metastasis. Because the ultrasound energy heats bone tissue faster than soft tissue with minimal energy penetration, MRgFUS is used clinically as a palliative treatment to reduce pain caused by bone metastases (Catane et al., 2007; Liberman et al., 2009). The results of 36 procedures in a multicenter study showed that MRgFUS treatment provided significant pain relief in 72% of the patients (18 out of 25 patients) with the reduction of the average visual analog pain score (VAS) from 5.9 before the treatment to 1.8 at 3-month follow-up (Liberman et al., 2009). In addition to uterine fibroids and bone metastases, multiple clinical trials for other disease sites are ongoing, such as, the clinical application of thermal ablation for prostate, liver, and brain cancer. MRgFUS is an attractive approach for focal treatment similar to image-guided stereotactic radiosurgery (SRS) in radiation oncology.

10.3.2 Components of MRgFUS System

For MRgFUS, several components (radio-frequency coils [RF coils] and a magnet with gradient coils) are added to a conventional MRI system to utilize a high-intensity focused ultrasound system (Abdullah et al., 2010; Haar and Coussios, 2007; Ghanouni et al., 2015). First, a customized MR-compatible table is equipped to support an ultrasound transducer and an acoustically transparent fluid bath (water or light oil) within a standard MR magnet bore. The table includes the powder modules for the transducer and the fluid bath cooling system, as shown in Figure 10.12. Second, the fluid bath located between the transducer and the patient is acoustically transparent, which provides an acoustic window for ultrasound and cooling of the patient's skin. Third, there is a multi-element phased array ultrasound transducer for generating high-intensity focused ultrasound in the 0.5–8 MHz frequency range. The high-intensity ultrasound energy is concentrated on a specific point (focal spot) using the transducer to attain a temperature of over 55°C (60°C–85°C) to disintegrate the tumor. Finally, a closed-loop feedback system combined with real-time MR thermometry is employed to monitor and adjust the treatment procedure in real time. During treatment, 16 thermometry with high spatial and temporal resolution within the targeted tumor volume is important to ensure treatment efficacy and patient safety.

10.3.3 MRI Guidance in FUS

Using MR image guidance has clear benefits compared to other imaging modalities in the FUS procedure: (1) radiation-free imaging, allowing frequent scans and real-time target localization during the procedure (Brown et al., 2014); (2) superior soft-tissue contrast compared to diagnostic ultrasound, which allows accurate target delineation and normal tissue sparing in planning and treatment; (3) endogenous contrast for multicontrast image acquisition by utilizing properties in human tissue; (4) multiparametric imaging to facilitate accurate tissue characterization and assessment of treatment; (5) flexible imaging-plane selection for accurate internal organ motion evaluation; and (6) real-time *in vivo* MR thermometry (Parker et al., 1983), which allows real-time energy deposition management through the feedback devices.

10.3.3.1 MRI in Treatment Planning and Assessment

Target volume as well as the surrounding healthy tissues should be defined precisely in treatment planning and assessment. Tumor location, size, and characteristics can be determined by using multicontrast MR imaging techniques with MR parameters such as T1 (longitudinal relaxation time), T2 (transverse relaxation time), and proton density (e.g., T1-weighted and T2-weighted images). For example, in the uterine fibroids, in T1- and T2-weighted images (displayed in Figure 10.13), fibroids showed homogeneous-lower intensity or heterogeneous-higher intensity compared to skeletal muscle on T2-weighted images (arrow). Due to insensitivity to the MRgFUS treatment, the fibroids with high intensity in the T2-weighted image were excluded in tumor definition during treatment planning (Funaki et al., 2007; Lénárd et al., 2008). Multicontrast MR images can be utilized to assess the presence of any hyperemia within healthy tissue right after the treatment. Furthermore, the diffusion-weighted imaging (DWI) technique can be useful to evaluate structural changes of the treated region with apparent diffusion coefficient (ADC) after the treatment (Abdullah et al., 2010).

Postprocedure verification can be achieved by post-treatment anatomic imaging. The contrast-enhanced MR imaging technique is often employed to identify the tumor extent and assessment of the treatment (e.g., gadolinium-enhanced T1-weighted MRI for fibroid). For example, Figure 10.13a shows a fibroid prior to a treatment in a T2-weighted image; Figure 10.13b shows the fibroid in a contrast-enhanced T1-weighted image with fat saturation right after the treatment. A series of contrast-enhanced T1-wighted images at follow-up can be used to determine NPV, corresponding to the treatment outcome.

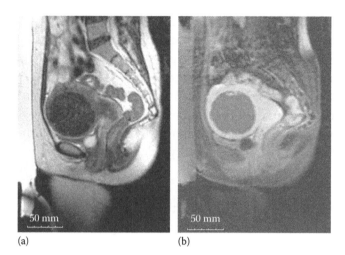

(a)   (b)

FIGURE 10.13   Uterine fibroid MR images. (a) T2-weighted sagittal MR image prior to the MRgFUS treatment. The fibroid shows very low intensity compared to skeletal muscle (arrow). (b) Contrast-enhanced T1-weighted sagittal MR image with fat saturation immediately after the treatment. (From Funaki, K. et al., *Am. J. Obstet. Gynecol.,* 196, 184. e1–e184. e62007.)

After the target and surrounding healthy tissue have been defined, paths of ultrasound should be optimized to improve focused ultrasound delivery efficiency and healthy tissue sparing. The ultrasound paths through the skull should be checked carefully because focused ultrasound can be blurred with intensive energy deposition on the bone (Aubry et al., 2003; Clement and Hynynen, 2002). In conventional MRgFUS treatments, information on a bony structure such as the location and thickness can be obtained from conventional CT images with suitable image contrast of bone and soft tissue for FUS treatments. Bone presents very low intensity (dark) on MR images with conventional MRI techniques, in contrast to CT imaging techniques, because of extremely short T2* value (range of the mean T2*: 0.42–0.50 ms) compared to tissue (range of the mean T2: 40–100 ms) on a 1.5T MR system (Reichert et al., 2005; Nishimura, 2010). However, CT-based treatment planning can introduce uncertainty into the treatment because the patient position in treatment may not be consistent with that on the CT images. In addition, MR images before and during the treatment need to be registered on planning CT images so additional uncertainty can be included into the treatment. Cortical bone imaging using ultrashort-TE (UTE) MR pulse sequences is now available (Reichert et al., 2005). These techniques utilize fast gradient systems to increase detectability of the signal from the bone. Signal from the tissue can be limited due to extremely short echo time acquisition compared to the long T2 value of the tissue. These techniques have not been utilized in clinical practice, and feasibility studies are ongoing (Miller et al., 2015).

10.3.3.2 MR-Thermometry
Real-time *in vivo* thermometry is important to control the thermal volume and to monitor *in vivo* temperature for ablation. In addition to anatomic imaging, MRI can be utilized

to measure *in vivo* temperature changes in real time because variations of MR parameters (such as $T1$, $T2$, proton density, and proton resonance frequency) are correlated with temperature change (Parker et al., 1983). The most common thermometry technique utilizes a proton-resonance frequency (PRF) shift based on proton chemical shifts in MRI (Ishihara et al., 1995). When the temperature increases, the proton chemical shift occurs at lower frequencies due to changes in the hydrogen bonds, which reduces the average PRF in the voxels. In MRI, the phase of gradient-echo images reflects the change of resonance frequency. Therefore, in MRgFUS, the phase of a baseline image is acquired, prior to the thermal ablation procedure, as the baseline temperature of the target volume. Then the temperature changes during the thermal ablation procedure can be measured from the phase differences between images acquired during the procedure and the baseline images (Holbrook et al., 2010). As shown in Figure 10.14, the *in vivo* thermal maps of patients provide quantitative temperature measurements in real time using MR thermometry. The range of temperature is sufficient to infer the range of the thermal ablation. The thermal consistency of the PRF thermometry to various types of tissue demonstrates reliable real-time thermometry for *in vivo* MRgFUS (Ghanouni et al., 2015; Peters et al., 1998). However, this

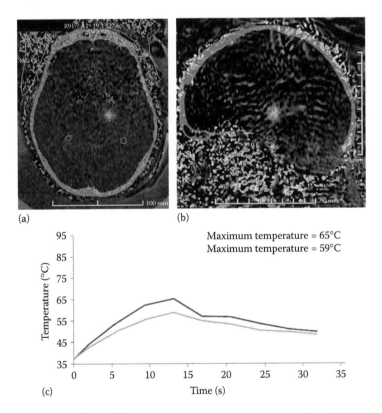

FIGURE 10.14 Example of MR thermometry: temperature maps using the PRF thermometry (a) axial, (b) sagittal planes. Skull image is overlaid on the temperature maps (light-gray shaded area). (c) Focal spot is indicated with red cross-hair (spotted white areas), showing maximum temperature of 3×3 ROI. (From Ghanouni, P., et al., *AJR Am. J. Roentgenol.*, 205, 150–159, 2015.)

technique relies on the phase difference between two image sets, which makes it very sensitive to patient motion. The positional variations due to inadvertent patient motion can cause image mismatch, inducing severe uncertainty in the use of thermometry. Additional care should be taken, such as using patient immobilization and respiratory motion management, to eliminate unexpected phase difference other than temperature changes during the treatment procedure.

10.4 TUMOR TREATING FIELDS

Cancer therapy using TTFs is a novel technology that disrupts tumor cell mitosis by alternating electric fields. A TTF technique (Novo TTF-100A System—Novocure, Portsmouth, New Hampshire) was initially approved in 2011 by the U.S. Food and Drug Administration (FDA) for the treatment of adults with recurrent GBM as an alternative to standard medical therapy after surgical and radiation treatment options were exhausted. On October 5, 2015, the U.S. FDA approved the same technique in combination with temozolomide (TMZ) for the treatment of adults with newly diagnosed, supratentorial GBM after maximal debulking surgery and radiotherapy. The TTF device contains electrically insulated surface transducer arrays and is placed on the patient's skin surface. It is operated by either a portable battery or power supply to produce alternating electric fields within the human body. Patients carry the device in a bag or backpack and receive continuous treatment with minimal interruption of daily life. In this section, the clinical data and mechanism of TTF will be described.

10.4.1 Background

The effect of electric fields on living tissue has been studied in both the laboratory and clinical environments. For instance, alternating electric fields stimulate excitable tissues through membrane depolarization at very low frequencies (under 1 kHz) (Polk, 1995). Stimulations of nerve, muscle, heart, and bone are examples of the practical use of such fields (Bassett, 1984; Palti, 1966; Polk, 1995). At very high frequencies (higher than many MHz), tissue heating becomes dominant due to dielectric losses, and such fields are often used for diathermy and radio-frequency tumor ablation (Chou, 1995; Elson, 1995). Contrarily, intermediate-frequency electric fields are known to be relatively ineffective in both nerve-muscle stimulation due to too-fast alternation and heating due to minute dielectric losses, and such fields of low to moderate intensities are commonly considered to have minimal biological effect (Elson, 1995).

Kirson et al. (et al.2004), however, reported that when low-intensity (less than 2 V/cm), intermediate-frequency (100–300 kHz), alternating electric fields were delivered by means of insulated electrodes for 24 h to cells undergoing mitosis that was oriented roughly along the field direction, they showed a profound inhibitory effect on the growth rate of a variety of human and rodent tumor cell lines (Patricia C, U-118, U-87, H-1299, MDA231, PC3, B16F1, F-98, C-6, RG2, and CT-26). Only dividing cells were affected by this nonthermal effect, while cells at rest were left intact. Two modes of action for these antitumor effects were observed: (1) arrest of cell proliferation, which was manifested by interference with the proper formation of the mitotic spindle, and (2) destruction of cells while undergoing division, which resulted in rapid disintegration of the dividing cells. Directional forces exerted by these specific fields on charges and dipoles within the dividing cells were calculated and found to be consistent with both effects. They also identified unique cellular processes as a result of TTF exposure by utilizing time-lapse microphotography of mouse melanoma cell cultures. Statistically significant prolongation of mitosis was observed in TTF-treated cells. In addition, one-quarter of the treated cells was destroyed. As mentioned, while cellular destruction was observed in mitotic cells, quiescent cells remained intact both functionally and morphologically. Nuclear rotation was also observed in TTF-treated cell cultures. In the form of spatially organized mitotic spindles in dividing cells, microtubules could be disoriented by TTF forces due to their very large electric dipole moments. In the control cell cultures, 95% of the mitotic spindles were intact and exhibited normal features in cells undergoing mitosis compared to 50% of abnormal cell activity in TTF treated cultures. In vivo application to mice having tumors resulted in significant tumor growth reduction and extensive destruction of tumor cells within 3 to 6 days. Based on such results, TTF could be an attractive candidate to serve as a novel treatment modality for cancer.

10.4.2 TTF for GBM and FDA Approval

Kirson *et al.* (2007) performed an industry-sponsored study and found TTF was effective to additional cell lines (human breast carcinoma [MDA-MB-231] and human non-small-cell lung carcinoma [H1299]) and animal tumor models (intradermal B16F1 melanoma and intracranial F-98 glioma). Then, based on such results, they initiated a pilot clinical trial of the effects of TTF in 10 patients with recurrent GBM. Median time to progression (TTP) and median overall survival (OS) in these patients were 26.1 weeks and 62.2 weeks, respectively. Progression-free survival at 6 months (PFS6) of 50% (23%–77%; 95% confidence interval [CI]) was also observed. It was noted by the authors that these TTP and OS values were more than double the reported medians of historical control patients. They did not observe device-related serious adverse events (SAEs) after more than 70 months of cumulative treatment in all of the patients. A mild-to-moderate contact dermatitis beneath the field delivering electrodes was the only but common device-related adverse event (AE) observed (reported in 9 out of 10 participants).

In addition to the study on recurrent GBM (rGBM), Kirson et al. (2007, 2009) also tested TTF for newly diagnosed GBM. While the group of rGBM was treated with TTF only, newly diagnosed GBM patients got TTF combined with TMZ. They were treated for an average of 1 year continuously. When compared to a matched group of 32 concurrent controls who received TMZ alone, progress-free survival (PFS) was significantly different (about 60% versus less than 5% at 100 weeks—hazard ratio (HR) 3.32; 95% CI, 1.9–5.9; $p = 0.0002$), and the difference in OS (about 75% versus about 20% at 30 months) was also significant ($p = 0.0018$). This study was promising and encouraged researchers to pursue prospective trials. Through this study, the authors concluded that combining chemotherapeutic cancer treatment with TTF might increase chemotherapeutic effectiveness and sensitivity without increasing treatment-related toxicity.

A phase III, multinational, and randomized controlled pivotal clinical trial, funded and sponsored by the device manufacturer (Novocure, Ltd), was performed, and results were reported as an abstract in 2010 (Stupp et al., 2010) and as a full paper in 2012 (Stupp et al., 2012). This study intended to have patients age 18 years or older with histologically confirmed GBM and with radiologically confirmed disease progression. Patients had adequate hematologic, renal, and hepatic function with a Karnofsky performance status greater than or equal to 70%. Prior therapy must have included radiotherapy with or without concomitant and/or adjuvant TMZ. Patients with infratentorial tumor were excluded. Patients with implanted electronic medical devices were also excluded. Randomization was made in a 1:1 ratio between TTF therapy and the physician's choice of active chemotherapy (active control). Uninterrupted treatment was recommended for TTF patients, although they were allowed to take treatment breaks of up to 1 h, twice per day. TTF patients were also able to take 2 to 3 days off treatment at the end of each 4-week period, which was the minimal TTF treatment period needed to reverse tumor growth. One full treatment course was a period of 28 days. OS was the primary end point and secondary end points were PFS rate at 6 months, median TTP, 1-year survival rate, quality of life, and radiological response. A total of 237 adult patients from 28 clinical centers were enrolled, with 117 subjects randomized to chemotherapy and 120 to TTF therapy. Four patients in the TTF group and 26 in the chemotherapy group never started on trial. For the 116 patients in the active TTF treatment cohort, an average of 4.2 months of treatment was completed. The median age of patients was 54 years (range 23–80) and Karnofsky performance status was 80% (range 50–100). Actual number of prior treatments ranged from 1 to 6. Marginally higher OS was observed, with the median OS of 6.6 months in the TTF group compared to 6 months in chemotherapy group. Survival rate at 1 year was 20% and 20%, PFS rate at 6 months was 21.4% and 15.1% ($p = 0.13$), respectively, in TTF and active control patients. Mild (14%) to moderate (2%) skin rash beneath the transducer arrays were observed as the TTF-related AEs. SAEs occurred in 6% and 16% ($p = 0.022$) of patients treated with TTF and chemotherapy, respectively. In the analysis of quality of life, TTF was favored in most domains. This trial was not successful in reaching to its primary goal, which was showing the superiority of TTF over chemotherapy. However, as commented by Debonis et al. (De Bonis et al., 2012), it has shown at least the equivalence of TTF to chemotherapy, with decreased toxicity and improved quality of life.

On April 15, 2011, based on the studies described above (i.e., [Stupp et al., 2010] and later [Stupp et al., 2012]), the US FDA approved the premarket approval application (PMA) for NovoTTF™100A System (NovoCure™ Ltd., Portsmouth, New Hampshire; Haifa, Israel). The device is now marketed as Optune™ (NovoCure Ltd., Portsmouth, New Hampshire; Haifa, Israel) which is a portable, noninvasive device designed for the delivery of TTF to the head. Optune (NovoTTF at that time) was approved as a solitary treatment for adults (22 years of age or older) with histologically confirmed, rGBM in the supratentorial region of the brain after receiving chemotherapy. The device was intended to be used as a monotherapy and is intended as an alternative to standard medical therapy for GBM after surgical and radiation options have been exhausted. The expedited PMA included a requirement for a postmarket clinical study in patients with rGBM. The primary question to be addressed by the study was whether the OS of patients treated with NovoTTF-100A would be noninferior to that of patients treated with the best standard of care.

A subsequent randomized clinical trial, supported by the manufacturer, was performed and interim analysis was reported (Stupp et al., 2015). In this study, the authors evaluated the safety and efficacy of TTF in individuals with newly diagnosed GBM following chemoradiation therapy. Between July 2009 and November 2014, a total of 695 subjects from multiple clinical sites were randomized to either TTF with TMZ or TMZ alone, in a 2:1 ratio. PFS in the intent-to-treat (ITT) population (significant threshold, $p \leq 0.01$) was identified as the primary end point, and OS in the per-protocol population ($n = 280$) was a powered secondary end point (significant threshold, $p \leq 0.006$). The interim analysis was based on the first 315 participants who had completed at least 18 months of follow-up. At interim analysis, significant difference of PFS was observed, with the median PFS of 7.1 months (5.9–8.2; 95% CI) in the TTF plus TMZ group and 4 months (3.3–5.2; 95% CI) in the TMZ group. HR was 0.62 (0.43–0.89; 98.7% CI; $p = 0.001$). OS in the per-protocol population, secondary end point, also showed significant improvement, with the median OS of 20.5 months (16.7–25.0; 95% CI) and 15.6 months (13.3–19.1; 95% CI), respectively, in the TTF group and the control group. HR was 0.64 (0.42–0.98; 99.4% CI; $p = 0.004$). Because of the significant difference found in the interim analysis, the study was terminated, and individuals in the control group were offered TTF in addition to TMZ. There were 11 patients who crossed over and began using TTF. The incidence, distribution, and severity of adverse events were similar across both treatment groups, but a higher incidence of localized skin reactions in the TTF plus TMZ group was observed. The investigators stated that the trial met its primary and main secondary end points.

Based on the study by Stupp et al. (2015), on October 5, 2015, the FDA approved an expansion of indication for TTF (i.e., Optune device this time) to treat patients with newly diagnosed, supratentorial GBM. It is administered along with TMZ following maximal debulking surgery and radiation therapy. With this expanded indication, Optune can be used as part of a standard treatment for GBM before the disease progresses. However, for newly diagnosed GBM, it is intended to be used as an adjunct therapy rather than as a substitute for standard treatments. The device is designed to be portable, and it can be powered by batteries or plugged directly into an electrical outlet. It can be used at home or work, allowing the patients to continue their normal daily activities.

On the other hand, Omar (2014) introduced an approach for combining TTF and bevacizumab for rGBM treatment. He stated that, prior to the approval of the TTF system, the only FDA-approved treatment for rGBM was bevacizumab, and it would be worth testing the combination of the two.

Wong et al. (2015) treated a series of patients with TTF and bevacizumab alone ($n = 34$) or in combination with a regimen consisting of 6-thioguanine, lomustine, capecitabine, and celecoxib (TCCC) ($n = 3$). They observed a trend of prolonged OS for the latter cohort compared to the former cohort, with the median of 10.3 (7.7–13.6) months versus 4.1 (0.3–22.7) months ($p = 0.0951$). This study illustrated that it is possible to improve survival and achieve a response in patients with end-stage rGBM.

10.4.3 Other Studies

Salzberg et al. (2008) performed a pilot study using the NovoTTF-100A device in which the safety, tolerability, and effectiveness of TTF treatment were evaluated for a total of 6 patients with locally advanced or metastatic solid tumors. Every patient involved was heavily pretreated with several lines of therapy and had no additional standard treatment option. Individual TTF exposure ranged from 13 to 46 days, and every patient tolerated the treatment well without related serious AEs. Greater than 80% of compliance was also observed. Through the study, 1 patient showed partial response of a treated skin metastasis from a primary breast cancer, tumor growth was arrested during treatment for 3 patients, and 1 patient experienced disease progression. The last patient, who had mesothelioma, showed lesion regression near the treatment field with simultaneous tumor stability or progression in distal areas.

A phase I/II trial of TTF therapy in combination with pemetrexed for advanced non-small-cell lung cancer (NSCLC) was performed by Pless et al. (2010, 2013). In this trial, pemetrexed 500 mg/m² iv q3w together with daily TTF therapy (approximately 12 h/day), applied to the chest and upper abdomen, were given to 42 inoperable stage IIIB (with pleural effusion) and IV NSCLC patients who had had tumor progression. This study was designed to evaluate in-field progression as the primary end point. The median age of the patient group was 63 years. Among all patients, 76% had stage IV disease, 78% had adeno-carcinoma, and 17% had performance status of 2. The median times for reaching in-field progression and systemic progression were 28 weeks and 22 weeks, respectively. A partial remission was observed in 6 (14.6%) patients, while 20 (48.8%) had stable disease. Survival rate at 1 year was 57%, and the median OS was 13.8 months. No TTF-related SAEs were observed. The only device-related AE was mild to moderate dermatitis under the electrodes, which usually improved with topical steroid use. The authors stated that the combination of TTF and pemetrexed as a second-line therapy for NSCLC is safe and potentially more effective than pemetrexed alone.

In 2014, Giladi et al. (2014) reported the results of a study on evaluating the effects of combining TTF with standard chemotherapeutic agents on several NSCLC cell lines, both *in vitro* and *in vivo*. Cell lines tested were Lewis lung carcinoma and KLN205 squamous cell carcinoma in mice. They observed enhanced treatment efficacy across all cell lines with the addition of TTF to chemotherapy (pemetrexed, cisplatin, or paclitaxel). They also found the inhibitory effects of TTF were maximal at 150 kHz for all NSCLC cell lines tested. This study suggests that combining TTF therapy with chemotherapy may provide an additive efficacy benefit in the management of NSCLC.

10.4.4 Mechanism

10.4.4.1 Cell Cycle

The mechanism of TTF effect has not been fully identified. However, observations show that TTF affects cells during their reproduction, especially the mitosis of cell cycle. The cell cycle, through which a cell produces two daughter cells, consists of three states: (1) quiescent, (2) interphase, and (3) cell division. Each state, its substages, and major events are summarized in Table 10.2.

TABLE 10.2   State, Description and Major Event in Cell Cycle

State	Description (Abbreviation)	Major Event
Quiescent	Gap 0 (G_0)	Resting phase
Interphase	Gap 1 (G_1)	Cell growth in size
	Synthesis (S)	DNA replication
	Gap 2 (G_2)	Cell growth in size
Cell division	Mitosis (M)	Orderly division

The cell division state is called mitosis, and it contains four phases: prophase, metaphase, anaphase, and telophase. Through mitosis, a eukaryotic cell separates the chromosomes in its cell nucleus into two identical sets in two nuclei. Errors in mitosis can cause either cell death through apoptosis or mutations that may lead to cancer. Cytokinesis is an event in which cytoplasm is divided into two daughter cells, and it follows mitosis directly. Major events happening in each mitosis phase and in cytokinesis are summarized in Table 10.3 and illustrated in Figure 10.15.

TABLE 10.3   Phase and Major Event in Mitosis and Cytokinesis

Phase	Major Events
Prophase	Chromatin condenses into chromosomes.
Metaphase	Chromosomes line up along metaphase plate.
Anaphase	Chromosomes break at centromeres, and sister chromatids move to opposite ends of the cell.
Telophase and cytokinesis	Nuclear membrane reforms, nucleoli reappear, chromosomes unwind into chromatin. Myosin II and actin filament ring contract to cleave cell in two.

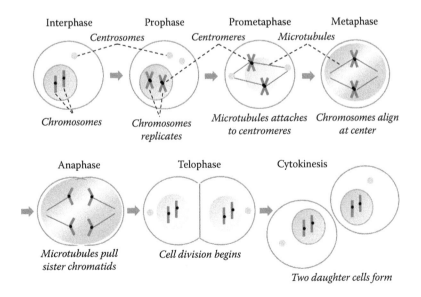

FIGURE 10.15   Illustration of mitosis and cytokinesis.

10.4.4.2 TTF Effect on Cells

TTF likely exerts forces or movement on definable molecular targets that have critical roles in a mitotic process or processes. A protein contains complex charge structures on its surface that are based on how much charges the surface amino acid side chains have. Depending on the arrangement of acidic and basic side chains, regional separations of surface charge of a protein can occur. The protein, with charge separation, develops a dipole moment, which is similar to that observed in bar magnets. Within an electric field, the protein having the dipole moment orients toward the oppositely charged pole of the field. Obviously, a realignment of the protein can be induced by the repolarization of the alternating field, and such proteins would experience rotational forces within the TTF (Kirson et al., 2004). Therefore, proteins that have high dipole moments and are important for mitotic process are considered to be the main targets of TTF perturbation.

It is not clear exactly what proteins are the targets, but there are two clearly observed TTF effects:

1. As illustrated in Figure 10.15, microtubules attach to centromeres and pull sister chromatids in opposite directions. They rely on their electric charges to form. When TTF is applied, however, they struggle to form normally and tend to fail at pulling the chromatids, resulting in either slowing or stopping cell division, which in turn leads to cell death.

2. Once chromatids are pulled in opposite directions, cell division begins, and it starts to take an hourglass shape. This geometry is susceptible to severe perturbation under TTF that can push chromatids back to the center of the dividing cell, resulting in structural damage and cell death.

10.5 SUMMARY

This chapter described three therapy technologies: MRT, MRI-guided focused ultrasound therapy, and TTFs. Although not recognized well, their potential can be significant, they are worth further investigation.

The concept of MRT was introduced almost half a century ago, and its biological effect has been demonstrated in various studies. The biggest technical obstacle in MRT is maintenance of the peak-and-valley dose distribution in the submillimeter range, including instances for patients' physiological motion. Therefore, synchrotron-generated X-rays that yield a huge output still remains the most realistic approach to the clinical application of MRT, despite its slow progress over the years. In seeking a more affordable and accessible microbeam system, however, a compact microbeam therapy system using CNT-based X-ray technology has been recently introduced and is actively under development. Another important task is to improve dose calculation/verification accuracy. MC simulation is an essential method in MRT study because experimental methods are expensive and not easily accessible. Biological mechanism of MRT should also be clearly identified before clinical application.

While the concept of MRT was introduced a long time ago, there are still many technical obstacles to be solved before clinical application. One of the most critical subjects for further research/development is how to make a radiation source/machine that can provide huge output, and is affordable and easy to use in a routine clinical environment. This task is extremely challenging and may require a revolutionary approach.

An active research group is trying to develop a compact microbeam therapy system using CNT-based X-ray technology. Another important area is the improvement of dose calculation/verification accuracy. As can be easily imagined, microbeam dosimetry is considerably more difficult and challenging than macrobeam dosimetry. MC simulation is considered necessary, and diverse MC codes are utilized by many research groups. The biological mechanism of MRT should also be clearly identified before clinical application.

As described, MRgFUS is an emerging technology that integrates MRI with therapeutically focused ultrasound. MRI produces anatomic images and *in vivo* thermometry; meanwhile high-intensity focused ultrasound yields local heating within the target volume. In clinical application, MRgFUS can be used in the combination with conventional treatment techniques and can even replace conventional invasive methods. For example, MRgFUS has been employed for the treatment of uterine fibroids; bone metastases; and prostate, liver, and brain cancer. Assuming such trends continue, the practice of MRgFUS is also likely to increase. Readers are thus encouraged to be more familiar with available hyperthermia outcome database and explore new opportunities for further enhancement.

TTF-based therapy is an interesting emerging technology that utilizes alternating electric fields to disrupt tumor cell mitosis. Several clinical trials have investigated the effectiveness of TTF. A trial for rGBM patients provided results supporting TTF's potential and brought FDA approval on a TTF device. Another clinical trial showed that TTF could be effective for newly diagnosed GBM as well. Multiple trials of applying TTF to various tumors are ongoing. Early results are encouraging, and the approach clearly deserves more scientific and clinical investigations.

Among the three technologies mentioned in this chapter, two are nonionizing radiation therapy methods. The authors intended to describe for readers diverse technologies that are outside the traditional delivery of radiation treatments.

REFERENCES

Abdullah, B., R. Subramaniam, S. Omar, et al. 2010. Magnetic resonance-guided focused ultrasound surgery (MRgFUS) treatment for uterine fibroids. *Biomed Imaging Interv J* 6 (2):e15.

Anderson, D., E. A. Siegbahn, B. G. Fallone, R. Serduc, and B. Warkentin. 2012. Evaluation of dose-volume metrics for microbeam radiation therapy dose distributions in head phantoms of various sizes using Monte Carlo simulations. *Phys Med Biol* 57 (10):3223–3248.

Aubry, J.-F., M. Tanter, M. Pernot, J.-L. Thomas, and M. Fink. 2003. Experimental demonstration of noninvasive transskull adaptive focusing based on prior computed tomography scans. *J Acoust Soc Am* 113 (1):84–93.

Bassett, C. 1984. The development and application of pulsed electromagnetic fields (PEMFs) for ununited fractures and arthrodeses. *Orthop Clin North Am* 15 (1):61–87.

Blattmann, H., J.-O. Gebbers, E. Bräuer-Krisch, et al. 2005. Applications of synchrotron X-rays to radiotherapy. *Nucl Instrum Meth Phys Res Sect A* 548 (1):17–22.

Bordelon, D. E., J. Zhang, S. Graboski, et al. 2008. A nanotube based electron microbeam cellular irradiator for radiobiology research. *Rev Sci Instrum* 79 (12):125102.

Bräuer-Krisch, E., A. Bravin, M. Lerch, et al. 2003. MOSFET dosimetry for microbeam radiation therapy at the european synchrotron radiation facility. *Med Phys* 30 (4):583–589.

Brauer-Krisch, E., H. Requardt, T. Brochard, et al. 2009. New technology enables high precision multislit collimators for microbeam radiation therapy. *Rev Sci Instrum* 80 (7):074301.

Brauer-Krisch, E., R. Serduc, E. A. Siegbahn, et al. 2010. Effects of pulsed, spatially fractionated, microscopic synchrotron X-ray beams on normal and tumoral brain tissue. *Mutat Res* 704 (1–3):160–166.

Brown, R. W., Y.-C. N. Cheng, E. M. Haacke, M. R. Thompson, and R. Venkatesan. 2014. *Magnetic resonance imaging: Physical principles and sequence design*: John Wiley & Sons, Hoboken, NJ.

Catane, R., A. Beck, Y. Inbar, et al. 2007. MR-guided focused ultrasound surgery (MRgFUS) for the palliation of pain in patients with bone metastases—Preliminary clinical experience. *Ann Oncol* 18 (1):163–167.

Chou, C. K. 1995. Radiofrequency hyperthermia in cancer therapy. In *Biologic Effects of Nonionizing Electromagnetic Fields*, pp. 1424–1428. City, ST: CRC Press.

Chtcheprov, P., L. Burk, H. Yuan, et al. 2014. Physiologically gated microbeam radiation using a field emission x-ray source array. *Med Phys* 41 (8):081705.

Clement, G., and K. Hynynen. 2002. A non-invasive method for focusing ultrasound through the human skull. *Phys Med Biol* 47 (8):1219–1236.

Cline, H. E., J. F. Schenck, K. Hynynen, et al. 1992. MR-guided focused ultrasound surgery. *J Comput Assist Tomogr* 16 (6):956–965.

Curtis, H. J. 1967. The use of deuteron microbeam for simulating the biological effects of heavy cosmic-ray particles. *Radiat Res Suppl* 7:250–257.

Dai, H., J. H. Hafner, A. G. Rinzler, D. T. Colbert, and R. E. Smalley. 1996. Nanotubes as nanoprobes in scanning probe microscopy. *Nature* 384 (6605):147–150.

De Bonis, P., F. Doglietto, C. Anile, A. Pompucci, and A. Mangiola. 2012. Electric fields for the treatment of glioblastoma. *Expert Rev Neurother* 12 (10):1181–1184.

De Felici, M., R. Felici, C. Ferrero, et al. 2007. Monte Carlo assessment of peak-to-valley dose ratio for MRT. *Nucl Instrum Methods Phys Res Sect A* 580 (1):489–492.

De Felici, M., R. Felici, M. S. del Rio, et al. 2005. Dose distribution from X-ray microbeam arrays applied to radiation therapy: An EGS4 Monte Carlo study. *Med Phys* 32 (8):2455–2463.

De Heer, W. A., A. Chatelain, and D. Ugarte. 1995. A carbon nanotube field-emission electron source. *Science* 270 (5239):1179–1180.

Dilmanian, F. A., G. M. Morris, N. Zhong, et al. 2003. Murine EMT-6 carcinoma: High therapeutic efficacy of microbeam radiation therapy. *Radiat Res* 159 (5):632–641.

Dilmanian, F. A., T. M. Button, G. Le Duc, et al. 2002. Response of rat intracranial 9L gliosarcoma to microbeam radiation therapy. *Neuro Oncol* 4 (1):26–38.

Elson, E. 1995. Biologic effects of radiofrequency and microwave fields. In *The Biomedical Engineering Handbook*, J.,D. Bronzino (Ed.), pp. 1417–1423. Boca Raton, FL: CRC Press.

Falvo, M., G. Clary, R. Taylor, et al. 1997. Bending and buckling of carbon nanotubes under large strain. *Nature* 389 (6651):582–584.

Fowler, R. H., and L. Nordheim. 1928. Electron emission in intense electric fields. *Proceedings of the Royal Society of London A: Mathematical*. London, UK: Physical and Engineering Sciences.

Funaki, K., H. Fukunishi, T. Funaki, et al. 2007. Magnetic resonance-guided focused ultrasound surgery for uterine fibroids: Relationship between the therapeutic effects and signal intensity of preexisting T2-weighted magnetic resonance images. *Am J Obstet Gynecol* 196 (2):184. e1–e184.

Ghanouni, P., K. B. Pauly, W. J. Elias, et al. 2015. Transcranial MR-guided focused ultrasound: A review of the technology and neuro applications. *AJR Am J Roentgenol* 205 (1):150–159.

Gianfelice, D., C. Gupta, W. Kucharczyk, et al. 2008. Palliative treatment of painful bone metastases with MR imaging–Guided focused ultrasound 1. *Radiology* 249 (1):355–363.

Giladi, M., U. Weinberg, R. S. Schneiderman, et al. 2014. Alternating electric fields (tumor-treating fields therapy) can improve chemotherapy treatment efficacy in non-small cell lung cancer both in vitro and in vivo. *Semin Oncol* 41:S35–S41.

Gomer, R. 1961. *Field Emission and Field Ionization*. Vol. 34. Cambridge, MA: Harvard University Press.

Gonzales, B., D. Spronk, Y. Cheng, et al. 2013. Rectangular computed tomography using a stationary array of CNT emitters: Initial experimental results. *SPIE Medical Imaging*. Proceedings Vol. 8668:86685K-1 to 86685K-8.

Grotzer, M. A., E. Schültke, E. Bräuer-Krisch, *et al.* 2015. Microbeam radiation therapy: Clinical perspectives. *Phys Medica* 31(6):564–567.

Haar, G. T., and C. Coussios. 2007. High intensity focused ultrasound: Physical principles and devices. *Int J Hyperthermia* 23 (2):89–104.

Hadsell, M., G. Cao, J. Zhang, et al. 2014. Pilot study for compact microbeam radiation therapy using a carbon nanotube field emission micro-CT scanner. *Med Phys* 41 (6):061710.

Hadsell, M., J. Zhang, P. Laganis, et al. 2013. A first generation compact microbeam radiation therapy system based on carbon nanotube X-ray technology. *Appl Phys Lett* 103 (18):183505.

Holbrook, A. B., J. M. Santos, E. Kaye, V. Rieke, and K. B. Pauly. 2010. Real-time MR thermometry for monitoring HIFU ablations of the liver. *Magn Reson Med* 63 (2):365–373.

Hynynen, K., A. Darkazanli, E. Unger, and J. Schenck. 1993. MRI-guided noninvasive ultrasound surgery. *Med Phys* 20 (1):107–115.

Iijima, S. 1991. Helical microtubules of graphitic carbon. *Nature* 354 (6348):56–58.

Ishihara, Y., A. Calderon, H. Watanabe, et al. 1995. A precise and fast temperature mapping using water proton chemical shift. *Magn Reson Med* 34 (6):814–823.

Jolesz, F. A., and N. McDannold. 2008. Current status and future potential of MRI-guided focused ultrasound surgery. *J Magn Reson Imaging* 27 (2):391–399.

Kirson, E. D., R. S. Schneiderman, V. Dbalý, et al. 2009. Chemotherapeutic treatment efficacy and sensitivity are increased by adjuvant alternating electric fields (TTFields). *BMC Med Phys* 9 (1):1.

Kirson, E. D., V. Dbalý, F. Tovaryš, et al. 2007. Alternating electric fields arrest cell proliferation in animal tumor models and human brain tumors. *Proc Natl Acad Sci* 104 (24):10152–10157.

Kirson, E. D., Z. Gurvich, R. Schneiderman, et al. 2004. Disruption of cancer cell replication by alternating electric fields. *Cancer Res* 64 (9):3288–3295.

Laissue, J. A., H. Blattmann, H. P. Wagner, M. A. Grotzer, and D. N. Slatkin. 2007. Prospects for microbeam radiation therapy of brain tumours in children to reduce neurological sequelae. *Dev Med Child Neurol* 49 (8):577–581.

Lénárd, Z. M., N. J. McDannold, F. M. Fennessy, et al. 2008. Uterine leiomyomas: MR imaging–guided focused ultrasound surgery—Imaging predictors of success 1. *Radiology* 249 (1):187–194.

Liberman, B., D. Gianfelice, Y. Inbar, et al. 2009. Pain palliation in patients with bone metastases using MR-guided focused ultrasound surgery: A multicenter study. *Ann Surg Oncol* 16 (1):140–146.

Maltz, J. S., F. Sprenger, J. Fuerst, et al. 2009. Fixed gantry tomosynthesis system for radiation therapy image guidance based on a multiple source X-ray tube with carbon nanotube cathodes. *Med Phys* 36 (5):1624–1636.

Martínez-Rovira, I., and Y. Prezado. 2011. Monte Carlo dose enhancement studies in microbeam radiation therapy. *Med Phys* 38 (7):4430–4439.

Martinez-Rovira, I., J. Sempau, and Y. Prezado. 2012a. Development and commissioning of a Monte Carlo photon beam model for the forthcoming clinical trials in microbeam radiation therapy. *Med Phys* 39 (1):119–131.

Martinez-Rovira, I., J. Sempau, and Y. Prezado. 2012b. Monte Carlo-based treatment planning system calculation engine for microbeam radiation therapy. *Med Phys* 39 (5):2829–2838.

Miller, G. W., M. Eames, J. Snell, and J.-F. Aubry. 2015. Ultrashort echo-time MRI versus CT for skull aberration correction in MR-guided transcranial focused ultrasound: In vitro comparison on human calvaria. *Med Phys* 42 (5):2223–2233.

Miura, M., H. Blattmann, E. Brauer-Krisch, et al. 2006. Radiosurgical palliation of aggressive murine SCCVII squamous cell carcinomas using synchrotron-generated X-ray microbeams. *Br J Radiol* 79 (937):71–75.

Nettelbeck, H., G. J. Takacs, M. L. F. Lerch, and A. B. Rosenfeld. 2009. Microbeam radiation therapy: A Monte Carlo study of the influence of the source, multislit collimator, and beam divergence on microbeams. *Med Phys* 36 (2):447–459.

Nishimura, D. G. 2010. *Principles of Magnetic Resonance Imaging.* Hoboken, NJ: Standford University.

Ohno, Y., M. Torikoshi, M. Suzuki, et al. 2008. Dose distribution of a 125 keV mean energy microplanar X-ray beam for basic studies on microbeam radiotherapy. *Med Phys* 35 (7):3252–3258.

Omar, A. I. 2014. Tumor treating field therapy in combination with bevacizumab for the treatment of recurrent glioblastoma. *J Vis Exp* (92):e51638.

Palti, Y. 1966. Stimulation of internal organs by means of externally applied electrodes. *J Appl Physiol* 21 (5):1619–1623.

Parker, D. L., V. Smith, P. Sheldon, L. E. Crooks, and L. Fussell. 1983. Temperature distribution measurements in two-dimensional NMR imaging. *Med Phys* 10 (3):321–325.

Peters, R. T., R. S. Hinks, and R. M. Henkelman. 1998. Ex vivo tissue-type independence in proton-resonance frequency shift MR thermometry. *Magn Reson Med* 40 (3):454–459.

Pless, M., C. Droege, R. von Moos, M. Salzberg, and D. Betticher. 2013. A phase I/II trial of tumor treating fields (TTFields) therapy in combination with pemetrexed for advanced non-small cell lung cancer. *Lung Cancer* 81 (3):445–450.

Pless, M., D. Betticher, M. Buess, et al. 2010. A phase II study of tumor treating fields (TTFields) in combination with pemetrexed for advanced non-small-cell lung cancer (NSCLC). *Ann Oncol.* Vol. 21, Sup. 8:viii125.

Polk, C. 1995. Therapeutic applications of low-frequency sinusoidal and pulsed electric and magnetic fields. In *The Biomedical Engineering Handbook, Second ed., Vol. 2 Set*, J.D. Bronzino (Ed.), pp. 1404–1416. Boca Raton, FL: CRC Press.

Prezado, Y., I. Martiinez-Rovira, and M. Saanchez. 2012. Scatter factors assessment in microbeam radiation therapy. *Med Phys* 39 (3):1234–1238.

Qian, X., A. Tucker, E. Gidcumb, et al. 2012. High resolution stationary digital breast tomosynthesis using distributed carbon nanotube x-ray source array. *Med Phys* 39 (4):2090–2099.

Reichert, I. L., M. D. Robson, P. D. Gatehouse, et al. 2005. Magnetic resonance imaging of cortical bone with ultrashort TE pulse sequences. *Magn Reson Imaging* 23 (5):611–618.

Rutkowski, S., U. Bode, F. Deinlein, et al. 2005. Treatment of early childhood medulloblastoma by postoperative chemotherapy alone. *N Engl J Med* 352 (10):978–986.

Salvetat, J.-P., G. A. D. Briggs, J.-M. Bonard, et al. 1999. Elastic and shear moduli of single-walled carbon nanotube ropes. *Phys Rev Lett* 82 (5):944–947.

Salzberg, M., E. Kirson, Y. Palti, and C. Rochlitz. 2008. A pilot study with very low-intensity, intermediate-frequency electric fields in patients with locally advanced and/or metastatic solid tumors. *Oncol Res Treat* 31 (7):362–365.

Schreiber, E. C., and S. X. Chang. 2012. Monte Carlo simulation of a compact microbeam radiotherapy system based on carbon nanotube field emission technology. *Med Phys* 39 (8):4669–4678.

Serduc, R., P. Verant, J. C. Vial, et al. 2006. In vivo two-photon microscopy study of short-term effects of microbeam irradiation on normal mouse brain microvasculature. *Int J Radiat Oncol Biol Phys* 64 (5):1519–1527.

Siegbahn, E. A., J. Stepanek, E. Bräuer-Krisch, and A. Bravin. 2006. Determination of dosimetrical quantities used in microbeam radiation therapy (MRT) with Monte Carlo simulations. *Med Phys* 33 (9):3248–3259.

Siegbahn, E., E. Bräuer-Krisch, A. Bravin, et al. 2009. MOSFET dosimetry with high spatial resolution in intense synchrotron-generated X-ray microbeams. *Med Phys* 36 (4):1128–1137.

Slatkin, D. N. 1992. Microbeam radiation therapy. *Med Phys* 19 (6):1395–1400.

Spiga, J., E. A. Siegbahn, E. Bräuer-Krisch, P. Randaccio, and A. Bravin. 2007. The GEANT4 toolkit for microdosimetry calculations: Application to microbeam radiation therapy (MRT). *Med Phys* 34 (11):4322–4330.

Stepanek, J., H. Blattmann, J. A. Laissue, et al. 2000. Physics study of microbeam radiation therapy with PSI-version of Monte Carlo code GEANT as a new computational tool. *Med Phys* 27 (7):1664–1675.

Straile, W. E., and H. B. Chase. 1963. The use of elongate microbeams of X-rays for simulating the effects of cosmic rays on tissues: A study of wound healing and hair follicle regeneration. *Radiat Res* 18 (1):65–75.

Stupp, R., A. Kanner, H. Engelhard, et al. 2010. A prospective, randomized, open-label, phase III clinical trial of NovoTTF-100A versus best standard of care chemotherapy in patients with recurrent glioblastoma. *J Clin Oncol* 28 (18_suppl):LBA2007–LBA2007.

Stupp, R., E. T. Wong, A. A. Kanner, et al. 2012. NovoTTF-100A versus physician's choice chemotherapy in recurrent glioblastoma: A randomised phase III trial of a novel treatment modality. *Eur J Cancer* 48 (14):2192–2202.

Stupp, R., S. Taillibert, A. A. Kanner, et al. 2015. Maintenance therapy with tumor-treating fields plus temozolomide vs temozolomide alone for glioblastoma: A randomized clinical trial. *JAMA* 314 (23):2535–2543.

Thomlinson, W., D. Chapman, N. Gmür, and N. Lazarz. 1988. The superconducting wiggler beamport at the national synchrotron light source. *Nucl Instrum Methods Phys Res Sect A* 266 (1):226–233.

Ticca, C., T. Bignardi, F. Zucconi, et al. 2014. Efficacy of MR-guided focused ultrasound surgery (MRgFUS) of uterine fibroids: Evaluation of non-perfused volume (NPV), fibroid shrinkage and clinical improvement at 6 months follow-up. *Transl Cancer Res* 3 (5):413–420.

Wang, S., X. Calderon, R. Peng, et al. 2011. A carbon nanotube field emission multipixel x-ray array source for microradiotherapy application. *Appl Phys Lett* 98 (21):213701.

Wang, S., Z. Liu, L. An, O. Zhou, and S. Chang. 2007. Fabrication and characterization of individually controlled multi-pixel carbon nanotube cathode array chip for micro-RT application for cancer research. *MRS Symp Proc.* 2007 October 1; 1065E: 1065-QQ04-08.

Wong, E. T., E. Lok, and K. D. Swanson. 2015. Clinical benefit in recurrent glioblastoma from adjuvant NovoTTF-100A and TCCC after temozolomide and bevacizumab failure: A preliminary observation. *Cancer Med* 4 (3):383–391.

Yuan, H., L. Zhang, J. E. Frank, et al. 2015. Treating brain tumor with microbeam radiation generated by a compact carbon-nanotube-based irradiator: Initial radiation efficacy study. *Radiat Res* 184 (3):322–333.

Yue, G., Q. Qiu, B. Gao, et al. 2002. Generation of continuous and pulsed diagnostic imaging x-ray radiation using a carbon-nanotube-based field-emission cathode. *Appl Phys Lett* 81 (2):355–357.

Zhang, J., G. Yang, Y. Cheng, et al. 2005. Stationary scanning x-ray source based on carbon nanotube field emitters. *Appl Phys Lett* 86 (18):184104.

Zhang, L., H. Yuan, L. M. Burk, et al. 2014. Image-guided microbeam irradiation to brain tumour bearing mice using a carbon nanotube x-ray source array. *Phys Med Biol* 59 (5):1283–1303.

CHAPTER 11

New Dosimetry Materials, Devices, and Systems

Hosang Jin and Daniel Johnson

CONTENTS

11.1 INTRODUCTION

In radiation therapy, sparing normal tissue and achieving effective tumor control rely on accurate dose delivery. The dose delivery depends on a quality dosimetric protocol, with the aim to deliver the dose within 5% of that prescribed, as recommended by the International Commission on Radiation Units and Measurements (ICRU) (ICRU, 1976).

Each new advance in dosimetry technology can have a significant impact on the routine practice of medical physics and, more important, on patient outcome. As data processing accelerates, so too does the development of new technology, enabling those not directly exposed to these methods and techniques to become luddites within their own field of practice. A prime example is the rapid evolution of patient-centric quality assurance (QA) methods because current implementations of newer QA methodologies vary widely across treatment centers.

Over the last decade, the precision delivery of radiation therapy has rapidly advanced via the introduction of several notable technologies: multimodality image guidance, highly conformal arc delivery, precise robotic gantry and couch control, unflattened photon beams, an enhanced understanding of small field dosimetry, and advanced developments in ion-beam therapies.

Operationally, practitioners adapt and achieve a balance between existing and emergent technologies via three fundamental approaches. First, existing conventional technologies are upgraded and adapted for new clinical applications. A prime example of this adaptive approach is the use of liquid-filled ionization chambers (LICs). Historically, these ionization chambers were utilized for electronic portal imaging devices (EPIDs) commonly used to verify patient positioning (van Elmpta et al., 2008). As technological demands have grown, the role of LICs has expanded to include dose verification. Currently, two-dimensional (2D) arrays of LICs have been adapted to verify dose in small-field intensity-modulated radiation therapy (IMRT) and stereotactic radiosurgery (SRS) treatments (Markovic et al., 2014; Poppe et al., 2013; Knill et al., 2016).

Beyond the adaptation of existing tools, a second method enhances the implementation of conventional dosimeters. Previously, dosimeters were often only capable of single-point measurements of dose distributions. Patient-specific QA has benefited through the use of efficient two- and three-dimensional (3D) diode detector arrays and thermoluminescent dosimeter (TLD) systems (Feygelman et al., 2009; Feygelman, et al. 2011; Gajewski et al., 2013).

Finally, new dosimetry materials and systems, not generally used in radiation therapy, have been developed to measure patient dose precisely. A recent implementation in dosimetry is the use of acoustic computed tomography (CT). The dose delivery plan is verified using acoustic and ultrasound wave measurements for pulsed particle beam delivery (Xiang et al., 2013; Assmann et al., 2015).

To function within clinical specifications, new treatment technologies demand evolution of dosimetry technologies. Each new technology may make a significant impact on the routine practice of medical physics. While many innovations have been developed for QA methods, especially those that are patient specific, few clinical physicists are familiar with methods and techniques to which they are not directly exposed. This chapter introduces new developments in dosimetric materials, devices, systems, and advanced QA methods.

11.2 IONIZATION CHAMBERS

The stability, dose linearity, independence of response from beam quality, and traceability to national calibration laboratories have made the ionization chamber the standard dosimeter in radiation therapy. While ionization chamber arrays are playing an important role in machine and patient-specific QA, their clinical applications are limited by their relatively large active chamber volume. These larger volumes routinely result in a lower accuracy with radiation fields of smaller size or in higher dose gradient regions. This makes the construction of high spatial resolution detector arrays challenging because avoiding partial-volume effects may lead to a decrease in response and precision as detector elements decrease in volume. Conventional 2D ionization chamber arrays do not have a spatial detector density sufficient to measure 3D projections of complex dose distributions accurately.

New developments in LICs offer solutions to the problems of air-filled detectors and advanced applications of 2D ionization chamber arrays for patient-specific QA. Chamber-based time-resolved dosimetry is also developed as a reasonable solution to measure the temporal nature of dose delivery.

11.2.1 Liquid-Filled Ionization Chambers

11.2.1.1 Point Liquid-Filled Ionization Chambers

When compared to gas-filled ionization chambers of the same volume, the LIC exhibits a greater response to dose because the higher density of the liquid greatly increases the ionization density. This allows for sufficiently large signals as reductions in cavity size can be achieved to improve spatial resolution. This has led to the increased utilization of LICs in the area of small-field dosimetry (Chung et al., 2012; Francescon et al., 2012). The liquids used within the LICs are of a density similar to that of water, thus reducing the perturbation effects on the fluence. Any energy dependence is also negligible because the collision mass stopping power ratio of water-to-liquid is nearly independent of the energy of the electrons typically available with medical accelerators.

LICs have a higher response rate, so they are also subject to much larger recombination effects. These effects result from both the recombination of the electrons with their ions of origin, and the recombination of two ions originating from separate ionization events. The potential applied across the chamber can greatly alter the effects of recombination, though it is only the latter recombination that is subject to dose rate dependence (Johansson et al., 1997; Tolli et al., 2010; Andersson and Tolli, 2011). Recombination effects in air-filled chambers are easily evaluated using a two-voltage method (Task Group 21, 1983). The two-voltage method cannot be directly applied to evaluate the collection efficiency of the LICs because no saturation of charge collection occurs at high voltage (Stewart et al., 2007). Alternatively, practitioners can vary the dose rate (dose per pulse) to determine the effect of general recombination (Tolli et al., 2010). The ion's reduced mobility within the nonpolar liquid filling the LICs increases the likelihood of charge recombination prior to collection, which enhances recombination losses. Dose per pulse can depend on field size, distance, and the depth of measurement, which can also alter the recombination effects and hence the collection efficiency.

While small, the volume effect is still a potential problem in LIC dosimetry. Like parallel plate chambers, these devices are generally designed to have a high spatial resolution along their smallest dimension. Results can strongly depend on the type of measurements conducted and the detector's orientation with respect to the beam. Perturbing effects of the liquid, chamber wall, and electrode may affect the detector response, and their influence can be difficult to evaluate.

The commercially available microLion LIC (a thin cylindrical cavity of 2.5 mm diameter and 0.35 mm height) from PTW (Freiburg, Germany) has been modeled with BEAMnrc to investigate the recombination effects of the iso-octane (2.2.4-trimethylpentane) liquid filling (Wagner et al., 2013). Results suggest more than 2% loss in signal between 0 and 1.58 mGy/pulse occurs by recombination, and the largest deviation on uncorrected output factors is 0.35%. Under small field conditions, the deviation of uncorrected output factors by the orientation of the detector can be 2.7%. In addition, these factors are device specific and depend on the size of the source focal spot. Temperature dependence of the iso-octane-filled ionization chamber was also investigated in an approximately $\pm10°C$ interval around the standard 20°C room temperature (Gomez et al., 2014). The chamber exhibited signal variations with temperature in the measured temperature range for two beam qualities, about 0.24%/K for ^{60}Co and about 0.20%/K for 50 kV X-rays.

11.2.1.2 Two-Dimensional Liquid-Filled Ionization Chamber Array

LIC arrays utilize an insulating fluid as the sensitive medium, the higher density of which provides a sensitivity increase as great as 1,000 times compared to air-filled chambers of equivalent volume. This increased sensitivity allows for a significant reduction in detector volumes and an increase in sampling frequency (Johansson et al., 1997).

An early application of the liquid-filled chamber was in a linear array (LA 48, PTW, Freiburg, Germany) of 47 detectors, each with a 4×4 cm^2 area of sensitivity. Introduced in the 1990s, detailed dosimetric analysis of the system concluded that the array was suitable for the measurements of both standard and intensity-modulated dose profiles (Martens et al., 2001).

The PTW Octavius 1000 SRS, a 2D LIC array consisting of 977 detector elements, is constructed in a fashion similar to its precursor, the PTW's LA48 linear array (Poppe et al., 2013). Investigations demonstrate that the Octavius 1000 SRS detector can be highly accurate when measuring clinical photon beams. Because the detector array offers a higher spatial resolution than an air-filled chamber array and minimal energy dependence, it excels in situations where the dosimetry involves steep changing dose gradients, as in IMRT or stereotactic modalities. Though this detector array is not significantly affected by ion-recombination for a standard 6 MV beam, it can be affected by the higher-rate dose pulses from unflattened beams of small aperture sizes.

11.2.2 Ionization Chamber Array Detectors for Patient-Specific QA

11.2.2.1 Two-Dimensional Dosimetry

The conventional 2D dosimetric QA in IMRT and volumetric modulated arc therapy (VMAT) compares a 2D dose map from a treatment planning system (TPS) to a dose distribution measured with 2D detectors such as film, ionization chamber or diode array, or EPID. Most recently, 2D transmission detectors have been introduced for patient-specific IMRT or VMAT QA.

One of the 2D transmission detectors is a flat, translucent, multiwire ionization chamber placed in a holder attached to or permanently installed on a linac (PTW DAVID) (Poppe et al., 2006). In this system, the number of detection wires matches that of Multileaf collimator (MLC) leaf pairs where each wire (tungsten: a diameter of 100 μm) is aligned with the central line of a leaf pair (effective collection volume: 0.03 cm^3/cm). The opening of each leaf pair is monitored by an individual wire, whose signal is proportional to the line integral of ionization density over its length. The wire does not have a spatial resolution; it detects only the integral dose for each leaf pair that correlated to an MLC error, as shown in Figure 11.1 (Poppe et al., 2010). For the daily monitoring of patient treatment, reference values are collected during pretreatment plan verification and compared with readings recorded during daily treatment. Limitations of the transmission detectors are the change of beam quality (attenuation of the beam) and the secondary electron contamination of the photon beam (increase of dose in the surface and buildup regions) (Venkataraman et al., 2009; Poppe et al., 2010).

Another ionization chamber-based transmission detector was developed for the verification of treatment delivery (Islam et al., 2009). This system consists of an area-integrating, energy-fluence monitoring sensor with a one-dimensional (1D) spatial variation in the chamber sensitivity proportional to the electrode plate separation. This "thickness gradient" records the variable response (a change in response of approximately 0.5%/mm) that depends on beam position. For verification of treatment delivery, a precalculated dose-area product using the jaw settings, beam segment shapes, and monitor units is compared to the measured signal. This detector reduces the intensity of a 6 MV beam by 7% and increases the surface dose from 19.5% to 22.5% for a 10 × 10 cm^2 field.

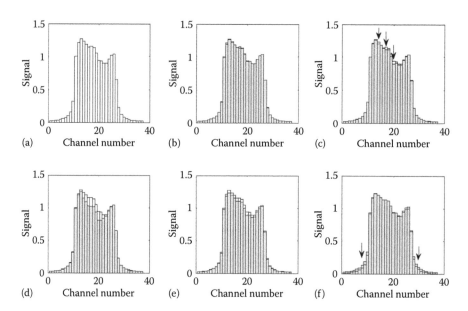

FIGURE 11.1   MLC errors detected by a 2D transmission detector. Opening of a leaf pair monitored by the detector wire for daily treatment is compared to the reference measurement (e.g., pretreatment dosimetric plan). (a) Reference measurement. (b) Third fraction. (c) Fifth fraction with artificial MLC error. (d) Ninth fraction with loss of segment. (e) Twenty-third fraction without reticle. (f) Twenty-fifth fraction with 10 MV. (From Poppe, B., et al., 2010, *Radiother. Oncol.*, 95, 158–165. With permission.)

11.2.2.2 Three-Dimensional Dosimetry

Generally, 2D dosimetric QA provides insufficient information on the highly complex IMRT/VMAT delivery. For in-depth evaluation, it is recommended that the clinical relevance of dose-volume values for targets and critical structures from the dose-volume histogram (DVH) curves be analyzed (Nelms et al., 2011). This can be achieved through dose reconstruction using a 3D volumetric CT scan of phantom or patient using the existing 2D measured dose maps. This method is more desirable than the conventional pretreatment QA techniques because dosimetric errors can be evaluated more thoroughly in 3D space.

To accomplish a 3D dose reconstruction using 2D measurements, 2D transmission detectors can be used to detect the fluence distribution during clinical treatment. A 2D pixel-segmented array of 1,600 air-vented plane parallel chambers (3.8 mm in diameter and 2 mm in height, 6.5 mm apart, which projects to 1 cm at the isocenter) directly mounted on a linac gantry was used as the transmission detector (COMPASS, IBA, Louvain-La-Neuve, Belgium) (Venkataraman et al., 2009; Nakaguchi et al., 2012). In this system, 3D dose distributions are computed with a collapsed cone convolution/superposition algorithm on the patient CT dataset imported in the DICOM RT format from a TPS based on the detected fluence map. The calculated 3D dose distributions are subsequently compared with those from the TPS using the gamma test and DVHs.

Another method of using conventional planar QA measurements is to reconstruct the 3D dose distribution (PTW OCTAVIUS 4D system and its accompanying software,

VERISOFT; Stathakis et al., 2013). Dose is measured with a PTW Seven29 2D ionization chamber array placed at the center of a cylindrical phantom (the Octavius 4D phantom). The phantom rotates synchronously with the gantry via digital inclinometer so that the detector panel is always orthogonal to the beam. The physical rotation of the phantom enables 3D dose reconstruction with the proprietary software for VMAT QA.

The 3D dose is reconstructed in the phantom through a ray segment, connecting a detector to the focus of the beam, through the conversion of percent depth doses (PDDs). The dose D(r) at the distance r from the detector along the ray is given by $D_0 \times PDD(r)/PDD_0$, where D_0 is the measured dose at the current detector, PDD(r) is the PDD at the distance r, and PDD_0 is the PDD at the position of the current detector. The agreement between measured and calculated dose distributions using a volumetric 3D gamma index and 3%/3 mm (dose difference/distance to agreement [DTA]) criteria ranged from 92.3% to 99.9% for all of the 10 patients tested (Stathakis et al., 2013).

To reconstruct the 3D dose in a patient, the water-equivalent depth (WED) is calculated with the patient CT data to account for inhomogeneity along the ray and the irregular shape of the patient surface (Allgaier et al., 2013). The dose in patient is given by $D_{Det} \cdot TPR(WED_{CT})/TPR(WED_{Det}) \cdot (a_{Det}/a_{CT})^2$, where D_{Det} is the dose of the current detector, TPR is a tissue phantom ratio, WED_{CT} is the WED of the current voxel in the patient, WED_{Det} is the WED of the current detector, a_{Det} is the distance between focus and current detector, and a_{CT} is distance between focus and current voxel in the patient. The accuracy of the DVHs determined by the OCTAVIUS 4D has been evaluated through comparison with those exported from TPS. The dose accuracy of DVHs in the patient geometry has been estimated to vary from ± 2% (relatively homogeneous plans) to ± 6% (lung plans).

The advantage of dynamic detector panel rotation is the decrease in the directional dependence of detector response. Main sources of uncertainties for the system include the coarse spatial resolution of the detector panel and the correction-based dose calculation algorithm, known for inaccurate calculations in the presence of lung volumes.

11.3 SOLID-STATE DOSIMETERS

A semiconductor detector can offer both high resolution and signal strength. The physics of free electron production and collection in diode semiconductors is relevant to their clinical use. The process governing radiation-induced charge mobility and collection is an indirect recombination effect. Unlike the ionization chambers, the semiconductor detector's minority charge carrier is captured by an interim recombination center prior to the recombination with a free majority carrier. This leads to a response that depends on instantaneous dose rate, accumulated dose, and temperature (Yorke et al., 2005). Dependencies on angle of incident radiation and average beam energy result from detector design and materials used in the construction of the semiconductor substrate, electrical connections and protective layers that can contain high Z materials (Saw et al., 1998; Jursinic, 2009). Additional effects arise due in part to the inherent shielding used to compensate for their overresponse to the photoelectric photons interacting through silicon (Yin et al., 2004; Griessbach et al., 2005). The observed field-size dependence is generally governed by the buildup used in the diode construction because both the semi-conducting silicon and surrounding materials contribute to the perturbation of the dose response (Rikner and Grusell, 1987).

New solid-state dosimeters aim to address these potential pitfalls, and novel 2D and 3D solid-state detector systems are rapidly being developed for clinical introduction.

11.3.1 New Solid-State Dosimeters

11.3.1.1 In Vivo Spherical Diodes

As *in vivo* dose monitoring and modulated arc therapy gain in popularity in the community, a weak angular dependence of response is a desirable feature for almost any dosimeter. To this end, spherically symmetric diodes have been introduced.

The Sphelar® One diode (Kyosemi Corporation, Kyoto, Japan) has been evaluated for use in radiotherapy (Broisman and Shani, 2011; Barbes et al., 2014). These low-cost diodes were originally designed for solar power collection and optimized for the visible light energy spectrum. This particular diode has a unique geometry, exhibiting an angular dependence lower than that of other diode detectors. At a diameter of nearly 2 mm, the collection efficiency of the Sphelar® One diode is expected to have sufficient sensitivity to allow both high spatial resolution and clinical viability. The variation of response with accumulated dose, accuracy, and reproducibility of the spherical diode is comparable to that of other diode detectors, generally negligible for relative dosimetry in typical clinical settings. Variation with dose rate is no more than $\pm$ 0.4%. When measurements of total scatter factors were normalized to the 10×10 cm^2 field, overresponse in large fields could be corrected empirically, while their utilization as a small field dosimeter demands additional Monte Carlo (MC) simulations of the dependence on field size.

Tests of the Sphelar® One diode with brachytherapy sources demonstrated its versatility in measuring low energy photons (10–40 keV) (Broisman and Shani, 2011). Results compared well with the dose distributions published for each source as determined by methods outlined by the American Association of Physicists in Medicine (AAPM) Task Group 43 (Meigooni, 1995).

11.3.1.2 Synthetic Diamond Detector

The recently available T60019 microDiamond detector (PTW, Freiburg, Germany) is specifically designed for small field dosimetry and marketed as tissue equivalent in megavoltage photon beams of field sizes ranging from 1×1 to 40×40 cm^2. The active layer is sandwiched between a thin aluminum foil electrode and a boron-doped p-type diamond conductor. The active volume of this 1 μ thick, 2.2 mm diameter synthetic diamond disc is 0.004 mm^3.

The manufacturing process of the synthetic diamond material reduces variation between detectors and provides a very small active element size (0.004 mm^3). While the mass energy absorption coefficient of diamond and water are similar, the diamond has a density 3.5 times higher than water, leading to an overresponse in small fields (Scott et al., 2012). The microDiamond detector measured relative output ratios in water for field sizes ranging between 2×2 to 3×3 cm^2 compared well with those determined with fiber-optic scintillation dosimeter, demonstrating a deviation of less than 0.5%. When volume-averaging corrections are not taken into account, the overresponse of the detectors is obfuscated. The synthetic diamond detector has a smaller and more accurately controlled active volume when compared to previous diamond detectors that used natural chips (Ralston et al., 2014).

A study of five microDiamond detectors shows that all displayed similar overresponse, and they varied by approximately 2% under identical, clinical beam conditions (Ralston et al., 2014). Published correction factors may be used in conjunction with volume averaging factors. Some studies of the microDiamond detector contradict these findings. Morales et al. (2014) and Chalkley and Heyes (2014) found that the necessary correction factors were within 1% down to field sizes of approximately 5×5 mm^2 at 6 MV when combining simulated and experimental results. On the other hand, Underwood et al. (2015) demonstrated experimentally that correction factors of almost 6% at 10 MV, and between 4% and 5% at 6 MV, were necessary, findings that aligned more closely with those of Ralston et al. (2014).

With the high-dose-per-pulse beams without flattening filters, collection efficiency drops with dose-per-pulse, a decrease of 2.2% at 2.2 mGy/pulse (Brualla-González et al., 2016). The ion recombination in diamond detectors is a more involved process than in ionization chambers, with impurities playing a significant role. As such, the collection efficiency versus dose-per-pulse is not a linear dependence in the near-saturation region as it is for ionization chambers. An empirical model (Fowler and Attix, 1966) does well to model the relationship. No significant dependence of collection efficiency on the pulse repetition frequency is observed because the decay time of charge carriers in shallow traps is only microseconds (Balducci et al., 2006), orders of magnitude below the millisecond range of the period of pulsed beams used clinically (Brualla-González et al., 2016).

11.3.1.3 Dose Magnifying Glass

The dose magnifying glass (DMG) strip detector consists of an array of 128 phosphor implanted n$^+$ strips on a p-type silicon wafer. The spatial resolution of the DMG is generally on the order of millimeters. A DMG with a 0.2-mm-high spatial resolution was designed and prototyped at the Centre for Medical Radiation Physics at the University of Wollongong (Australia) and proved capable of utilization in patient specific IMRT verification (Wong et al., 2010). This system was subsequently evaluated for dose rate and angular position dependence. A custom phantom was designed to enable accurate and reproducible irradiation conditions that model SRS.

Dose rate changes of 750% and 2,200% over reference conditions lead to a change in response of the DMG that was less than 5% and 11%, respectively. The variations in beam angle were seen to change the response by as much as 15.3%. At isocenter, positioning uncertainty in the DMG was determined to be $\pm$ 0.8 mm, that is, 2% for the smallest SRS cone diameter. SRS cone output factors, S_{cp} scatter factors, profiles, and penumbrae were also measured and agreed well with established data. The maximum deviation between DMG and EBT2 Gafchromic film measurements for the S_{cp} values was 3.8% for a photon SRS cone of 5 mm diameter. An SRS arc delivery of four noncoplanar beams exhibited a maximum difference in profiles of less than 2.5% when compared to radiochromic film. In addition, the real-time data acquisition of DMG arrays allows for timely analysis of small field radiotherapy (Wong et al., 2011).

11.3.1.4 Gallium Nitride–Based Direct-Gap Semiconductor

Gallium nitride (GaN)–based devices have been designed and prototyped for radiation detection (Duboz et al., 2008; Duboz et al., 2009). These detectors consist of multiple layers of nanometer thick GaN and AlGaN. The fixed space charge at the interface of the AlGaN/GaN heterostructures occurs due to the difference in polarization of the materials, resulting in the production of a 2D electron gas (Dimitrov et al., 2000).

Advantages of these devices over retrospective devices such as TLDs include a faster and more convenient analysis procedure. In addition, the active volume of these devices is on the order of 0.001 mm^3 (Hofstetter et al., 2011). GaN is not water equivalent, so there is demonstrable energy dependence in its response. The mass attenuation coefficient of GaN is much higher than for water for low-energy photons, increasing the likelihood of photoelectric interactions. This overresponse to scattered radiation has been studied and addressed through the introduction of a system of multiple detector elements (Pittet et al., 2013; Wang et al., 2013).

A characterization study of GaN dosimeters included testing reproducibility and dependence on dose rate, field size, source-to-surface distance (SSD), hard wedges, incident beam angle, total dose, and temperature. Observed factors requiring correction were changes in field size and temperature. The trend of the response to decrease with increasing temperatures, within the range expected for *in vivo* readings, is related to a change in the GaN luminescence properties as a function of temperature (Reshchikov et al., 2005). To improve the accuracy of measurements, temperature sensitivity may require compensation (Chaikh et al., 2014). As an implantable dosimetry system, the real-time readability allows for the GaN system to be a prime candidate for independent dose verifications of radiotherapy patients under treatment. The in-house prototypes tested exhibit a linear response and negligible angular dependence (Ismail et al., 2011).

11.3.2 Two-Dimensional RADPOS Dosimetry

The ideal *in vivo* dosimeter is both easy to localize and very small. Metal oxide semiconductor field-effect transistors (MOSFETs) have been widely used as such a dosimeter for patient dose verification (Soubra et al., 1994; Marcie et al., 2005; Beyer et al., 2007; Bloemen-Van Gurp et al., 2007). The problem of measurement errors is unavoidable in *in vivo* dosimetry due to the difficulties in accurately measuring point doses (Kohno et al., 2012). To reduce these potential errors, one approach is to monitor the relative position of the MOSFET detector. The radiation positioning system (RADPOS) consists of a MOSFET dosimeter and electromagnetic positioning device (Best Medical Canada, Ottawa, Ontario, Canada), as shown in Figure 11.2 (Kohno et al., 2015). Able to provide position and dose measurements simultaneously, the average absolute deviation of the measured RADPOS position from the actual position is on the order of 1 mm (Cherpak et al., 2009; Cherpak et al., 2011; Cherpak et al., 2012).

Metallic objects near the RADPOS transmitter or sensors can reduce this positioning accuracy. Magnetic and/or electrically conductive materials can distort the transmitted

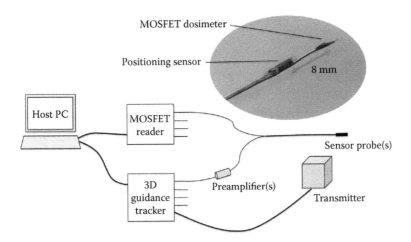

FIGURE 11.2 Schematic of RADPOS patient dosimetry system. (From Kohno, R., et al., 2015, *Int. J. Med. Phys. Clin. Eng. Radiat. Oncol.,* 4, 318–325. With permission.)

signal, which may result in systematic shifts in the measured position. The magnitude of the shift artifact depends on the physical shape of the obstruction, its ferromagnetism, and its electrical conductivity. The highest accuracy is achievable when interfering metals are no closer than 20 cm to the system.

The utilization of the RADPOS system in proton therapy presented the challenge of the close proximity to the patient of the metallic nozzle compensator components (Kohno et al., 2015). It has been demonstrated that, while an increased attention to system setup is necessary when using the RADPOS in proton therapy, RADPOS positions can be measured in real-time during irradiation, and the average absolute deviation of the RADPOS position is within 1.0 mm when following the manufacture's guidelines. The RADPOS system has also found applications in high-dose rate (HDR) gynecological brachytherapy (Reniers et al., 2012). Through the comparisons of the RADPOS positions at the time of imaging and those just prior to treatment, one may assess potential problems in the applicator geometry, allowing for replanning if the changes are deemed clinically relevant. The RADPOS can also measure the dose during delivery and can be compared to treatment planning predictions, allowing for discrepancies to be accounted for in adaptive treatments.

11.3.3 Solid-State Dosimeters for Patient-Specific QA

11.3.3.1 Solid-State EPID-Based Dose Reconstruction

The present generation of EPIDs is based on the active matrix flat panel imager (AMFPI) using noncrystalline, amorphous silicon (aSi). The technique of using exit dosimetry to verify *in vivo* dosimetry with EPIDs has been extensively investigated for patient-specific QA (van Elmpta et al., 2008; Slosarek et al., 2010; Wendling et al., 2012; Mijnheer et al., 2013; Camilleri et al., 2014; Hanson et al., 2014). In this approach, transmitted radiation is detected with EPID through a patient during an actual treatment and dose reconstructed through a back-projection algorithm. The reconstructed dose is compared to distributions from TPS in 3D space.

This technique has a high potential for dose-guided radiation therapy (DGRT) when combined with online cone-beam CT images used for patient setup verification. The EPID is an ideal detector for this type of transit dosimetry because of its high spatial resolution, large detection area, wide dynamic range, and real-time imaging. Challenges persist: aSi EPID-based portal imagers suffer from optical glare and asymmetrical radiation backscatter from the EPID support structure (Ko et al., 2004). Fluence distributions measured during treatment cannot completely resolve the contributions from patient setup, machine output, or other error sources.

In a study, EPID-based *in vivo* dosimetry methods showed a mean deviation between reconstructed (D_{REC}) and calculated (D_{TPS}) dose for 25 pelvic modulated fields delivered of $0.1 \pm 1.0\%$ (Camilleri et al., 2014). The *in vivo* dose reconstructions also showed that mean deviations between D_{REC} and D_{TPS} were $1.0 \pm 2.2\%$ for the 53 patients treated by 3D conformal therapy and $0.3 \pm 2.6\%$ for the 92 patients treated by IMRT. In another study, EPID-based *in vivo* dosimetry showed better agreements between D_{REC} and D_{TPS} than TLD-based *in vivo* dosimetry for all five prostate patients tested (Hanson et al., 2014). The in-house EPID-based *in vivo* dosimetry is being used as the primary method of patient-specific QA at the Royal Marsden Hospital Sutton, United Kingdom.

11.3.3.2 Two-Dimensional Diode Array-Based Dose Reconstruction

A planned dose perturbation (PDP) algorithm was developed to reconstruct 3D dose in patients using the conventional 2D pretreatment QA measurements obtained by the MapCHECK diode detector array (Sun Nuclear Corporation, Melbourne, Florida) (Zhen et al., 2011). Per-beam planar dose QA measurements, DICOM RT plan, structure set, patient CT scan, and 3D planned dose from a TPS are entered into the 3D dose perturbation software (Sun Nuclear 3DVH). PDP uses an absolute dose difference matrix (2D error mask) between a calculated dose distribution in a phantom and that measured during the pretreatment QA. The 3D planned dose in the patient CT scan is perturbed by back-projection of the 2D error masks for all beams that are modified by a contribution modifying function (CMF). The CMF converts the error from the specific phantom geometry to that of the patient using total energy released per unit mass (TERMA) parameters and MC dose kernels designed specifically for Compton effects of high-energy photons. PDP adjusts the 3D dose in the patient plan through the back-projection only if and where dose differences are detected in dosimetry array systems.

To evaluate the accuracy of PDP in patient DVH predictions, 96 head and neck IMRT plans were generated for 24 patients using four error-induced beam models: double MLC transmission, half MLC transmission, shallow MLC penumbra, and very shallow MLC penumbra (Zhen et al., 2011). The error-free plans were used as "virtually delivered" plans. PDP-estimated DVH values using the error mask between error-induced and error-free plans were compared to DVH values from the "true" 3D error-induced planned dose. A median value of difference between PDP estimations and true DVH metrics for six different organs was close to 0 Gy (prescribed target dose of 75 Gy) with a tight interquartile range (≤ 0.3 Gy), indicating that the predicted DVH metrics are very close to the actual value.

This method strongly relies on accuracy of the per-beam planar QA measurement and detector density. Keeling et al. (2014) reported that minimizing setup errors of the detector array and high resolution (doubling the diode detector resolution by shift) for small targets (i.e., PTV < 5 cc) are essential for more accurate evaluation.

11.3.3.3 Three-Dimensional Diode Array Dosimeters

3D dosimeters are ideal instruments to detect dosimetric errors in the highly conformal fields of IMRT and VMAT. The 3D dosimetry systems for IMRT and VMAT QA have been constructed by arranging conventional point detectors about a 3D space. To date, cross-sectional "X" and "O" diode array geometries have been developed (Feygelman et al., 2011).

The "X" detector arrangement (Delta^4, ScandiDos, Uppsala, Sweden) consists of 1,069 p-type silicon diodes (active volume: 1 mm in diameter and 0.05 mm thick) in a 22-cm-diameter cylindrical polymethyl methacrylate (PMMA) phantom (Figure 11.3) (Feygelman et al., 2009). The detectors are arranged on two orthogonal planes: the measurement area of 20 × 20 cm^2 ("main board") on the first plane and two halves ("wings") covering 20 × 10 cm^2 each on the second plane. The detector spacing is 0.5 cm in the central 6 × 6 cm^2 area, and 1.0 cm apart elsewhere.

The "O" detector arrangement (ArcCHECK, Sun Nuclear corporation, Melbourne, Florida) consists of 1,386 n-type diodes with 0.8 × 0.8 mm^2 active measuring area embedded in a 3D cylindrical phantom shell (PMMA with a length of 32.4 cm, an outer diameter of 26.6 cm, and an inner hole diameter of 15.1 cm), as shown in Figure 11.4 (Feygelman et al., 2011). The diodes are positioned at a distance of 10.4 cm from the center with buildup

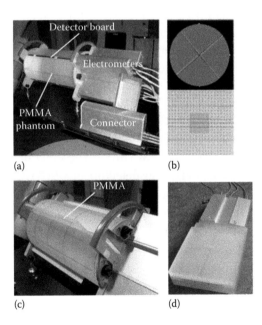

(a)   (b)

(c)   (d)

FIGURE 11.3   A 3D dosimeter of "X" detector arrangement consisting of 1069 p-type cylindrical silicone diodes: (a) measurement setup, (b) CT scan of the device with an axial slice (top) and an oblique reconstruction through the midplane (bottom), (c) the phantom with film along the main board and the wings removed with PMMA slab shown half-inserted, and (d) assembled calibration phantom with the main detector board inserted. (From Feygelman, V., et al., 2009, *J. Appl. Clin. Med. Phys.*, 10, 64–77. With permission.)

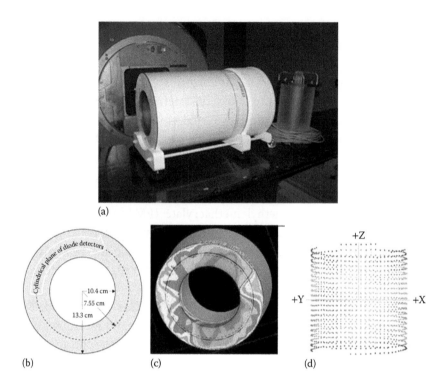

FIGURE 11.4   A 3D dosimeter of "O" detector arrangement consisting of 1386 n-type diodes: (a) overview with an optional PMMA plug on the right, (b) axial cross section of phantom and detector geometry, (c) rotational plan calculated on the virtual model of the device, and (d) 3D view of the diode detector positions. (From Feygelman, V., et al., 2011, *J. Appl. Clin. Med. Phys.*, 12, 146–168. With permission.)

and backscatter of approximately 2.9 cm PMMA each. The diode is 1 cm apart along both the cylindrical length and circumference in a helical pattern. This arrangement increases the detector density in the beam's eye view (BEV). These 3D detectors have comparable gamma pass-rates to conventional 2D array detectors in VMAT QA and can be used effectively for helical Tomotherapy QA (Feygelman et al., 2010; Masi et al., 2011).

Fundamentally, these diode-based dosimeters intrinsically share disadvantages of conventional diodes: angular response dependence, long-term change of sensitivity, dependence of dose rate and temperature, and insufficient spatial resolution. The detector density is also not great enough to adequately cover all clinical cases of highly conformal dose delivery in 3D space. Currently available 3D dosimeter arrangements do not meet the spatial resolution requirements defined by the resolution-time-accuracy-precision (RTAP) criteria (≤ 1 mm^3 spatial resolution, ≤ 1 h imaging time, accurate to within 3%, and within 1% precision) for stringent radiosurgery QA (Oldham et al., 2001). In addition, error sensitivity is substantially dependent on setup errors of the dosimetry systems. The sensitivity of the Delta^4 detector is strongly reduced along the lateral direction (Masi et al., 2011). Another study found that a translational/rotational setup accuracy of 2 mm/2° and 2 mm/1° was required for the "X" and "O" detector arrangement, respectively, to achieve a mean pass rate reduction of less than 3% for VMAT QA with the 3%/3 mm criteria (Li et al., 2013). The 2 mm translational and 1° rotational errors reduced the mean gamma pass rates by 2.4% and 3.8% for "X" and "O" detector arrangements, respectively.

Clinical relevance of measured discrepancies is evaluated by full volumetric dose reconstruction from the phantom measurement to the patient's anatomy. DVH-based metrics in the patient can be calculated from measurements made by the Delta[4] detector (Hauri et al., 2014). An independent pencil beam algorithm commissioned using the 3D dose distributions of open squared fields from a TPS is used to calculate the dose delivered to the patient geometry. The energy fluence used in the pencil beam algorithm is approximated from the measured dose in the detector array. The ArcCHECK detector array is combined with PDP to provide clinically relevant DVH-based patient-specific QA for VMAT or tomotherapy (Infusino et al., 2014).

11.3.4 Time-Resolved Dosimetry

One research group adopted the ArcCHECK detector array with PDP for time-resolved dosimetry (Nelms et al., 2012). The dose information acquired at 50 ms intervals is stored with gantry angle information (incremented between control points) determined by a virtual inclinometer and is used to synchronize all dynamic machine parameters to absolute time because gantry speed may be variable. The time-stamped dose measurement for each diode is used to create a 3D dose correction grid (treatment plan versus interpolated high-density measurement) after obtaining a high-density (2 mm voxel) volumetric dose distribution on the solid cylindrical phantom via convolution of TERMA with 3D dose scatter kernels. The dose correction grid based on the phantom is directly mapped to the TPS dose on the patient dataset using a simple ratio between the plan and the morphed measurement. The 3D cumulative PDP mapping was in good agreement with the four-dimensional (4D) PDP mapping summed for the entire delivery time for individual sub-beams (99.5% gamma passing rate with 0.5% local/1 mm). This work demonstrates potential to resolve errors in dose delivery at each control point for VMAT delivery.

The Delta[4] detector array was also used for the time-resolved measurement with data collected at a rate of 72 Hz (Ravkilde et al., 2013). The phantom is placed on a programmable motion stage, reproducing eight patient-measured tumor trajectories (4 lungs and 4 prostates), and is tracked via electromagnetic tracking system. The dose delivered in each treatment is reconstructed using tracking log files of gantry angle and real-time target position and Dynalog files of MLC positions, gantry angle, and beam-on flag for divisions of VMAT fields as sub-arcs (e.g., 10 sub-arcs of 35.8° for the 358° VMAT field). MLC position information stored in Dynalog files are combined with sub-beams with isocenter shifts (via motion tracking) and imported into a TPS to calculate dose. The measured dose of the phantom is synchronized with the treatment delivery by means of the beam-on flag in the Dynalog files. This can be subsequently compared with the reconstructed 3D dose for each sub-arc and the accumulated dose of the complete arc field. This time-resolved technique makes possible the evaluation of robustness of treatment plans with organ motion.

The time-resolved dosimetry can be also performed using EPID (Yeo et al., 2013). Researchers studied optimal operating conditions (dose rates and time resolution) of EPID for time-resolved dosimetry of IMRT and VMAT using test patterns of an open field, sliding window IM beam, step-and-shoot IM beam, and VMAT beam with various monitor units (MUs) for two dose rates (300 and 600 MU/min). The study revealed that sliding-window IMRT required minimal dose rate fluctuations; irradiation durations for step-and-shoot IMRT should be greater than 8 and 12 s/shot for 300 and 600 MU/min, respectively. In addition, frame averaging (0.95 s for an image) is required to resolve nonuniformity of EPID images for VMAT delivery. Validation of 4D beam delivery may be accomplished by combining time-resolved EPID imaging with the 3D dose reconstruction.

11.4 PLASTIC SCINTILLATION DOSIMETERS

Plastic scintillators manufactured from low-Z materials have been demonstrated to exhibit a nearly water-equivalent response to radiotherapeutic photon and electron beams, matching more closely than air, lithium fluoride, or silicon. Plastic scintillators display minimal temperature dependence; an acceptable resistance to radiation damage (Beddar et al., 1992a); and a stable, reproducible, and linear response to dose (Beddar et al., 1992b). Currently, the light collection efficiency of plastic scintillators is low, and the generation of Cerenkov light within the optical fibers is a significant source of noise for these systems (Archambault et al., 2005). The Cerenkov light generated is estimated to be 0.1% of the magnitude of the scintillation light (Teymurazyan and Pang, 2012), though directly coupling the scintillator fibers to the photodetector can drastically reduce the noise contribution of Cerenkov light by eliminating the production of Cerenkov light in the optical light-guide fiber.

Many research efforts have focused on methods of calibration and correction of optical stem effects for plastic scintillation detectors (PSDs). Established methods used for the correction of optical stem effects in PSD have been through the use of air-core light guides (Liu et al., 2011) and chromatic removal (Fontbonne et al., 2002; Guillot et al., 2011). The process of chromatic removal involves discriminating between the total acquired light signal, comprised of both Cerenkov light and fluorescence generated within the optical fiber, and the component signal, being primarily comprised of light from the scintillator.

Scintillation dosimeters are deemed suitable for use in planar array configurations (Aoyama et al., 1997; Archambault et al., 2007; Bartesaghi et al., 2007; Cartwright et al., 2009; Lacroix et al., 2008; Lee et al., 2008). The sensitive volume of each detector element can be on the order of 1 mm³, allowing for an increased packing density of detector elements and thus an increased spatial resolution. Both MC simulations and experimental studies have confirmed that densely packed scintillator elements do not lead to the perturbation of dose between neighboring scintillators (Naseri et al., 2010; Wang and Beddar, 2011).

New developments of PSDs include *in vivo* PSD, plastic scintillating fiber for small field dosimetry, new PSD arrays, and applications of PSDs in ion-beam therapy.

11.4.1 Point Plastic Scintillation Dosimeters

The OARtrac® (RadiaDyne, LLC., Houston, Texas) is an *in vivo* scintillation dosimetry system developed to measure rectal wall dose during prostate radiotherapy, as shown in Figure 11.5 (Klawikowski et al., 2014). This system utilizes two independent PSDs that, together with a duplex fiber-optical connector, constitute a PSD embedded

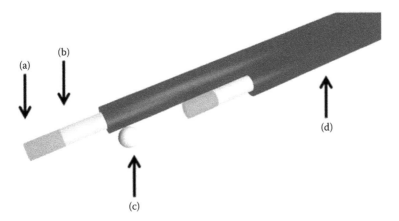

FIGURE 11.5 Scale model of OARtrac probe detector illustrating scintillator fibers and tantalum fiducial placement. (a) Dual 2 mm scintillating fibers with 15 mm separation. (b) Plastic optical fibers transmit emitted light to CCD camera. (c) A 1.0 mm diameter tantalum fiducial for visualization on computed tomographic images. (d) Polyethylene jackets prevent external light contamination. (From Klawikowski, S. J., et al., 2014, *Phys. Med. Biol.*, 59, N27–N36. With permission.)

within a disposable endorectal balloon assembly. A fiber-optic cable transmits detector signals out of the treatment vault, where a charge-coupled device (CCD) camera measures the scintillator output signal.

An accredited dosimetry calibration laboratory (ADCL) independently calibrates each probe as an integral part of the manufacturing process. These probes are sealed within an endorectal balloon post calibration. The disposable nature of the PSD assemblies increases convenience and clinical hygiene. The OARtrac® system has the average dose deviation of only 0.01% from the expected value at dose levels of 200 cGy within a clinical setting and an accompanying overall error of ± 5.4% at the 95% confidence interval (± 2.7% at the 1σ interval).

The Exradin W1 (Standard Imaging Inc., Middleton, Wisconsin) is another commercial dosimetry system based on a fiber-coupled organic PSD, as shown in Figure 11.6 (Beierholm et al., 2014). Provided as a solution for small field dosimetry and routine QA applications, characterization of the W1 has been performed by only a few selected groups (Bourgouin et al., 2012).

A comparison study between the W1 and a custom-built PSD system showed similarities in design, composition, and chromatic-removal-based stem effect correction. Differences were seen in the methods employed for optical spectrum discrimination and the acquisition of data. The characterization of the commercial Exradin W1 through this comparison demonstrated reproducibility and intra-use variability both within 0.4% and a dose rate independence within 0.5%. Variations of the correction coefficients with beam quality stress the need for a stem-effect baseline correction for any change in beam quality.

A dose-to-water calibration must be performed each time a PSD system is used for any absolute dose measurements because gain-correction coefficient variability can be as great as 3.0%. Experimental results demonstrate conflicting findings when investigating the comparison between fiber-coupled organic PSDs and ionization chambers when

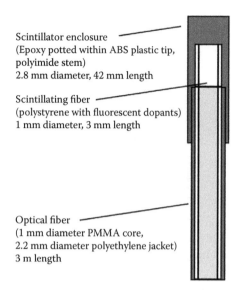

Scintillator enclosure
(Epoxy potted within ABS plastic tip,
polyimide stem)
2.8 mm diameter, 42 mm length

Scintillating fiber
(polystyrene with fluorescent dopants)
1 mm diameter, 3 mm length

Optical fiber
(1 mm diameter PMMA core,
2.2 mm diameter polyethylene jacket)
3 m length

FIGURE 11.6 Cross-sectional schematic of Exradin W1. (From Beierholm, A. R., et al., 2014, *Radiat. Meas.*, 69, 50–56. With permission.)

measuring output factors in large fields (Guillot et al., 2011). These studies emphasized the necessity of minimizing the exposure of optical fiber to the field. The performance between detection techniques is more consistent for field sizes below 10×10 cm. The W1 and Farmer chamber may differ by as much as 1.5% when measuring reference dose to water in different beam qualities, possibly indicating a systematic uncertainty in response with beam quality. This brings into question the value of PSDs to measure k_Q factors accurately for ionization chambers.

The Exradin W1 scintillator exhibits dosimetric characteristics that make it a reasonable alternative to diodes for relative dosimetry. Its small dimensions make the W1 especially appropriate for small field sizes. The W1 scintillator has an edge over diodes in its water equivalency, resulting in a response independent from energy for the higher energy X-rays and electron beams. Temperature correction factors keep uncertainties around 2% when applied during measurements. While commercially available, the W1 has yet to be adapted for use within a scanning water tank. In measuring small fields, the optical cable experiences relatively rapid changes in dose. These scenarios are likely to require unique Cerenkov correction factors that may be additionally dependent on the way that scanning is performed and the field size being scanned (Beierholm et al., 2014).

11.4.2 Plastic Scintillation Dosimeter Arrays

11.4.2.1 Plastic Scintillator-Based EPIDs

The typical EPID indirectly detects radiation via a copper plate and phosphor (Gd_2O_2S:Tb) screen, where photon and electron interactions deposit energy, resulting in the emission of scintillation photons, a fraction of which are subsequently detected by an underlying array of optically sensitive a-Si photodiodes. This design fundamentally limits the EPID to a detective quantum efficiency (DQE) of approximately 1% (El-Mohri et al., 2001).

While removal of the metal plate and phosphor layers allows for the direct detection of the incident radiation with a near water-equivalent response (Sabet et al., 2010), the detection efficiency decreases by nearly 90% in the DEQ of approximately 1% (Vial et al., 2008). Replacing the phosphor layer with segmented high-Z scintillators may increase the DQE by greater than 20% for MV photons, exhibiting virtually zero spatial frequency (Sawant et al., 2006; Wang et al., 2009). Through the use of thicker scintillators, one can increase the DQE, and segmentation allows for high spatial resolution; yet the high-Z scintillators incorporated into these designs results in a fundamentally non-water-equivalent detector.

Prototype EPIDs under investigation utilize plastic scintillator fiber arrays coupled directly to photodiodes (PSA-EPIDs), providing simultaneous imaging and dosimetry information (Blake et al., 2013). In initial testing, the imaging abilities of the standard EPID, employing a copper plate and Gd_2O_2S:Tb phosphor screen, are superior to those of the scintillation prototype. MC simulations have demonstrated the use of plastic scintillating fibers may make it feasible to construct a water-equivalent EPID with both better energy response and higher detection efficiency than the flat-panel-based EPIDs currently available (Blake et al., 2014). The increase in the detection efficiency of the PSA-EPIDs as a function of optical fiber length is partially counteracted through a concurrent reduction in sensitivity to X-rays. Because the energy of peak efficiency increases with fiber length, it is likely that self-absorption of optical photons within the fibers is a leading factor for decreased efficiency for lengths approaching 50 mm. When using thicker scintillators, improving detection efficiency comes at the cost of decreasing spatial resolution. An extramural absorber to prevent optical crosstalk, in addition to a scintillator core with an increased Rayleigh scattering length, is recommended to maximize both detection efficiency and spatial resolution. Simulated optical transport parameters demonstrated no significant influence in the PSA–EPID's water-equivalent dose response, allowing for an optimization of the imaging capabilities of the PSA–EPID.

11.4.2.2 Photomultiplier Tube Array Dosimetry

Modern fiber-optic-based scintillation arrays may utilize photodiodes (Letourneau et al., 1999; Therriault-Proulx et al., 2011) or CCD elements (Cartwright et al., 2009; Lacroix et al., 2009) for the detection of the optical signal. Photodiode arrays assign a single photodiode to each optical channel and have been seen to exhibit responses that are highly temperature dependent (Letourneau et al., 1999). The greater detector area of the CCD chip allows for multiple optical fibers to be interfaced via a single objective lens. Both photodiode and CCD detector exhibit a relatively low sensitivity, requiring acquisition times of over 5 s to obtain an adequate signal. This renders these technologies a poor choice when detecting rapidly changing signals (Archambault et al., 2006).

Photomultiplier tubes (PMTs) are adept at accurately measuring low-intensity optical signals and exhibit both higher signal-to-noise ratios (SNRs) and readout speeds than either photodiodes or CCD chip systems. Their use in array-type systems has been limited because each array element demands an individual photocathode/anode and electrometer to measure the signal current. This setback has been potentially alleviated with the commercial introduction of fast-response multichannel electrometers (Vertilon Corporation,

Westford, Massachusetts), and the release of multi-anode PMTs, in which multiple photocathode/anode pairs are located within a single envelope (Hamamatsu Photonics K.K., Shizuoka, Japan).

Recent developments in coupling these novel technologies have renewed the interest in the creation of PMT-based array detectors. Work has been presented outlining the implementation and testing of a readout system based on a multi-anode PMT and a multichannel electrometer (Liu et al., 2012). Employed on an existing fiber-optic dosimetry array, the expressed goal of such work is to improve the system's dosimetric accuracy and acquisition speed. This study revealed that while interchannel optical crosstalk did occur, it could be corrected. The array was able to achieve a greater SNR and precision than comparable CCD-based systems, especially at higher sampling rates. The fast sampling rate and lack of dead time between acquisitions allow for accurate measurements of rapidly changing dose rates. The PMT array system had a measurement uncertainty of 0.54% for a 0.1 s acquisition, while an electron-multiplying CCD-based system showed a measurement uncertainty of near 3.0%. Confirmation of the result was performed through testing the device during source retraction in HDR brachytherapy and through measurements of individual beam pulses produced from a linear accelerator system. The enhanced sensitivity of the system allows for the use of smaller scintillators, thus achieving a higher packing density within the array. PMT array systems may provide desired improvements in both the spatial and temporal resolution for dosimetry in the clinical setting.

11.4.2.3 Three-Dimensional Plastic Scintillation Dosimeter

The plastic scintillator is an effective, real-time 3D dosimeter with high spatial resolution. New detectors based on optical tomography within scintillator blocks were studied to determine 3D dose distributions for bremsstrahlung (6–21 MV), electrons (6–21 MeV), and protons (68 MeV) (Kroll et al., 2013). This portable detector was based on a plastic scintillator block (BC-408) and four digital CCD cameras. An iterative reconstruction algorithm was used to reconstruct distributions of scintillation light detected by the CCD cameras. The tomographic reconstruction was achieved by imaging all light emitted along a line parallel to the CCD camera axis, a technique analogous to the parallel projections of the 3D distribution on a 2D plane. Dose profiles generated by proton and electron beams were in agreement with reference measurements after further refinements and corrections were applied to the data. The drawbacks of this scintillator-based system include the nonlinear light output for high stopping-power radiation and imaging artifacts.

11.4.3 Plastic Scintillation Dosimeters for Ion-Beam Therapy

In ion therapy, PSDs suffer significantly from the quenching effect, the degradation of light production efficiency due to the high ionization density by high linear energy transfer (LET) charged particles, when the detector is close to a Bragg peak. Wang et al. (2012) hence investigated a quenching correction factor (QCF) for a polystyrene-based PSD (BCF-12) in depth-dose measurements for passively scattered proton beams. A linear relationship between QCF (the ratio of a normalized depth-dose curve measured by a Markus chamber to that measured by the PSD) and LET (calculated by using the MCNPX MC code) was expressed as an empirical formula: QCF = 0.881 + 0.00796 × LET. The QCF-corrected PSD measurement agreed to within 5% of the chamber-based measurement. In another study, entrance dosimetry was performed using a PSD for passively scattered proton beams (Wootton et al., 2015). It was demonstrated that correction factors were required due to a 7% (250 MeV) to 10% (140 MeV) PSD signal loss resulting from the quenching effect.

Paradoxically, the Cerenkov radiation produced by interactions between charged particles and plastic optical fiber can be used to measure the Bragg peaks and spread-out Bragg peaks (SOBPs) of therapeutic proton beams without the quenching effect (Jang et al., 2012). The Bragg peak and SOBP measurements of proton beams in a water phantom by Cerenkov radiation using multimode, step index optical fibers with a core/cladding structure (the outer diameter: 1.0 mm; the cladding thickness: 0.01 mm) were in good agreement with those made via Marcus chamber. This concept was further developed into an optical fiber array incorporating multi-cathode/anode PMT and was subsequently tested under clinical proton beam conditions. The measurement of SOBPs in real time is a prominent quality of the optical Cerenkov detection array because the majority of commercially available equipment cannot provide this measurement (Jakel et al., 2000; Lomax et al., 2004; Paganetti et al., 2004). During system testing with clinical proton treatment plans, measurement results agreed well with those obtained through ionization-chamber array, differing by less than 0.5 mm. Future developments of this Cerenkov system may be a scaling into 2D and 3D arrays because the optical signal for each coordinate axis can be measured independently through separate PMTs, and both beam position and intensity can be determined with great precision (Son et al., 2015).

Cerenkov radiation detection systems may be used for prompt gamma imaging. Prompt gamma imaging is among the more promising indirect, online 2D techniques to verify the range of charged particles; however, the direct detection of the relatively high energy (approximately a few MeV) prompt gammas photons is extremely challenging. Cerenkov detectors convert MeV gammas to ultraviolet or visible light to be detected easily by a plastic optical fiber. A feasibility study using Cerenkov detector material (7.2 g of In_2O_3 + 90 g cladding, density of 2.82 g/cm^3, and Z_{eff} = 33.7) for prompt gamma detection as range verification in proton therapy was performed using MC simulation (Lau et al., 2015). The Cerenkov material investigated was able to estimate the proton range with timescales on the order of tens of nanoseconds by detecting Cerenkov photons emitted by the energy range of the deexcitation of ^{16}O photons (approximately 6 MeV).

11.5 LUMINESCENCE DOSIMETERS

Luminescence-based ionizing radiation detectors are applied extensively in individual dosimetry and clinical dosimetry, including IMRT and ion-beam therapy (Olko, 2010). Three groups of luminescence detectors are most common in radiation dosimetry: thermoluminescence detectors (TLDs), optically stimulated luminescence (OSL) detectors, and radiophotoluminescence (RPL) glasses. In luminescence materials, ionizing radiation creates electron-hole pairs. Electrons trapped within defects escape through external thermal stimulation (TLD), optical excitation (OSL), or by ultraviolet (UV) light (RPL). These electrons pair again with holes at the recombination centers and result in the emission of light. The intensity of light emitted is detected for dosimetric purposes. The advantages of luminescence dosimeters include high sensitivity, linear dose response, tissue-equivalent Z_{eff}, and reusability. Recent developments include 2D TLD-based dosimetry, and advances in OSL and RPL glass dosimetry.

11.5.1 Two-Dimensional Thermoluminescence Dosimetry

Clinically, TLDs are used routinely in the form of chips, rods, powders, and microcubes, for phantom-based measurements and *in vivo* dosimetry. The most commonly used TLD material available, (LiF:Mg. Ti) has a spatial resolution limited to the physical dimensions of the TLD, ranging between 6 mm (rods) and 1 mm (microcubes). TLD-based 2D dosimetry systems were generally created using surface-deposited TL powder and subsequently scanned and heated with laser beam (Iwata et al., 1992). The most current 2D TLD techniques utilize CCD cameras that are sensitive enough to detect the TL signal, pinpoint its source, and reconstruct the information as a 2D digital image. One such prototype planar TLD-foil reader consists of a combination of a 78 mm heater and 12-bit CCD camera (resolution of 640 × 480 pixels). The TLD foils utilized are of a 0.3 mm thickness and can vary in diameter (up to 70 mm), the active component of which is a mixture of LiF:Mg, Cu, P powder and ethylene tetrafluoroethylene (ETFE) polymer. This marriage of detector and reader allows for a spatial resolution better than 0.5 mm. These systems, having the same advantages of TLDs, exhibit a flat energy response and reusability, and have been tested under a number of radiation systems: ^{60}Co, 60 MeV protons, 6 MV photons, and microbeam irradiation (Gajewski et al., 2013).

Some difficulties inherent to TLDs are also found in 2D TLDs, including fading, nonlinear dose response (above 1 Gy), and variable energy/LET response in keV ranges. An overall uncertainty better than 2%–3% requires a reproducible preparation and annealing process in addition to corrections for energy response, dose response, fading, and systemic uncertainties of the TLD reader system. Additional problems have arisen from the 2D nature of the system, including nonuniform TLD-foil sensitivity, pixel corrections for the CCD camera, and optical distortions of the lens system. While troublesome, these problems can be accounted for most efficiently through the use of task specific software (Olko et al., 2006).

11.5.2 Optically Stimulated Luminescence Dosimetry

Optically stimulated luminescence dosimeters (OSLDs) have many advantages as dosimeters: a compact size (1 × 1 × 0.2 mm³)", the ability to provide rapid results (in approximately 2 s), and an exhibited stability of several days (Jursinic, 2007; Viamonte et al., 2008). A low intrinsic buildup also allows OSLDs to perform surface dosimetry accurately. OSLDs are beginning to replace both diodes and TLDs not only for *in vivo* dosimetry but also for routine clinical dose measurements.

To determine the absorbed dose received by an OSLD, the dosimeter, comprised of carbon doped sapphire (Al_2O_3:C), is illuminated by light-emitting diodes (LEDs). The luminescence resulting from this stimulation is captured with PMT. The dose absorbed by the dosimeter is determined as it is proportional to the integral of recorded counts as a function of time (Yukihara et al., 2008).

Commercial OSLD systems are user friendly, can provide repeated readings in seconds, and are designed to be individually identifiable for clinical use. While these features are well known in the business of personal dosimetry, Landauer Inc. (Glenwood, Illinois, United States) has expanded the use of OLSDs into the clinic with the introduction of the *nanoDot* dosimeter and the *InLight* System.

The adoption of the OSLD in the clinic, unfortunately, has been slow. Recently, the Imaging and Radiation Oncology Core (IROC) has transitioned from TLDs to OSLDs in their program to perform postal dosimetry audits independently. When the IROC adopted the OSLD for use in monitoring the clinical beams of participating clinics nationwide, they had to determine the validity of multiple characteristics of the OSLD technology. The linearity, fade characteristics, energy dependence, signal depletion per read, reproducibility following bleaching, uncertainty, batch characteristics, and comparison to the accuracy of ionization chamber and TLD measurements all had to be determined. Although well established for the related TLD system (Attix, 1975), the OSLD behavior under shipping and other conditions of irradiation also had to be established (Kry et al., 2015). While the OSLD response is invariant to many clinically relevant conditions, practitioners need to make angular corrections related to the setup and design of buildup materials (Mrcela et al., 2011). Practitioners should also take into account the OSLD dependence on accumulated dose because small, long-term fading effects occur when using OSLD in certain applications.

The OSLD luminescence signal is enhanced at higher doses. Deep electron traps are filled and additional electrons are trapped within the luminescence traps. As such, OSLDs tend to exhibit a supralinear response to doses above 3 Gy (Jursinic, 2007). Irradiation of OSLDs with therapeutic clinical photon beams of energy 6 and 18 MV, as well as electron energies ranging between 6 and 20 MeV, demonstrate the dosimeters' energy independence. At lower energies, their response relative to the 6 MV photons has been found to range between factors of 1.05 for ^{60}Co to 3.5 for X-rays with an average energy of 35 keV (Yukihara and McKeever, 2006).

In particle therapy, OSLDs exhibit a strong dependence on LET. While their relative response under proton irradiation displays an increase by a factor of 1.06 when compared to a 6 MV photon beam, the known supralinearity has been observed as subdued. Under beams of carbon ions of LET ranging between 78 and 140 keV/µm, the response decreased (relative to the 6 MV photons) to values between 0.40 and 0.50, respectively. No supralinearity occurs under carbon irradiations of up to 4 Gy and is presumed to be related to the increase in LET (Reft, 2009).

OSLDs can be utilized to determine dose within, and adjacent to, the treatment field. It has been observed that the accuracy of measurements made outside the reference field suffers from an inconsistent energy spectrum. For out-of-field measurements, correction factors can adjust for overresponse that has been observed to be over 30% (Scarboro et al., 2012).

11.5.3 Radiophotoluminescent Glass Dosimeters

Radiophotoluminescent glass dosimeters (RGDs) are a solid-state class dosimeter, well established for measuring dose deposited by photons of energy from 30 to 1.3 MeV (Yamamoto et al., 2011). RDGs are composed of a silver-activated phosphate glass and emit a dose-proportional signal of photons of approximately 650 nm when exposed to UV light. The range of linear response and high sensitivity to dose spans from 0.1 to 1,000 cGy (Sato et al., 2013).

The chemically stable silver ions doped into the phosphate glass of RGDs are uniformly distributed. When irradiated, the ionization places electrons into the conduction band, where they become trapped by the silver ions, forming stable islands for later luminescence. These islands can form through the migration of holes through the glass, eventually forming doubly ionized silver. The silver islands release a luminescence signal of intensity that is proportional to absorbed dose when exposed to UV light. As the silver islands of luminescence return to a ground state only through thermal annealing, multiple readouts of these detectors are possible without detriment to the dose information recorded.

The most common RGDs used in medical dosimetry are GD-300 series small glass rods (Knežević et al., 2013). Composed of 31.55% P, 51.16% O, 6.12% Al, 11.0% Na, and 0.17% Ag by weight, the rods have a density of 2.61 g/cm^3, an effective atomic number of 11.04 (Piesch, 1972), and an active volume located 0.7 mm from their 1-mm-diameter and 0.6-mm-thick tip. The length of the RGD rod element is 8.5 mm (GD-301 and GD-351) or 12.0 mm (GD-302M and GD-352M). This GD series dosimeter is surrounded by a 0.3–0.5-mm-thick plastic holder. Due to the exceedingly small variability in their response, corrections of sensitivity between individual detectors are unwarranted (Perks et al., 2005).

Recently, RGDs have been used for both postal audits (Perks et al., 2005; Rah et al., 2009) and small field dosimetry (Araki et al., 2003; Araki et al., 2004; Rah et al., 2008) of photon beams used in radiotherapy. While RGDs can be read repeatedly and demonstrate a strong stability over time (Hoshi et al., 2000; Tsuda, 2000), the use of high-Z materials in their construction results in a response dependence on photon energy. Investigators have observed a decrease in response with increasing energy as great as 3%. RGDs of neodymium-doped lithium–lead–boron–silver and lithium–bismuth–boron–silver glass have been studied for accurate, real-time dosimetry in HDR brachytherapy (Correia et al., 2013). The response of RGD can depend on field size by as much as 2% because dose contributions from scattered low-energy photons increase with increasing field size. A depth dependence of 2% has also been observed beyond the buildup region (Mizuno et al., 2008).

RGDs have been fabricated via gas-particle jet flame system in the form of 50 μ-diameter spherical beads, as shown in Figure 11.7 (Sato et al., 2013). Subsequent testing demonstrated the viability of the microbeads as radiation dosimeters. Flexible 2D glass-type sheet dosimeters were formed by bonding RGD glass beads to aluminum sheets. The readout system for the RGD sheets was comprised of a digital camera, supplementary optical lenses and filter, and a UV light source (Zushi et al., 2014).

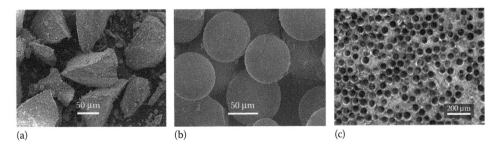

FIGURE 11.7   RPL glass particles. SEM: scanning electron microscopy. OM: optical microscopy. (a) Pulverized glass particle (SEM). (b) Spherical glass particle (SEM). (c) Spherical glass particle (OM). (From Sato, F., et al., 2013, *Radiat. Meas.*, 55, 68–71. With permission.)

11.6 POINT ALANINE DOSIMETERS

Alanine dosimetry systems, often referred to as electron spin resonance (ESR), electron paramagnetic resonance (EPR), or electron magnetic resonance (EMR) dosimetry, exhibit sensitivity over a wide dose range, a high degree of precision, linear dose dependence, energy and dose rate independence, and tissue equivalence. Alanine is among the simplest of amino acids (2-aminopropanoic acid). Consisting of a methyl (CH_3), carboxylic (COOH), and amino (NH_2) groups bound to a central carbon atom, in addition to a hydrogen atom, the methyl is the distinguishing feature of alanine from other amino acids. Alanine is pressed in the form of rods, pellets, films, and cables.

Existing in two isomeric forms, D-alanine and L-alanine, the L-alanine and a racemic mixture (DL-alanine) are commonly used for dosimetry (Baffa and Kinoshita, 2014). As the radiation ionizes alanine, it induces the creation of free radicals and the liberation of the NH_2 group bound, giving rise to the paramagnetic properties from the carbon atom and presence of an unpaired electron (Miyagawa and Gordy, 1960; Heydari et al., 2002). The resulting characteristic ESR spectrum of alanine following irradiation consists of a dominant central line and four smaller peaks, each corresponding to a hydrogen atom in the molecule. The radiation-induced paramagnetic centers by the presence of an unpaired electron with the central carbon create a central line of greatest amplitude in the ESR spectrum. The dose information is effectively obtained by determining the concentration of unpaired electrons produced by ionizing radiation (Miyagawa and Gordy, 1960). Procedures for using alanine/ESR dosimetry were established in the International Organization for Standardization/American Society for Testing and Materials (ISO/ASTM 51607:2013), standard practice for use of the alanine–EPR dosimetry system protocol, additionally describing a means of achieving ISO/ASTM52628 (Standard Practice for Dosimetry in Radiation Processing) compliance (ISO/ASTM International, 2013).

As the field of spectroscopy advanced in the 1990s, alanine dosimetry for the therapeutic range became affordable and thus accessible (Sharpe et al., 1996; Nagy et al., 2002; Anton et al., 2009). The uncertainty of the system, and resultant dose calibration curve, stems from two aspects of the measurement: the calibration source and ESR readout system

(Bergstrand et al., 1998; Anton, 2006). For dosimeters irradiated with ^{60}Co, precision has been achieved as low as 0.5% for doses between 5 and 25 Gy (Anton, 2006).

External factors influencing the ESR spectral peaks are the inherent background signal, environmental temperature, and humidity effects in ESR signal measurement (Alexandre et al., 1992). An early application of the ESR dosimeter was *in vivo* patient measurements. In testing alanine dosimeters with ^{60}Co and ^{192}Ir sources, low absorbed dose (60 cGy) measurements were found to exhibit an uncertainty of less than 5%, decreasing to less than 1% for doses greater than 10 Gy (Kuntz et al., 1996; Schaeken and Scalliet, 1996). Applying alanine ESR dosimetry in the QA of 3D conformal external beam radiotherapy of prostate cancer (dosimeter probe $0.5 \times 0.5 \times 0.2$ cm^3), anterior and posterior rectal wall dose measurements agreed with calculated dose to within 1.5% and 3.5%, respectively (Wagner et al., 2008).

Total body irradiation (TBI) dose calculations have also been validated *in vivo* with an ESR dosimetry-based system (Schaeken et al., 2010). Tomotherapy equipment has been successfully calibrated using ESR/EPR dosimetry (Perichon et al., 2011), as has the Gamma Knife Perfexion (Elekta, Stockholm, Sweden) SRS system (Hornbeck et al., 2014). EPR measurements agreed with ionization chamber results within 1% when phantom attenuations were corrected by MC simulations.

Advancing the potential of EPR/ESR dosimetry has often been through the addition of alanine into materials, or the infusion of materials into alanine. With the goal of creating a 3D dosimeter analogous to *Bang-Gel*, aggregated alanine has been infused into an agarose gel (Olsson et al., 1996). Improvements in sensitivity have been attempted by doping alanine with high-Z materials. The detriment of this is a decrease in tissue equivalence. Introducing materials of high atomic number increases the number of secondary free electrons produced during irradiation and therefore the production of free radicals within the alanine. Potassium iodide (KI) doping at less than 15% increased detector sensitivity through the production of free radicals when irradiated with diagnostic X-rays of energy below 100 kVp (Chen et al., 2008; Chen et al., 2010). Improvements in the sensitivity were also observed for gadolinium doping under 6-MV clinical photon doses (Marrale et al. 2011). Beyond chemical doping, the introduction of high-Z nano-colloidal materials also increases the alanine dosimeter sensitivity. Sensitivity was optimized with 30 nm nanoparticles homogeneously distributed inside the alanine matrix. Likewise, gold nanoparticles improved sensitivity thresfold at a 3% (w/w) particle concentration (Guidelli et al., 2012a, 2012b). Improvements in sensitivity, introduction of minidosimeters, and higher RF spectrometers all enhance the clinical potential for EPR applications in IMRT, VMAT, and radiosurgery (Baffa and Kinoshita, 2014).

Alanine also responds well to neutron, proton, and ion beam irradiation, a feature useful during the rapid adoption of particle therapy techniques (Onori et al., 1997; Trompier et al., 2004; Herrmann et al., 2011). A novel verification technique based on EPR/alanine dosimetry was developed for proton eye radiotherapy (Michalec et al., 2014). Because the dependence between entrance dose/isocenter dose ratio and modulation width is linear, the isocenter (tumor) dose can be predicted by measuring the entrance dose using alanine pellets (4.8 mm in diameter, 3.0 mm height; 96% alanine and 4% binder by weight). During irradiation, the detector pellets were placed in the beam, on the inner side of the final collimator, where dose interference was negligible. Results showed good agreement between the alanine dosimetry and on-line transmission chamber dose measurements in ocular proton radiotherapy (0.3%–2.6%).

11.7 POTENTIAL DOSIMETRY MATERIALS, DEVICES, AND SYSTEMS

Some dosimetry technologies are gaining appreciation in the research community and are being developed commercially. These dosimetry devices and systems are increasing the number of options available to the practice of medical radiation dosimetry.

11.7.1 Calorimetry

Calorimetry is one of the truly absolute dosimetry techniques because its calibration can be determined independent of irradiation, dependent only on the physical properties of the system (Attix, 1986). Because dose is fundamentally defined as energy deposited per unit mass, calorimetry measures the rise in temperature in an absorbing medium, resulting from the energy deposited. The large size of typical calorimetry systems can hinder their ease of use, so ionization chambers have been the preferred reference dosimeter in clinical use. Recently developed and miniaturized calorimeters may offer a direct and accurate means of determining absorbed dose in small and nonstandard radiation fields, eliminating the calibration factors currently needed for different radiation beam quality (Duane et al., 2012).

Dose calibration for nonstandard fields has become a challenge with the emergence of compact linear accelerators, advanced IMRT, and SRS (Alfonso et al., 2008). With an estimated accuracy of ±1.2%, the calorimetry-based dosimetry of clinical beams shows great promise. A smaller graphite calorimeter, of a scale similar to traditional Farmer chambers, may be used as a clinical reference dosimeter with no need of third-party calibration (Renaud et al., 2013). This graphite calorimeter (US provisional patent 61/652,540) utilizes an aerogel-type material as thermal insulator, providing an increased strength necessary for clinical duty. The difference in dose-to-water measurements using the TG-51 protocol and the calorimeter at 400 and 1,000 MU/min for a 6 MV photon beam was 0.2% and 1.2%, respectively.

With graphite calorimeters, low-energy electron contamination can lead to discrepancies in PDD measurements when compared to MC calculation. Dose measurements of a 1.6 MeV electron beam were higher than MC simulations at a depth of 0.15 cm. Both measurements and MC calculations prove that energy absorption is linear up to 3 mm thicknesses of graphite. Isothermal operation can shield the graphite calorimeter from thermal fluctuations of the surrounding environment, and provides a stable background for accurate dose measurement with an overall uncertainty of less than 1% (Benny et al., 2013).

The graphite calorimeter is also a viable dosimeter for ion-beam therapy. In a study with a graphite calorimeter for a 80-MeV/A carbon-ion beam (Rossomme et al., 2014), the experimental w_{air}-value (35.5 ± 0.9 J/C) was higher than the value recommended by the TRS-398

(IAEA 2000) (34.5 ± 0.52 J/C). Additional investigations of heat transfer kinetics within the components of the calorimeter, as well as ion recombination corrections, are necessary.

Whereas graphite calorimeters utilize thermistors to determine temperature rise, optical interferometry has been developed recently to accomplish the same goal. An interferometer can determine the change in refractive index in water to infer change in temperature without influencing the radiation beam (Cavan and Meyer, 2014). Never-before-realized dosimetric precision can be achieved through the application of optical interferometry to calorimetric dosimetry. Interferometry resolves changes in the interference pattern resulting from differences in the phase of the two light waves reflecting from the irradiated and unirradiated states of the object, respectively.

With advances in modern imaging sensors, complementary metal oxide semiconductors (CMOS), and CCDs, the development of the digital interferometer and numerical algorithms allow for the direct reconstruction of the hologram (or interferogram), a recoded reproduction of the optical conditions at a moment in time. The application of these technologies allows the interferograms to be determined accurately and in real time, and reveals optical phase information at any position within the image.

Experimentally, digital holographic interferometry (DHI) as an optical calorimeter has demonstrated its potential as a radiation dosimeter in water for HDR brachytherapy (Cavan and Meyer, 2014). A prototype DHI system was able to obtain relative dose distributions within 1 cm of the source, agreeing well with calculated isodose lines. The most promising aspect of DHI is the ability to measure an absorbed dose to water at high resolution. The measurements are absolute dose measurements, independent of beam modality, energy, or dose rate. The noninvasive aspect of DHI lends itself well to challenging applications in microbeam therapy, proton and light-ion beam therapy, and surface dosimetry.

11.7.2 Germanium-Doped Optical Fibers

Thermoluminescent (TL) silica dioxide (SiO_2) optical fiber is a promising material for the detection of radiation, having a physical diameter of just a few tens of microns and a polymer coating providing resistance to water and corrosion (Abdulla et al., 2001; Hashim et al., 2009; Bradley et al., 2012). TL fibers are also reusable without detriment to their dose response, exhibiting a low residual signal and high degree of reproducibility (Abdulla et al., 2001; Espinosa et al., 2006). Germanium (Ge) doping, as shown in Figure 11.8, has been shown to yield a significant increase in TL response (Begum et al., 2015). While other dopants have been investigated, including Al, Nd, Yb, Er, and Sm, Ge doping provides the greatest benefit (Hashim et al., 2009; Noor et al., 2011). Dosimetric characteristics of TL fibers have been established as similar to those seen in TLD dosimetry: reproducibility; linearity; fading; and dependencies on energy, dose rate, angle of incidence, and temperature (Noor et al., 2010). Ge-doped fibers have been found to have an effective atomic number between 11.5 and 13.4, implying that the dosimeters are not soft-tissue equivalent but are more analogous to bone (Hashim et al., 2013).

Reproducibility testing of 9 μ Ge-doped optical fibers reveals no statistically significant difference between five repeat measurements for megavoltage photon beams. The response is linear over a dose range between 5 and 1,000 cGy. Ge-doped optical fibers exhibit energy

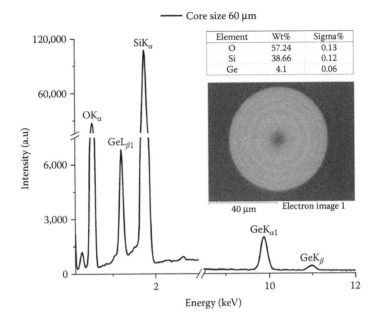

FIGURE 11.8   Scanning electron microscopy and energy dispersive X-ray spectroscopy analysis of 60 μm core Ge doped optical fiber. Concentric ring formation is a result of the modified chemical vapor deposition fabrication process, where a layer-by-layer style deposition occurs. (From Begum, M., et al., 2015, *Appl. Radiat. Isot.*, 100, 79–83. With permission.)

dependencies as great as 6% and 8% at 10 and 15 MV, respectively, compared to those of 6 MV. This under-response correlates well to the mass energy absorption coefficient at 6 MV being greater than that for 10 or 15 MV (Noor et al., 2014).

Ge-doped fiber dosimeters studied for use in HDR brachytherapy dosimetry were able to measure doses at distances from 2 mm to 2 cm from the source. MC simulations were found to agree well with measured doses from ^{133}Ba and ^{60}Co sources, 3% and 1%, respectively (Issa et al., 2012). It has been surmised that the microcrystalline structures may induce orientation effects similar to asymmetrical doping (Noor et al., 2014).

11.7.3 X-Ray Acoustics

When a pulse of radiation emitted from a linear accelerator strikes a material, an acoustic wave is induced within the media via the photoacoustic effect (Bowen et al., 1991). The amplitude of this acoustic wave is proportional to the amount of dose deposited and, as such, an image of the resulting dose distribution may be constructed through the measurement of the wave with ultrasonic transducers. This imaging technique, X-ray acoustic computed tomography (XACT), is an analogue of photoacoustic tomography (PAT), in which a pulsed laser system is used to induce the acoustic waves (Xiang et al., 2013).

Through a combination of MC and photoacoustic simulation methods, a photoacoustic wave simulation model has been developed and evaluated through comparison with experimental measurements (Hickling et al., 2014). This simulation technique was able to accurately predict the relative amplitude and the frequency spectrum of the photoacoustic waves produced following the pulsed irradiation of a lead brick in water. The environment in which radiotherapy occurs is inherently filled with acoustic noise, so signal filtering is required for the detection of the photoacoustic signals. Successful implementation of a band-pass filter has been based on the signal frequency spectrum predicted via simulation. The photoacoustic effect is a thermal process, and noninvasive tomographic ultrasound systems have also been utilized to obtain real-time measurements of the spatial distribution of absorbed dose in terms of temperature changes in micro-Kelvin degrees at a spatial resolution of 5 mm (Malyarenko et al., 2010). It has also been suggested that a further enhancement of the sensitivity of this technique may be achieved through the use of an optical system (Tosh, 2013).

The range of charge particle beams can also be verified by thermoacoustic signals (iono-acoustics) (Assmann et al., 2015). The thermoacoustic signals are generated due to localized energy loss of ion beams in tissue and can be detected by high-frequency ultrasonic transducers. A study under idealized condition was performed with a pulsed 20 MeV proton beam in a water phantom. The 20 MeV proton beam raises temperature by nearly 0.3 mK through local energy deposition in the Bragg peak region, translating to a pressure increase of 2.5 hPa. This pressure signal is captured by piezo-composite- (PZT-) based ultrasound immersion transducers. Contrary to conventional *in vivo* range verification based on positron emitter or prompt gamma ray, this system can detect the range with sub-millimeter (below 100 μm) accuracy and provide a simpler, cost-effective, and direct Bragg peak position measurement. This technique is limited to only pulsed beams (preferred frequencies in the range 100 kHz–10 MHz). It is considered unsuitable for the cyclotron-generated proton beams of pulsed frequencies between 50 and 100 MHz because the very high frequencies are rapidly damped within short distances from the Bragg peak.

11.7.4 Tomodosimetry

A high-resolution 2D tomodosimetry was developed for IMRT QA using an array of scintillation fibers (Goulet et al., 2012). A weighted projection of the incident dose distribution is obtained by aligning 50 parallel scintillating fibers (BCF60; 1 mm diameter) on a single imaging plane in a 30-cm-diameter, 15-cm-thick Masonite phantom. Both ends of each scintillating fiber are coupled to a photodetector to collect light signals proportional to the integral dose. Multiple dose projections are acquired at different angles about the beam axis, between the scintillating fibers and the dose distribution. Tomographic information between fiber angle and fiber number are acquired, information analogous to the sinogram used in tomography, to reconstruct the dose distribution using an iterative reconstruction algorithm. This tomodosimetry method provides a relatively high resolution (1×1 mm^2) dosimetry. Conversely, this technique exhibits comparatively higher levels of local variations due to the iterative reconstruction algorithm, and multiple data acquisitions at different angles are required to obtain dose distributions for each radiation field.

The concept of tomodosimetry can also be adapted to 3D dosimetry by placing scintillating fibers of variable length on multiple concentric cylindrical surfaces of known radii inside a phantom (Goulet et al., 2013). This phantom is rotating around the superior–inferior (SI) axis (parallel to the linac gantry axis of rotation) during light signal collection.

The light is acquired by a scintillating fiber arranged at a unique angle (ϕ) on the 2D cylindrical surface of the phantom. Weighted-dose integrals ($\vec{p}$; light signals collected by the scintillating fiber) from different cylinder rotation angles are correlated to the 2D dose distribution ($\vec{D}$) using a projection matrix A: $\vec{p} = A \cdot \vec{D}$. The matrix $A_{j,i}$ represents the contribution of the ith dose pixel to the jth weight dose integral and is calculated using a geometric factor and the optical attenuation. The 2D dose distribution is reconstructed with the collected signal $\vec{p}$ and the calculated matrix A through an expectation-maximization algorithm (iteration). The 3D dose distribution within the phantom is interpolated from multiple reconstructed dose surfaces, taking into consideration the beam divergence from a source position.

This technique is limited to a stable, static radiation field because the tomographic reconstruction requires a relatively fast signal acquisition and rotational speed (100 acquisitions/s and 100°/s; 0.1 MU/acquisition in this simulation study). Reconstruction artifacts can be induced by dynamically changing field shape, dose rate, and gantry speed, conditions commonly seen in VMAT. Increasing rotation speed, partial angle reconstruction, and incorporation of a temporal reconstruction are each proposed to alleviate this restriction.

In another study, a novel type of high-resolution 3D dosimeter was developed based on the real-time light acquisition of a plastic scintillator volume (Goulet et al., 2014). Using a plenoptic camera, a single-lens 3D camera with an array of microlenses covering several sensor pixels, light rays passing through the camera's main lens are separated into smaller images. Spatial and directional information of incident photons are recorded as the light rays specific over a small portion of incident angles are captured on each image pixel. This information is used to reconstruct the 3D light pattern emitted from the plastic scintillator by back-projecting the light rays using the iterative reconstruction algorithm. This reconstruction is analogous to tomodosimetry ($\vec{p} = A \cdot \vec{D}$) (Goulet et al., 2012, 2013). The dose reconstruction is constrained by a BEV projection acquired by the EPID to improve the spatial resolution in the direction of the main optical axis of the camera. In this study, a plastic scintillator cube ($10 \times 10 \times 10$ cm^3) embedded in the center of a cylindrical acrylic phantom with a plenoptic color camera (acquisition rate of one image/sec) was used to verify the step-and-shoot IMRT and VMAT delivery. The agreement between the reconstructed 3D dose distribution (spatial resolution of $2 \times 2 \times 2$ mm^3) and TPS calculation was within 3% of the maximum dose in low-gradient regions. This dosimetric technique is mainly restricted by optical scattering, inaccurate prediction of out-of-field low-dose regions in BEV projection, and an inadequate image acquisition rate required for dynamic delivery.

11.7.5 Liquid Scintillator for Proton Therapy

An intensity-modulated proton therapy (IMPT) delivers a large number of pencil beams to prescribed spots within a target to generate highly conformal dose distribution and thus requires high-resolution, small-field 3D dosimetry. Real-time 3D liquid scintillator (LS) detector systems have the potential to verify the range and dose of IMPT. Beddar et al. (2009) investigated LSs for IMPT beams. An organic LS (BC-531) was irradiated by a 120 MeV passively scattered proton beam with a field size of 2×2 cm^2. The scintillation light was measured with a high-sensitivity CCD camera with 658×496 pixels at acquisition rates of 20 and 10 fps. The quenching effect was simulated by GEANT4 MC toolkit, and the LS measurements were compared with Markus chamber-based measurements and MC simulations. The LS detector system could acquire multiple images within a single proton pulse with a submillimeter image resolution.

In a recent study by the same group, the LS detector system was used for verification of proton spot position in a scanning beam system (Archambault et al., 2012). The detector system consisted of a 20 cc LS volume. The light signal was continuously monitored by CCD camera with a field of view of 25.7 by 19.3 cm^2 and a pixel size of 0.4 mm at an acquisition rate of 50 ms/frame. Because only one camera was used, the image was a 2D projection of the 3D light distribution in crossline profiles and depth profiles of spot-scanned proton beams. The researchers concluded that the LS system provided sufficient submillimeter accuracy for verification of range, position, and intensity of intensity-modulated proton beams after a proper calibration of the system (e.g., proper quenching correction). A full 3D verification of IMPT with a two-camera system was considered possible.

11.8 CONCLUSION

Radiation therapy is rapidly evolving, and the demand for new dosimetry methods, materials, and systems is high. New developments in conventional detectors focus on decreasing the size of systems to increase spatial resolutions, increasing robustness for new environments, and increasing the ease of use, allowing for the improvement of clinical workflow. Clear understanding of the pros and cons of each detector and dosimetry system will ensure the proper application of new technologies.

Novel systems have been developed and tested by academic institutes and commercial vendors. Techniques common in high-energy and health physics may make their way into the medical field. The advancements in dosimetry systems for IMRT and VMAT will help ensure the benefits of radiation therapy. Dosimetry systems for ion-beam therapy require unique characteristics dissimilar to those used for photon or electron detection. These systems are being translated from the high-energy laboratories of experimental physics. New detector materials may not be realized commercially, yet the new information will provide the basis for future advancements.

REFERENCES

Abdulla, Y. A., Y. M. Amin, and D. A. Bradley. 2001. The thermoluminescence response of Ge-doped optical fibre subjected to photon irradiation. *Radiat Phys Chem* 61:409–410.

Alexandre, A. C., O. Baffa, and O. R. Nascimento. 1992. The influence of measurement and storage-conditions on alanine esr dosimeters. *App Radiat Isot* 43:1407–1411.

Alfonso, R., P. Andreo, R. Capote, et al. 2008. A new formalism for reference dosimetry of small and nonstandard fields. *Med Phys* 35:5179–5186.

Allgaier, B., E. Schüle, and J. Würfel. 2013. *Dose Reconstruction in the OCTAVIUS 4D Phantom and in the PatOngient Without Using Dose Information from the TPS*, Freiburg, Germany: PTW-Freiburg.

Andersson, J., and H. Tolli. 2011. Application of the two-dose-rate method for general recombination correction for liquid ionization chambers in continuous beams. *Phys Med Biol* 56:299–314.

Anton, M., R. P. Kapsch, and T. Hackel. 2009. Is there an influence of the surrounding material on the response of the alanine dosimetry system? *Phys Med Biol* 54:2029–2035.

Aoyama, T., S. Koyama, M. Tsuzaka, and H. Maekoshi. 1997. A depth-dose measuring device using a multichannel scintillating fiber array for electron beam therapy. *Med Phys* 24:1235–1239.

Araki, F., H. Kubo, T. Ikegami, T. Ishidoya, and C. Yang. 2003. Measurements of gamma-knife helmet output factors using a radiophotoluminescent glass rod dosimeter and a diode detector. *Med Phys* 30:1435–1435.

Araki, F., T. Ikegami, T. Ishidoya, et al. 2004. Development of a radiophotoluminescent glass plate detector for small field. *Int J Radiat Oncol Biol Phys* 60:S586.

Archambault, L., A. S. Beddar, L. Gingras, et al. 2007. Water-equivalent dosimeter array for small-field external beam radiotherapy. *Med Phys* 34:1583–1592.

Archambault, L., A. S. Beddar, L. Gingras, R. Roy, and L. Beaulieu. 2006. Measurement accuracy and Cerenkov removal for high performance, high spatial resolution scintillation dosimetry. *Med Phys* 33:128–135.

Archambault, L., F. Poenisch, N. Sahoo, et al. 2012. Verification of proton range, position, and intensity in IMPT with a 3D liquid scintillator detector system. *Med Phys* 39:1239–1246.

Archambault, L., J. Arsenault, L. Gingras, et al. 2005. Plastic scintillation dosimetry: Optimal selection of scintillating fibers and scintillators. *Med Phys* 32:2271–2278.

Assmann, W., S. Kellnberger, S. Reinhardt, et al. 2015. Ionoacoustic characterization of the proton Bragg peak with submillimeter accuracy. *Med Phys* 42:567.

Attix, F. H. 1975. Further consideration of track-interaction model for thermoluminescence in LiF(Tld-100). *J Appl Phys* 46:81–88.

Attix, F. H. 1986. *Introduction to Radiological Physics and Radiation Dosimetry*. New York, NY: Wiley.

Baffa, O., and A. Kinoshita. 2014. Clinical applications of alanine/electron spin resonance dosimetry. *Radiat Environ Biophys* 53:233–240.

Balducci, A., M. Chiorboli, M. G. Donato, et al. 2006. Analysis of trapping-detrapping defects in high quality single crystal diamond films grown by chemical vapor deposition. *Diam Relat Mater* 15:1878–1881.

Barbes, B., J. D. Azcona, J. Burguete, and J. M. Marti-Climent. 2014. Application of spherical diodes for megavoltage photon beams dosimetry. *Med Phys* 41:012102.

Bartesaghi, G., V. Conti, D. Bolognini, et al. 2007. Scintillating fiber dosimeter for radiotherapy. *Nucl Instrum Meth A* 581:80–83.

Beddar, A. S., T. R. Mackie, and F. H. Attix. 1992a. Water-equivalent plastic scintillation detectors for high-energy beam dosimetry. 1. Physical characteristics and theoretical considerations. *Phys Med Biol* 37:1883–1900.

Beddar, A. S., T. R. Mackie, and F. H. Attix. 1992b. Water-equivalent plastic scintillation detectors for high-energy beam dosimetry. 2. Properties and measurements. *Phys Med Biol* 37:1901–1913.

Beddar, S., L. Archambault, N. Sahoo, et al. 2009. Exploration of the potential of liquid scintillators for real-time 3D dosimetry of intensity modulated proton beams. *Med Phys* 36:1736–1743.

Begum, M., A.K.M. MizanurRahman, H.A. Abdul-Rashid, et al. 2015. Thermoluminescence characteristics of Ge-dopedoptical fibers with different dimensions for radiation dosimetry. *Appl Radiat Isot* 100:79–83.

Beierholm, A. R., C. F. Behrens, and C. E. Andersen. 2014. Dosimetric characterization of the Exradin W1 plastic scintillator detector through comparison with an in-house developed scintillator system. *Radiat Meas* 69:50–56.

Benny, P. G., T. Palani Selvam, and K. S. S. Sarma. 2013. Comparison of graphite calorimeter dosimetry system with Monte Carlo simulation at an industrial electron beam accelerator. *Nucl Instrum Meth A* 703:98–101.

Bergstrand, E. S., E. O. Hole, and E. Sagstuen. 1998. A simple method for estimating dose uncertainty in ESR/alanine dosimetry. *Appl Radiat Isot* 49:845–854.

Beyer, G. P., J. Pursley, G. Mann, et al. 2007. Radiation characteristics and hypo-fractionation dose response for the DVS implantable MOSFET dosimeter. *Med Phys* 34:2610.

Blake, S. J., A. L. McNamara, P. Vial, L. Holloway, and Z. Kuncic. 2014. Optimisation of the imaging and dosimetric characteristics of an electronic portal imaging device employing plastic scintillating fibres using Monte Carlo simulations. *Phys Med Biol* 59:6827–6840.

Blake, S. J., A. L. McNamara, S. Deshpande, et al. 2013. Characterization of a novel EPID designed for simultaneous imaging and dose verification in radiotherapy. *Med Phys* 40:091902.

Bloemen-Van Gurp, E. J., B. J. Munheer, T. A. M. Verschueren, and P. Lambin. 2007. Total body irradiation, toward optimal individual delivery: Dose evaluation with metal oxide field effect transistors, thermoluminescence detectors, and a treatment planning system. *Int J Radiat Oncol Biol Phys* 69:1297–1304.

Bourgouin, A., C. Bonde, E. Adams, et al. 2012. PO-0810 characterization and validation of a commercial plastic scintillation detector prototype. *Radiother Oncol* 103:S315.

Bowen, T., C. X. Chen, S. C. Liew, W. R. Lutz, and R. L. Nasoni. 1991. Observation of ultrasonic emission from edges of therapeutic X-ray-beams. *Phys Med Biol* 36:537–539.

Bradley, D. A., R. P. Hugtenburg, A. Nisbet, et al. 2012. Review of doped silica glass optical fibre: Their TL properties and potential applications in radiation therapy dosimetry. *Appl Radiat Isot* 71:2–11.

Broisman, A., and G. Shani. 2011. Application of spherical micro diodes for brachytherapy dosimetry. *Radiat Meas* 46:334–339.

Brualla-González, L., F. Gomez, M. Pombar, and J. Pardo-Montero. 2016. Dose rate dependence of the PTW 60019 microDiamond detector in high dose-per-pulse pulsed beams. *Phys Med Biol* 61:N11–N19.

Camilleri, J., J. Mazurier, D. Franck, et al. 2014. Clinical results of an EPID-based *in vivo* dosimetry method for pelvic cancers treated by intensity-modulated radiation therapy. *Phys Medica* 30:690–695.

Cartwright, L. E., J. Lambert, D. R. McKenzie, and N. Suchowerska. 2009. The angular dependence and effective point of measurement of a cylindrical scintillation dosimeter with and without a radio-opaque marker for brachytherapy. *Phys Med Biol* 54:2217–2227.

Cavan, A., and J. Meyer. 2014. Digital holographic interferometry: A novel optical calorimetry technique for radiation dosimetry. *Med Phys* 41:022102.

Chaikh, A., J. Balosso, J. Y. Giraud, et al. 2014. Characterization of GaN dosimetry for 6 MV photon beam in clinical conditions. *Radiat Meas* 71:392–395.

Chalkley, A., and G. Heyes. 2014. Evaluation of a synthetic single-crystal diamond detector for relative dosimetry measurements on a CyberKnife (TM). *Brit J Radiol* 87: 20130768.

Chen, F., J. V. Ramirez, P. Nicolucci, and O. Baffa. 2010. Sensitivity comparison of two L-Alanine doped blends to different photon energies. *Health Phys* 98:383–387.

Chen, F., P. Nicolucci, and O. Baffa. 2008. Enhanced sensitivity of alanine dosimeters to low-energy X-rays: Preliminary results. *Radiat Meas* 43:467–470.

Cherpak, A., G. Kertzscher, and J. Cygler. 2012. Application of RADPOS *in vivo* dosimetry for QA of high dose rate brachytherapy. *Med Phys* 39:3968.

Cherpak, A., M. Serban, J. Seuntjens, and J. E. Cygler. 2011. 4D dose-position verification in radiation therapy using the RADPOS system in a deformable lung phantom. *Med Phys* 38:179–187.

Cherpak, A., W. Ding, A. Hallil, and J. E. Cygler. 2009. Evaluation of a novel 4D *in vivo* dosimetry system. *Med Phys* 36:1672–1679.

Chung, E., E. Soisson, and J. Seuntjens. 2012. Dose homogeneity specification for reference dosimetry of nonstandard fields. *Med Phys* 39:407–414.

Correia, A., S. Chiquita, N. S. Hussain, R. Pirraco, and C. Rosa. 2013. A multi-sensor dosimeter for brachytherapy based on radioluminescent fiber sensors. In *Fifth European Workshop on Optical Fibre Sensors*, Jaroszewicz, L. R. (Ed.), May 19–22. Warsaw, Poland: Krakow.

Dimitrov, R., M. Murphy, J. Smart, et al. 2000. Two-dimensional electron gases in Ga-face and N-face AlGaN/GaN heterostructures grown by plasma-induced molecular beam epitaxy and metalorganic chemical vapor deposition on sapphire. *J Appl Phys* 87:3375–3380.

Duane, S., M. Aldehaybes, M. Bailey, et al. 2012. An absorbed dose calorimeter for IMRT dosimetry. *Metrologia* 49:S168–S173.

Duboz, J. Y., B. Beaumont, J. L. Reverchon, and A. D. Wieck. 2009. Anomalous photoresponse of GaN X-ray Schottky detectors. *J Appl Phys* 105:114512.

Duboz, J. Y., M. Lauegt, D. Schenk, et al. 2008. GaN for X-ray detection. *Appl Phys Lett* 92:263501.

El-Mohri, Y., K. W. Jee, L. E. Antonuk, M. Maolinbay, and Q. H. Zhao. 2001. Determination of the detective quantum efficiency of a prototype, megavoltage indirect detection, active matrix flat-panel imager. *Med Phys* 28:2538–2550.

Espinosa, G., J. I. Golzarri, J. Bogard, and J. Garcia-Macedo. 2006. Commercial optical fibre as TLD material. *Radiat Prot Dosim* 119:197–200.

Feygelman, V., D. Opp, K. Javedan, A. J. Saini, and G. Zhang. 2010. Evaluation of a 3D diode array dosimeter for helical tomotherapy delivery QA. *Med Dosim* 35:324–329.

Feygelman, V., G. Zhang, C. Stevens, and B. E. Nelms. 2011. Evaluation of a new VMAT QA device, or the "X" and "O" array geometries. *J Appl Clin Med Phys* 12:146–168.

Feygelman, V., K. Forster, D. Opp, and G. Nilsson. 2009. Evaluation of a biplanar diode array dosimeter for quality assurance of step-and-shoot IMRT. *J Appl Clin Med Phys* 10:64–77.

Fontbonne, J. M., G. Iltis, G. Ban, et al. 2002. Scintillating fiber dosimeter for radiation therapy accelerator. *IEEE T Nucl Sci* 49:2223–2227.

Fowler, J., and F. Attix. 1966. *Solid State Integrating Dosimeters*. New York, NY: Academic Press, pp. 241–290.

Francescon, P., W. Kilby, N. Satariano, and S. Cora. 2012. Monte Carlo simulated correction factors for machine specific reference field dose calibration and output factor measurement using fixed and iris collimators on the CyberKnife system. *Phys Med Biol* 57:3741–3758.

Gajewski, J., M. Kłosowski, P. Olko, and M. Waligórski. 2013. The response of 2D TL foils after doses of Co-60 Gamma-ray, 6 MV X-ray and 60 MeV Proton beams applied in radiotherapy. *Acta Phys Pol B Proc Suppl* 6:1021.

Gomez, F., D. Gonzalez-Castano, P. Diaz-Botana, and J. Pardo-Montero. 2014. Study of the PTW microLion chamber temperature dependence. *Phys Med Biol* 59:2705–2712.

Goulet, M., L. Archambault, L. Beaulieu, and L. Gingras. 2013. 3D tomodosimetry using long scintillating fibers: A feasibility study. *Med Phys* 40:101703.

Goulet, M., L. Archambault, L. Beaulieu, and L. Gingrasb. 2012. High resolution 2D dose measurement device based on a few long scintillating fibers and tomographic reconstruction. *Med Phys* 39:4840–4849.

Goulet, M., M. Rilling, L. Gingras et al. 2014. Novel, full 3D scintillation dosimetry using a static plenoptic camera. *Med Phys* 41:082101.

Griessbach, I., M. Lapp, J. Bohsung, G. Gademann, and D. Harder. 2005. Dosimetric characteristics of a new unshielded silicon diode and its application in clinical photon and electron beams. *Med Phys* 32:3750–3754.

Guidelli, E. J., A. P. Ramos, M. E. D. Zaniquelli, P. Nicolucci, and O. Baffa. 2012a. Synthesis and characterization of silver/alanine nanocomposites for radiation detection in medical applications: The influence of particle size on the detection properties. *Nanoscale* 4:2884–2893.

Guidelli, E. J., A. P. Ramos, M. E. D. Zaniquelli, P. Nicolucci, and O. Baffa. 2012b. Synthesis and characterization of gold/alanine nanocomposites with potential properties for medical application as radiation sensors. *ACS Appl Mater Interfaces* 4:5844–5851.

Guillot, M., L. Beaulieu, L. Archambault, S. Beddar, and L. Gingras. 2011. A new water-equivalent 2D plastic scintillation detectors array for the dosimetry of megavoltage energy photon beams in radiation therapy. *Med Phys* 38:6763–6774.

Hanson, I. M., V. N. Hansen, I. Olaciregui-Ruiz, and M. van Herk. 2014. Clinical implementation and rapid commissioning of an EPID based *in vivo* dosimetry system. *Phys Med Biol* 59:N171–N179.

Hashim, S., M. I. Saripan, A. T. A. Rahman, et al. 2013. Effective atomic number of Ge-doped and Al-doped optical fibers for radiation dosimetry purposes. *IEEE T Nucl Sci* 60:555–559.

Hashim, S., S. Al-Ahbabi, D. A. Bradley, et al. 2009. The thermoluminescence response of doped SiO₂ optical fibres subjected to photon and electron irradiations. *Appl Radiat Isot* 67:423–427.

Hauri, P., S. Verlaan, S. Graydon, et al. 2014. Clinical evaluation of an anatomy-based patient specific quality assurance system. *J Appl Clin Med Phys* 15:181–190.

Herrmann, R., O. Jakel, H. Palmans, P. Sharpe, and N. Bassler. 2011. Dose response of alanine detectors irradiated with carbon ion beams. *Med Phys* 38:1859–1866.

Heydari, M. Z., E. Malinen, E. O. Hole, and E. Sagstuen. 2002. Alanine radicals. 2. The composite polycrystalline alanine EPR spectrum studied by ENDOR, thermal annealing, and spectrum simulations. *J Phys Chem A* 106:8971–8977.

Hickling, S., M. Hobson, and I. El Naqa. 2014. Feasibility of X-ray acoustic computed tomography as a tool for noninvasive volumetric *in vivo* dosimetry. *Int J Radiat Oncol Biol Phys* 90:S843–S843.

Hofstetter, M., J. Howgate, I. D. Sharp, M. Stutzmann, and S. Thalhammer. 2011. Development and evaluation of gallium nitride-based thin films for X-ray dosimetry. *Phys Med Biol* 56:3215–3231.

Hornbeck, A., T. Garcia, M. Cuttat, and C. Jenny. 2014. Absolute calibration of the Gamma Knife(R) Perfexion and delivered dose verification using EPR/alanine dosimetry. *Med Phys* 41:061708.

Hoshi, Y., T. Nomura, T. Oda, et al. 2000. Application of a newly developed photoluminescence glass dosimeter for measuring the absorbed dose in individual mice exposed to low-dose rate Cs-137 gamma-rays. *J Radiat Res* 41:129–137.

IAEA. 2000. Absorbed dose determination in external beam radiotherapy: An international code of practice for dosimetry based on standards of absorbed dose to water (Technical reports series No. 398). Vienna, Austria: International Atomic Energy Agency.

ICRU. 1976. *ICRU Report 24: Determination of Absorbed Dose in a Patient ISrradiated by Beams of x or Gamma Rays in Radiotherapy Procedures.* Bethesda, MD: International Commission on Radiation Units and Measurements.

Infusino, E., A. Mameli, R. Conti, et al. 2014. Initial experience of ArcCHECK and 3DVH software for RapidArc treatment plan verification. *Med Dosim* 39:276–281.

Islam, M. K., B. D. Norrlinger, J. R. Smale, et al. 2009. An integral quality monitoring system for real-time verification of intensity modulated radiation therapy. *Med Phys* 36:5420–5428.

Ismail, A., P. Pittet, G. N. Lu, et al. 2011. *In vivo* dosimetric system based on Gallium Nitride radioluminescence. *Radiat Meas* 46:1960–1962.

ISO/ASTM International. 2013. *ISO/ASTM 51607:2013 Practice for Use of an Alanine-EPRdosimetry System.* Geneva, Switzerland: Author.

Issa, F., A. T. A. Rahman, R. P. Hugtenburg, D. A. Bradley, and A. Nisbet. 2012. Establishment of Ge-doped optical fibres as thermoluminescence dosimeters for brachytherapy. *Appl Radiat Isot* 70:1158–1161.

Iwata, K., H. Yoshimura, Y. Tsuji, et al. 1992. *In vivo* measurement of spatial dose distribution with thermoluminescent sheet around high dose-rate intracavitary source—Application to rectal-cancer. *Int J Radiat Oncol Biol Phys* 22:1109–1115.

Jakel, O., G. H. Hartmann, C. P. Karger, P. Heeg, and J. Rassow. 2000. Quality assurance for a treatment planning system in scanned ion beam therapy. *Med Phys* 27:1588–1600.

Jang, K. W., W. J. Yoo, S. H. Shin, D. Shin, and B. Lee. 2012. Fiber-optic Cerenkov radiation sensor for proton therapy dosimetry. *Opt Express* 20:13907–13914.

Johansson, B., G. Wickman, and J. BaharGogani. 1997. General collection efficiency for liquid iso-octane and tetramethylsilane in pulsed radiation. *Phys Med Biol* 42:1929–1938.

Jursinic, P. A. 2007. Characterization of optically stimulated luminescent dosimeters, OSLDs, for clinical dosimetric measurements. *Med Phys* 34:4594–4604.

Jursinic, P. A. 2009. Angular dependence of dose sensitivity of surface diodes. *Med Phys* 36:2165–2171.

Keeling, V. P., S. Ahmad, O. Algan, and H. Jin. 2014. Dependency of planned dose perturbation (PDP) on the spatial resolution of MapCHECK 2 detectors. *J Appl Clin Med Phys* 15:100–117.

Klawikowski, S. J., C. Zeringue, L. S. Wootton, G. S. Ibbott, and S. Beddar. 2014. Preliminary evaluation of the dosimetric accuracy of the *in vivo* plastic scintillation detector OARtrac system for prostate cancer treatments. *Phys Med Biol* 59:N27–N36.

Knežević, Ž., L. Stolarczyk, I. Bessieres, et al. 2013. Photon dosimetry methods outside the target volume in radiation therapy: Optically stimulated luminescence (OSL), thermoluminescence (TL) and radiophotoluminescence (RPL) dosimetry. *Radiat Meas* 57:9–18.

Knill, C., M. Snyder, J. T. Rakowski, et al. 2016. Investigating ion recombination effects in a liquid-filled ionization chamber array used for IMRT QA measurements. *Med Phys* 43:4946822.

Ko, L., J. Kim, and J. Siebers, 2004. Investigation of the optimal back-scatter for an aSi electronic portal imaging device. *Phys Med Biol* 49:1723–1738.

Kohno, R., H. Yamaguchi, K. Motegi, et al. 2015. Position verification of the RADPOS 4-D *in vivo* dosimetry system. *IJMPCERO* 4:318–325.

Kohno, R., K. Hotta, K. Matsubara, et al. 2012. *In vivo* proton dosimetry using a MOSFET detector in an anthropomorphic phantom with tissue inhomogeneity. *J Appl Clin Med Phys* 13:159–167.

Kroll, F., J. Pawelke, and L. Karsch. 2013. Preliminary investigations on the determination of three-dimensional dose distributions using scintillator blocks and optical tomography. *Med Phys* 40:082104.

Kry, S., L. Dromgoole, P. Alvarez, et al. 2015. IROC Houston on-site audits and parameters that affect performance. *Med Phys* 42:3692–3692.

Kuntz, F., J. Y. Pabst, J. P. Delpech, J. P. Wagner, and E. Marchioni. 1996. Alanine-ESR *in vivo* dosimetry: A feasibility study and possible applications. *Appl Radiat Isot* 47:1183–1188.

Lacroix, F., A. S. Beddar, M. Guillot, L. Beaulieu, and L. Gingras. 2009. A design methodology using signal-to-noise ratio for plastic scintillation detectors design and performance optimization. *Med Phys* 36:5214–5220.

Lacroix, F., L. Archambault, L. Gingras, et al. 2008. Clinical prototype of a plastic water-equivalent scintillating fiber dosimeter array for QA applications. *Med Phys* 35:3682–3690.

Lau, A., S. Ahmad, and Y. Chen. 2015. A feasibility study of using a Cherenkov detector material with the prompt gamma range verification technique in proton therapy (Abstract). *Med Phys* 42:3567.

Lee, B., K. W. Jang, D. H. Cho, et al. 2008. Measurement of two-dimensional photon beam distributions using a fiber-optic radiation sensor for small field radiation therapy. *IEEE T Nucl Sci* 55:2632–2636.

Letourneau, D., J. Pouliot, and R. Roy. 1999. Miniature scintillating detector for small field radiation therapy. *Med Phys* 26:2555–2561.

Li, G., S. Bai, N. Chen, et al. 2013. Evaluation of the sensitivity of two 3D diode array dosimetry systems to setup error for quality assurance (QA) of volumetric-modulated arc therapy (VMAT). *J Appl Clin Med Phys* 14:13–24.

Liu, P. Z. Y., N. Suchowerska, J. Lambert, P. Abolfathi, and D. R. McKenzie. 2011. Plastic scintillation dosimetry: Comparison of three solutions for the Cerenkov challenge. *Phys Med Biol* 56:5805–5821.

Liu, P. Z., N. Suchowerska, P. Abolfathi, and D. R. McKenzie. 2012. Real-time scintillation array dosimetry for radiotherapy: The advantages of photomultiplier detectors. *Med Phys* 39:1688–1695.

Lomax, A. J., T. Bohringer, A. Bolsi, et al. 2004. Treatment planning and verification of proton therapy using spot scanning: Initial experiences. *Med Phys* 31:3150–3157.

Malyarenko, E. V., J. S. Heyman, H. H. Chen-Mayer, and R. E. Tosh. 2010. Time-resolved radiation beam profiles in water obtained by ultrasonic tomography. *Metrologia* 47:208–218.

Marcie, S., E. Charpiot, R. J. Bensadoun, et al. 2005. *In vivo* measurements with MOSFET detectors in oropharynx and nasopharynx intensity-modulated radiation therapy. *Int J Radiat Oncol Biol Phys* 61:1603–1606.

Markovic, M., S. Stathakis, P. Mavroidis, I. A. Jurkovic, and N. Papanikolaou. 2014. Characterization of a two-dimensional liquid-filled ion chamber detector array used for verification of the treatments in radiotherapy. *Med Phys* 41:051704.

Marrale, M., A. Longo, M. Spano, et al. 2011. Sensitivity of alanine dosimeters with Gadolinium exposed to 6 MV photons at clinical doses. *Radiat Res* 176:821–826.

Martens, C., C. De Wagter, and W. De Neve. 2001. The value of the LA48 linear ion chamber array for characterization of intensity-modulated beams. *Phys Med Biol* 46:1131–1148.

Masi, L., F. Casamassima, R. Doro, and P. Francescon. 2011. Quality assurance of volumetric modulated arc therapy: Evaluation and comparison of different dosimetric systems. *Med Phys* 38:612–621.

Meigooni, A. S. 1995. Dosimetry of interstitial brachytherapy sources: recommendations. *Med Phys* 22:2.

Michalec, B., G. Mierzwinska, M. Ptaszkiewicz, et al. 2014. Applicability of EPR/alanine dosimetry for quality assurance in proton eye radiotherapy. *Radiat Prot Dosim* 159:137–140.

Mijnheer, B., S. Beddar, J. Izewska, and C. Reft. 2013. *In vivo* dosimetry in external beam radiotherapy. *Med Phys* 40:070903.

Miyagawa, I., and W. Gordy. 1960. Electron spin resonance of an irradiated single crystal of alanine: 2nd-order effects in free radical resonances. *J Chem Phys* 32:255–263.

Mizuno, H., T. Kanai, Y. Kusano, et al. 2008. Feasibility study of glass dosimeter postal dosimetry audit of high-energy radiotherapy photon beams. *Radiother Oncol* 86:258–263.

Morales, J. E., S. B. Crowe, R. Hill, N. Freeman, and J. V. Trapp. 2014. Dosimetry of cone-defined stereotactic radiosurgery fields with a commercial synthetic diamond detector. *Med Phys* 41:111702.

Mrcela, I., T. Bokulic, J. Izewska, et al. 2011. Optically stimulated luminescence *in vivo* dosimetry for radiotherapy: Physical characterization and clinical measurements in (60)Co beams. *Phys Med Biol* 56:6065–6082.

Nagy, V., S. V. Sholom, V. V. Chumak, and M. F. Desrosiers. 2002. Uncertainties in alanine dosimetry in the therapeutic dose range. *Appl Radiat Isot* 56:917–929.

Nakaguchi, Y., F. Araki, M. Maruyama, and S. Saiga. 2012. Dose verification of IMRT by use of a COMPASS transmission detector. *Radiol Phys Technol* 5:63–70.

Naseri, P., N. Suchowerska, and D. R. McKenzie. 2010. Scintillation dosimeter arrays using air core light guides: Simulation and experiment. *Phys Med Biol* 55:3401–3415.

Nelms, B. E., D. Opp, J. Robinson, et al. 2012. VMAT QA: Measurement-guided 4D dose reconstruction on a patient. *Med Phys* 39:4228–4238.

Nelms, B. E., H. Zhen, and W. A. Tomé. 2011. Per-beam, planar IMRT QA passing rates do not predict clinically relevant patient dose errors. *Med Phys* 38:1037–1044.

Noor, N. M., M. Hussein, D. A. Bradley, and A. Nisbet. 2010. The potential of Ge-doped optical fibre TL dosimetry for 3D verification of high energy IMRT photon beams. *Nucl Instrum Meth A* 619:157–162.

Noor, N. M., M. Hussein, D. A. Bradley, and A. Nisbet. 2011. Investigation of the use of Ge-doped optical fibre for in vitro IMRT prostate dosimetry. *Nucl Instrum Meth A* 652:819–823.

Noor, N. M., M. Hussein, T. Kadni, D. A. Bradley, and A. Nisbet. 2014. Characterization of Ge-doped optical fibres for MV radiotherapy dosimetry. *Radiat Phys Chem* 98:33–41.

Oldham, M., J. H. Siewerdsen, A. Shetty, and D. A. Jaffray. 2001. High resolution gel-dosimetry by optical-CT and MR scanning. *Med Phys* 28:1436–1445.

Olko, 2010. Advantages and disadvantages of luminescence dosimetry. *Radiat Meas* 45:506–511.

Olko, P., B. Marczewska, L. Czopyk, et al. 2006. New 2-D dosimetric technique for radiotherapy based on planar thermoluminescent detectors. *Radiat Prot Dosimetry* 118:213–218.

Olsson, S., E. Lund, and R. Erickson. 1996. Dose response and fading characteristics of an alanine-agarose gel. *Appl Radiat Isot* 47:1211–1217.

Onori, S., F. dErrico, C. DeAngelis, et al. 1997. Alanine dosimetry of proton therapy beams. *Med Phys* 24:447–453.

Paganetti, H., H. Jiang, S. Y. Lee, and H. M. Kooy. 2004. Accurate Monte Carlo simulations for nozzle design, commissioning and quality assurance for a proton radiation therapy facility. *Med Phys* 31:2107–2118.

Perichon, N., T. Garcia, P. Francois, et al. 2011. Calibration of helical tomotherapy machine using EPR/alanine dosimetry. *Med Phys* 38:1168–1177.

Perks, J., M. Gao, V. Smith, S. Skubic, and S. Goetsch. 2005. Glass rod detectors for small field, stereotactic radiosurgery dosimetric audit. *Med Phys* 32:726–732.

Piesch, E., 1972. Developments in radio photoluminescence dosimetry, *Topics in Radiat Dosim* 1:461–532.

Pittet, P., A. Ismail, J. Ribouton, et al. 2013. Fiber background rejection and crystal over-response compensation for GaN based *in vivo* dosimetry. *Phys Med* 29:487–492.

Poppe, B., C. Thieke, D. Beyer, et al. 2006. DAVID-a translucent multi-wire transmission ionization chamber for *in vivo* verification of IMRT and conformal irradiation techniques. *Phys Med Biol* 51:1237–1248.

Poppe, B., H. K. Looe, N. Chofor, et al. 2010. Clinical performance of a transmission detector array for the permanent supervision of IMRT deliveries. *Radiother Oncol* 95:158–165.

Poppe, B., T. S. Stelljes, H. K. Looe, et al. 2013. Performance parameters of a liquid filled ionization chamber array. *Med Phys* 40:082106.

Rah, J. E., D. O. Shin, J. S. Jang, et al. 2008. Application of a glass rod detector for the output factor measurement in the CyberKnife. *Appl Radiat Isot* 66:1980–1985.

Rah, J. E., S. Kim, K. H. Cheong, et al. 2009. Feasibility study of radiophotoluminescent glass rod dosimeter postal dose intercomparison for high energy photon beam. *Appl Radiat Isot* 67:324–328.

Ralston, A., M. Tyler, P. Liu, D. McKenzie, and N. Suchowerska. 2014. Over-response of synthetic microDiamond detectors in small radiation fields. *Phys Med Biol* 59:5873–5881.

Ravkilde, T., P. J. Keall, C. Graua, M. Høyer, and P. R. Poulsen. 2013. Time-resolved dose reconstruction by motion encoding of volumetric modulated arc therapy fields delivered with and without dynamic multi-leaf collimator tracking. *Acta Oncol* 52:1497–1503.

Reft, C. S. 2009. The energy dependence and dose response of a commercial optically stimulated luminescent detector for kilovoltage photon, megavoltage photon, and electron, proton, and carbon beams. *Med Phys* 36:1690–1699.

Renaud, J., D. Marchington, J. Seuntjens, and A. Sarfehnia. 2013. Development of a graphite probe calorimeter for absolute clinical dosimetry. *Med Phys* 40:020701.

Reniers, B., G. Landry, R. Eichner, A. Hallil, and F. Verhaegen. 2012. *In vivo* dosimetry for gynaecological brachytherapy using a novel position sensitive radiation detector: Feasibility study. *Med Phys* 39:1925–1935.

Reshchikov, M. A., X. Gu, B. Nemeth, J. Nause, and H. Morkoc. 2005. High quantum efficiency of photoluminescence in GaN and ZnO. In *Materials Research Society Symposium Proceedings*, Kuball, M., T. Myers, J. Redwing, and T. Mukai (Eds.), 0892-FF23-11. Boston, MA: Cambridge University Press.

Rikner, G., and E. Grusell. 1987. General specifications for silicon semiconductors for use in radiation-dosimetry. *Phys Med Biol* 32:1109–1117.

Rossomme, S., H. Palmans, R. Thomas, et al. 2014. Reference dosimetry for light-ion beams based on graphite calorimetry. *Radiat Prot Dosim* 161:92–95.

Sabet, M., F. W. Menk, and P. B. Greer. 2010. Evaluation of an a-Si EPID in direct detection configuration as a water-equivalent dosimeter for transit dosimetry. *Med Phys* 37:1459–1467.

Sato, F., Y. Toyota, D. Maki, et al. 2013. Development of bead-type radiophotoluminescence glass dosimeter applicable to various purposes. *Radiat Meas* 55:68–71.

Saw, C. B., J. Shi, and D. H. Hussey. 1998. Energy dependence of a new solid state diode for low energy photon beam dosimetry. *Med Dosim* 23:95–97.

Sawant, A., L. E. Antonuk, Y. El-Mohri, et al. 2006. Segmented crystalline scintillators: Empirical and theoretical investigation of a high quantum efficiency EPID based on an initial engineering prototype CsI(Tl) detector. *Med Phys* 33:1053–1066.

Scarboro, S. B., D. S. Followill, J. R. Kerns, R. A. White, and S. F. Kry. 2012. Energy response of optically stimulated luminescent dosimeters for non-reference measurement locations in a 6 MV photon beam. *Phys Med Biol* 57:2505–2515.

Schaeken, B., and P. Scalliet. 1996. One year of experience with alanine dosimetry in radiotherapy. *Appl Radiat Isot* 47:1177–1182.

Schaeken, B., S. Lelie, P. Meijnders, et al. 2010. Alanine/EPR dosimetry applied to the verification of a total body irradiation protocol and treatment planning dose calculation using a humanoid phantom. *Med Phys* 37:6292–6299.

Scott, A. J. D., S. Kumar, A. E. Nahum, and J. D. Fenwick. 2012. Characterizing the influence of detector density on dosimeter response in non-equilibrium small photon fields. *Phys Med Biol* 57:4461–4476.

Sharpe, P. H. G., K. Rajendran, and J. P. Sephton. 1996. Progress towards an alanine/ESR therapy level reference dosimetry service at NPL. *Appl Radiat Isot* 47:1171–1175.

Slosarek, K., M. Szlag, B. Bekman, and A. Grzadziel. 2010. EPID *in vivo* dosimetry in RapidArc technique. *Rep Prac Oncol Radiother* 15:8–14.

Son, J., M. Kim, D. Shin, et al. 2015. Development of a novel proton dosimetry system using an array of fiber-optic Cerenkov radiation sensors. *Radiother Oncol* 117:501–504.

Soubra, M., J. Cygler, and G. Mackay. 1994. Evaluation of a dual bias dual metal-oxide-silicon semiconductor field-effect transistor detector as radiation dosimeter. *Med Phys* 21:567–572.

Stathakis, S., P. Myers, C. Esquivel, P. Mavroidis, and N. Papanikolaou. 2013. Characterization of a novel 2D array dosimeter for patient specific quality assurance with volumetric arc therapy. *Med Phys* 40:071731.

Stewart, K. J., A. Elliott, and J. P. Seuntjens. 2007. Development of a guarded liquid ionization chamber for clinical dosimetry. *Phys Med Biol* 52:3089–3104.

Task Group 21. 1983. A protocol for the determination of absorbed dose from high-energy photon and electron-beams. *Med Phys* 10:741–771.

Teymurazyan, A., and G. Pang. 2012. Monte Carlo simulation of a novel water-equivalent electronic portal imaging device using plastic scintillating fibers. *Med Phys* 39:1518–1529.

Therriault-Proulx, F., T. M. Briere, F. Mourtada, et al. 2011. A phantom study of an *in vivo* dosimetry system using plastic scintillation detectors for real-time verification of Ir-192 HDR brachytherapy. *Med Phys* 38:2542–2551.

Tolli, H., R. Sjogren, and M. Wendelsten. 2010. A two-dose-rate method for general recombination correction for liquid ionization chambers in pulsed beams. *Phys Med Biol* 55:4247–4260.

Tosh, R. 2013. Interferometry for detection of temperature rise in a water phantom (Abstract). *Med Phys* 40:82–83.

Trompier, F., P. Fattibene, D. Tikunov, et al. 2004. EPR dosimetry in a mixed neutron and gamma radiation field. *Radiat Prot Dosim* 110:437–442.

Tsuda, M. 2000. A few remarks on photoluminescence dosimetry with high energy X-rays. *Jpn J Med Phys* 20:131–139.

Underwood, T. S., B. C. Rowland, R. Ferrand, and L. Vieillevigne. 2015. Application of the exradin W1 scintillator to determine ediode 60017 and microDiamond 60019 correction factors for relative dosimetry within small MV and FFF fields. *Phys Med Biol* 60:6669–6683.

van Elmpta, W., L. McDermottb, S. Nijstena, et al. 2008. A literature review of electronic portal imaging for radiotherapy dosimetry. *Radiat Oncol* 88:289–309.

Venkataraman, S., K. E. Malkoske, M. Jensen, et al. 2009. The influence of a novel transmission detector on 6 MV X-ray beam characteristics. *Phys Med Biol* 54:3173–3183.

Vial, P., P. B. Greer, L. Oliver, and C. Baldock. 2008. Initial evaluation of a commercial EPID modified to a novel direct-detection configuration for radiotherapy dosimetry. *Med Phys* 35:4362–4374.

Viamonte, A., L. A. R. da Rosa, L. A. Buckley, A. Cherpak, and J. E. Cygler. 2008. Radiotherapy dosimetry using a commercial OSL system. *Med Phys* 35:1261–1266.

Wagner, A., F. Crop, T. Lacornerie, F. Vandevelde, and N. Reynaert. 2013. Use of a liquid ionization chamber for stereotactic radiotherapy dosimetry. *Phys Med Biol* 58:2445–2459.

Wagner, D., M. Anton, H. Vorwerk, et al. 2008. *In vivo* alanine/electron spin resonance (ESR) dosimetry in radiotherapy of prostate cancer: A feasibility study. *Radiat Oncol* 88:140–147.

Wang, L. L. W., and S. Beddar. 2011. Study of the response of plastic scintillation detectors in small-field 6 MV photon beams by Monte Carlo simulations. *Med Phys* 38:1596–1599.

Wang, L. L. W., L. A. Perles, L. Archambault, et al. 2012. Determination of the quenching correction factors for plastic scintillation detectors in therapeutic high-energy proton beams. *Phys Med Biol* 57:7767–7781.

Wang, R., P. Pittet, J. Ribouton, et al. 2013. Implementation and validation of a fluence pencil kernels model for GaN-based dosimetry in photon beam radiotherapy. *Phys Med Biol* 58:6701–6712.

Wang, S., J. K. Gardner, J. J. Gordon, et al. 2009. Monte Carlo-based adaptive EPID dose kernel accounting for different field size responses of imagers. *Med Phys* 36:3582–3595.

Wendling, M., L. N. McDermott, A. Mans, et al. 2012. In aqua vivo EPID dosimetry. *Med Phys* 39:367–377.

Wong, J. H., M. Carolan, M. L. Lerch, et al. 2010. A silicon strip detector dose magnifying glass for IMRT dosimetry. *Med Phys* 37:427–439.

Wong, J. H., T. Knittel, S. Downes, et al. 2011. The use of a silicon strip detector dose magnifying glass in stereotactic radiotherapy QA and dosimetry. *Med Phys* 38:1226–1238.

Wootton, L., C. Holmes, N. Sahoo, and S. Beddar. 2015. Passively scattered proton beam entrance dosimetry with a plastic scintillation detector. *Phys Med Biol* 60:1185–1198.

Xiang, L. Z., B. Han, C. Carpenter, et al. 2013. X-ray acoustic computed tomography with pulsed X-ray beam from a medical linear accelerator. *Med Phys* 40:010701.

Yamamoto, T., D. Maki, F. Sato, et al. 2011. The recent investigations of radiophotoluminescence and its application. *Radiat Meas* 46:1554–1559.

Yeo, I. J., J. W. Jung, B. Patyal, et al. 2013. Conditions for reliable time-resolved dosimetry of electronic portal imaging devices for fixed-gantry IMRT and VMAT. *Med Phys* 40:072102.

Yin, Z., R. P. Hugtenburg, and A. H. Beddoe. 2004. Response corrections for solid-state detectors in megavoltage photon dosimetry. *Phys Med Biol* 49:3691–3702.

Yorke, E., R. Alecu, L. Ding, et al. 2005. Diode in vivo dosimetry for patients receiving external beam radiation therapy: Report of task group 62 of the radiation therapy committee. Madison, WI: Medical Physics Publishing.

Yukihara, E. G., and S. W. S. McKeever. 2006. Ionisation density dependence of the optically and thermally stimulated luminescence from Al_2O_3:C. *Radiat Prot Dosim* 119:206–217.

Yukihara, E. G., G. Mardirossian, M. Mirzasadeghi, S. Guduru, and S. Ahmad. 2008. Evaluation of Al_2O_3:C optically stimulated luminescence (OSL) dosimeters for passive dosimetry of high-energy photon and electron beams in radiotherapy. *Med Phys* 35:260–269.

Zhen, H., B. E. Nelms, and W. A. Tome. 2011. Moving from gamma passing rates to patient DVH-based QA metrics in pretreatment dose QA. *Med Phys* 38:5477–5489.

Zushi, N., F. Sato, Y. Kato, T. Yamamoto, and T. Iida. 2014. Flexible sheet with radiophotoluminescence glass beads for remotely monitoring high beta-surface-contamination. *Radiat Meas* 71:217–219.

CHAPTER 12

Safety Aspects, Failure Mode and Effect Analysis, and Safety Enhancement Technologies

Julian Perks and David Hoffman

CONTENTS

12.1 INTRODUCTION

The main emphasis of this chapter is on how safety and quality within an institution should not be maintained by only following guidelines but also by analysis of the practice at that specific institution. The chapter opens with some background on errors and public awareness of safety, first on a general scale of medical errors and then specific to radiation oncology.

Safety within the medical care system was brought into sharp and widespread public focus by the 1999 U.S. Institute of Medicine report "To Err Is Human: Building a Safer Health System." The report triggered a number of responses from health agencies, most notably with President Bill Clinton signing Senate Bill 580, "Healthcare Research and

Quality Act of 1999" (Stelfox et al., 2006). A number of professional bodies also published their responses and generated quality guidelines.

In 2008 a British-based consortium produced "Towards Safer Radiotherapy." The consortium consisted of the British Institute of Radiology, the Institute of Physics and Engineering in Medicine (IPEM), representatives from the National Health Service, the College of Radiographers, and the Royal College of Radiologists. This publication described the nature of human errors in radiotherapy, a methodology for safe radiation delivery, and ways to learn from and deal with errors.

Two years later the American Society for Radiation Oncology (ASTRO) published "Safety Is No Accident." As with the British publication, the ASTRO document was developed and endorsed by a large consortium—American Association of Medical Dosimetrists, American Association of Physicists in Medicine (AAPM), American Board of Radiology, American Brachytherapy Society, American College of Radiology, American Radium Society, American Society of Radiologic Technologists, Society of Chairmen of Academic Radiation Oncology Programs, and Society for Radiation Oncology Administrators.

"Safety Is No Accident" takes a somewhat different tack to the previous safety publications. The document opens with a discussion on the flow of patient care, expanding the safety concept to the patient planning process. There is a greater focus on the concept of "safety culture" and the responsibility of the team approach. Included in the team concept are definitions of the roles and responsibilities and a guide to staffing. Notably, the list of responsibilities shows the radiation oncologist as responsible for all aspects of care. Writing about staffing levels is a bold move for a professional society because it encroaches into the arena of clinic and personnel management, going as far as to specify numbers on full-time equivalent (FTE) staff required for given clinical practices, for example, intensity-modulated radiotherapy (IMRT) and brachytherapy.

ASTRO followed "Safety Is No Accident" with a number of white papers on safety concerns in specific areas of radiation oncology practice, including stereotactic body radiation and brachytherapy. The white papers allow ASTRO experts to give more in-depth recommendations for what constitutes good practice in specialized aspects of clinical radiation oncology.

12.2 SHIFT FROM PRESCRIPTIVE QA TO PRACTICE BASED QA

This section discusses AAPM task group (TG) reports, mainly TG40 and TG142 (Kutcher et al., 1994, Klein et al., 2009), along with other special procedure-based reports and how they prescribe tests to be performed and the tolerances regardless of the local practice. The discussion will allude to the strengths and weaknesses of this approach.

There is a class of quality assurance (QA) procedures and recommendations that are directly prescriptive. The logic is that one performs the tests described to measure a given quantity and if the measurement falls within a set tolerance, then all is well. The AAPM has a very strong compendium of such procedures in the form of task group reports. Each report is produced by a small committee formed from AAPM members known for their expertise and experience in the subject. The task group reports can be relatively broad, such as TG 142 (Quality assurance of medical accelerators), or very specific, such as TG 148 (QA for helical Tomotherapy) (Langen et al. 2010).

The AAPM task group reports form a substantial and extremely strong quality assurance compendium. They are meticulously constructed by current experts and renowned medical physicists practicing in specific areas. These TG reports clearly have a strong place in the quality of care paradigm but there is a more recent movement to add more individualized programs to institutions. The thought process here is not to detract from the TG reports but that local practice or patient mix could yield safety improvements by focusing on the smaller scale. The concept is that local clinical practice can vary, and considering these variations along with global QA practice should provide the patient with a safer treatment route. One of the simplest processes to begin with is a continuous quality improvement (CQI) project.

A CQI project usually focuses on a relatively narrow subject, but one that can have an impact on patient safety or comfort. An example is monitoring the time taken at each stage of planning and delivery for gynecological high-dose-rate (HDR) brachytherapy. With HDR, patients often receive planning and treatment in a single day, often involving sedation, and many team members are involved, and communication and patient hand-off are critical. A CQI project would show if any of the stages could be streamlined, and hence whether the patient experience could be smoother and more comfortable.

The example of a CQI project based on HDR brachytherapy shows how local practice should be considered in addition to published guidelines (ASTRO white papers and AAPM TG reports). Clearly, the published AAPM reports are essential for the quality assurance of the HDR source calibration, but there are many other aspects to be considered in terms of the practice—including patient sedation, patient transport, who performs morning quality assurance checks, how the patient (data) is handed off during planning, and so on. The safest and most efficient patient experience will be generated by considering both task group reports and local regulations, and how they integrate into local, clinical practice.

12.3 A SAFETY CULTURE

This section will detail the notion of a safety culture and what it would mean for a clinical department.

A number of publications (Potters and Kapur 2012, Ross 2015) have talked about a "safety culture" and what it means for the healthcare industry. A strong and positive safety culture within a department indicates a state of openness and communication between staff members. A state of awareness over safety and a departmental willingness to solve problems in a blame-free way has been shown to reduce accidents.

Conversely, there is evidence that an environment in which potential errors cannot be brought to light will result in those errors occurring. Establishing and maintaining a safety culture is the responsibility of both management and staff, and the balance is often difficult—a number of awkward situations that can arise. There may be a feeling of "that is the way we have always done things" or "why say anything, nothing changes." Additionally, different industries show marked differences in safety culture—a sign depicting "20 days without an accident" may be excusable on a construction site but it would be wholly inappropriate in the clinical setting.

12.4 THE FAILURE MODES AND EFFECTS ANALYSIS PROCESS

This section details the move to institution-specific QA practices and improvements with a review of the initial publications, a worked example, the range of applications, and the advantages and disadvantages.

The failure modes and effects analysis (FMEA) process was first employed by the aerospace industry in the 1960s and has been a staple of industrial safety analysis since. The term "failure modes" refers to the potential ways in which a failure might occur. The "effect analysis" attempts to understand the consequences of those failures. The premise of FMEA is that safety improvements are implemented in the most cost-effective manner. The first stage of a successful FMEA requires the entire team that is involved to assemble and lay out a process map. In the specific FMEA of radiation oncology, the representatives of the treatment team, including radiation oncologists, medical physicists, dosimetrists, nurses, and radiation therapists, should gather and design the process map to include each encounter that the patient passes through. Some points in the patient pathway may be repeated and some may occur only if certain conditions are met; hence, a process map needs to be created. The individual steps in the process map then form the "modes" of the analysis. To follow through with the FMEA, three risk factors are estimated for each mode.

For each mode in the patient treatment process, three risk factors form the risk priority number (RPN) and are estimated by the team performing the FMEA; the factors are the likelihood that an error could occur, the likelihood that the error could be detected, and the severity of the event should it reach the patient. The RPN of each mode is generated by ranking the three risks—(occurrence [O], detectability (inversely) (D), and severity [S], respectively) from 1 to 10, then multiplying the three numbers to determine the RPN. The RPNs are multiplied together so that any of the three factors can have full bearing on the final outcome. A certain risk may have a low rate of occurrence (2–3) but may have a catastrophic severity (9–10), so the multiple will strongly bear this out. The final stage of the FMEA is to take the highest RPNs and address them through changes in the workflow, new safety checks, or the implementation of new protocols. In this way, the highest risk modes of the entire chain of the treatment process receive the most attention. Although the FMEA process is rigorous, it is also labor intensive, and may be only that the top two or three RPN modes are addressed; this is logical, however, because resources to change practices are limited and should be spent in the most cost-effective manner, that is, addressing the worst modes.

As an example, consider the FMEA performed by the National Centre of Oncological Hadrontherapy (CNAO) Foundation in Pavia, Italy (Cantone et al., 2013) for a scanned beam proton treatment facility: One of the modes they considered was a manual correction of the patient's external contour. For this mode, the group identified that human error could yield an incorrect external contour and lead to errors in the dose distribution, with risk factors of $O = 4$, $D = 4$, and $S = 5$ for an RPN of 80. This team ranked 23 modes with RPNs ranging from 80 to 196. One important point to remember when embarking on an FMEA within a department is that the rankings are a personnel/group decision. One team may give a certain probability of occurrence for an event (determining patient identity at each treatment fraction, for example) as a 2 and another may give it a 5. The key to the FMEA is that the highest overall RPNs are tackled after all modes are assessed to the same (local) scale.

As a further illustration of the FMEA process, consider the nodes analyzed in an SBRT review from UC Davis (Perks et al. 2012), Figure 12.1.Most of the flow is linear, with one node leading to the next, but there are loops and switchbacks when the tolerance in a node

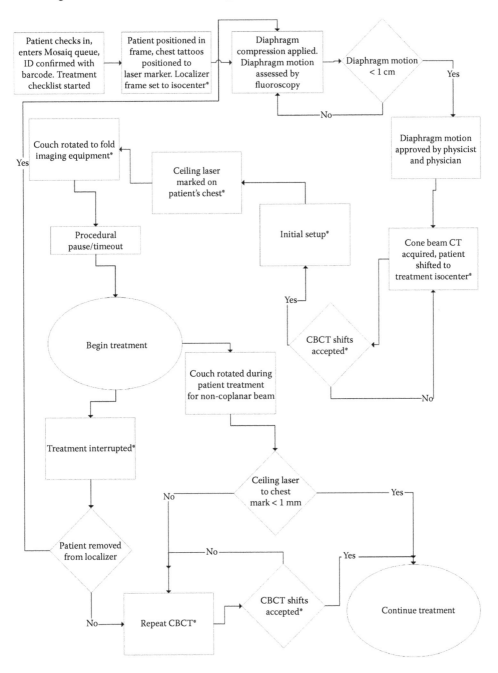

FIGURE 12.1 An illustration of the FMEA process for an SBRT (a stereotactic body radiation therapy treatment plan).

is not achieved. The FMEA assessment takes each node and the analyzing team discusses the risk factors associated with it. For example, take the node where the patient is positioned according to the shifts determined by the cone beam CT registration; the three contributing factors to the risk for this node would be: occurrence (how likely or how often would we expect a misalignment when applying cone beam CT derived shifts), detectability (if a misalignment occurred, how likely is it that the treatment team catches the error), and severity (what magnitude of error in applying the required shift could occur, and what the clinical impact is to the patient).

12.5 OTHER QUALITY METRICS AND ANALYSIS TOOLS

This section will detail the other available tools clinics may employ, including checklists; error, near-miss, and misadministration recording; root cause analysis; lean six sigma; and event learning.

Checklists may be the simplest but most effective and efficient method of ensuring a consistent and presumably safe treatment. The key to using a checklist is that it contains sufficient, salient information to act as an *aid memoir* without being overly cumbersome and therefore leading to noncompliance. For an extensive review of checklists in medicine, refer to *The Checklist Manifesto* (Gawande, 2011).

A clinic's definition of a medical event/misadministration should be based on both institutional policy and local and federal laws. In the case of radiation oncology, the standard factors associated with medical events (wrong patient, wrong site, wrong procedure) are augmented by considerations of the delivered dose differing from the prescription, as a percentage of the fractional and overall course dose.

In addition to recording and reporting medical events, a clinic may consider recording and auditing its "near misses." The term *near miss* is a little more nebulous compared to *medical event*, but one useful definition is "an error which is not caught at the expected level of checking but found before it reaches the patient." An example in the radiation oncology setting is a transposition of setup coordinates (left for right) that is not caught during plan second check but is found by the treating therapists as the patient is being set for the first fraction if treatment.

12.5.1 Root Cause Analysis Report

Following a medical event, a clinic may consider a formal investigation to determine contributing factors and prevent further occurrence. A root cause analysis is a process that details what happened, and aims to find the primary cause of the error by linking and tracing faults. A primer for a root cause analysis (RCA) is shown in the following:

1. What happened?

2. How did it happen?

3. Why did it happen?

    a. Human error?

    b. Systems issue?

    c. External factors?

    d. Was staffing an issue?

4. What was done (and by whom) to prevent recurrence?

5. Outcome: What was the outcome?

6. Any personnel trends involved?

7. Any departmental trends involved?

8. Risk reduction strategies:

9. What is to be monitored for the next six months and by whom?

A root cause analysis is reactive and is used after an event, whereas the FMEA is prospective, aiming to discover weaknesses or sources of error before they have their effect.

12.5.2 Lean Six Sigma

Lean six sigma is an analysis process that combines the six sigma (6σ) philosophy with the lean approach. The concept of 6σ may not seem immediately applicable in the healthcare setting. The aim in industries that apply 6σ is that a process produces 99.99966% of items without error, which is clearly desirable in manufacturing electronic components, but the number of patients seen in radiation oncology and the clinical variability make this a difficult concept to apply. Hence, the analysis system of lean six sigma is to consider the stages a patient progresses through in isolation and analyze them for efficiency (waste). The lean six sigma concept can be implemented and analyzed by following the acronym DMAIC—define, measure, analyze, improve, and control. When a logistical problem within a department is discovered (e.g., the flow of brachytherapy patients on their treatment day), each of the points of DMAIC can be spelled out, a process map can be drawn up, and bottlenecks in the system should become obvious and can be rectified. An example of a lean six sigma for surgical wait times has been carried out for the Veterans Affairs system (Valsangkar et al., 2017).

12.5.3 The Concept of Class Solutions

The term "class solution" is rarely used with regard to clinical radiation oncology, but it is a powerful concept that could show a significant safety benefit. The premise is that the planning of a particular diagnosis is governed by a fixed set of rules, so the checking of the plan should be more straightforward and errors should stand out. This principle has been applied to fractionated linear-accelerator-based treatments of pituitary adenoma (Perks et al., 1999) and inversely planned prostate implants (Lessard et al. 2006). Despite the power of using a fixed set of parameters for planning and checking treatment plans, the difficulties with the class solutions approach are deriving the plan parameters (linac gantry and couch angles, dose volume histogram points) by which plans will be fixed. Some dosimetrists and physicians may feel that class-solution-derived-plans can be further optimized, negating the "rules" by which the plan has been created.

12.6 SAFETY IN RADIATION ONCOLOGY RESEARCH

This section details additional safety considerations that the individual researcher and the institution must consider when engaging in research.

This section is specific to research conducted in the United States and the pertinent regulations. Outside the United States, researchers must familiarize themselves with the requisite practices and ethical codes.

Research is the acquisition of generalizable knowledge, and when human subjects are involved, specific requirements must be met. The terms "acquisition" and "generalizable" are key; it is not the publication of the research that invokes the regulations, but the conduct, where identifiable data on the human subject(s) are involved. All research involving human subjects is subject to institutional review board (IRB) review, approval, and monitoring.

An IRB is convened by the institution and should consist of scientific, nonscientific, and community members. They may review protocols based on scientific validity or the details of the science may be previewed by a more specific committee—one with more experience and in-depth knowledge. The IRB committee reviews protocols on the grounds of respect for persons, beneficence, and justice. Respect for persons deals with the informed consent aspects of the research, beneficence deals with the risk/benefit ratio, and justice ensures that the selection of research subjects is fair and equitable.

The IRB approval of a protocol is a regulatory step that must be adhered to, but it does not ensure safety. Researchers must, at all times, consider that they are departing from the clinical norm, and they may be running untested and unfamiliar techniques on human subjects (patients). A full working knowledge of the research protocol, whether it is written in house or the product of a national collaborative group trial, is essential.

12.7 INCIDENT REPORTING SYSTEMS

An incident reporting system is a digital registry where incidents can be logged with the aim of learning, distinguishing patterns, and ultimately preventing future occurrences. The main power of a reporting system comes from the input from the people with access to it. To be effective, the reporting system must be confidential and nonpunitive; this ensures that all incidents are reported without any fear of recriminations.

An excellent example of an incident reporting (and hence learning) system, which is often purported as the model system, is run by the airline industry (it is currently hosted by NASA). The Aviation Safety Reporting System offers online or paper submission of reports contains a search function, and offers collated sets of reports for a number of commonly reported categories.

For radiation oncology an online recording system exists in the form of the Radiation Oncology Incident Learning System (RO-ILS), presented by ASTRO. Originally established in 2011, the mission of RO-ILS is to facilitate safer and higher-quality care in radiation oncology. A strong feature of the system is that ASTRO membership is not required, bringing access to a wider audience. Quarterly reports are available; for example, the report from the third quarter of 2015 lists 197 logged incidents of which 68 were patient incidents, 101 were near misses, and 28 were unsafe conditions.

12.8 SAFETY ENHANCEMENT TECHNOLOGIES

A number of linear-accelerator-based features add to the safety of patient treatment, including a digital record-and-verify system (RVS); integrated portal dosimetry; and on board imaging, such as MV portal imaging and kV cone beam CT.

A digital RVS holds the patient demographics and their plan details. By acting as a conduit between the treatment planning system and the treatment delivery device, the RVS should eliminate transcription errors. The RVS takes the digital data directly from the planning system, holding it for each fraction and even allowing updates such as patient couch settings as they are acquired. The advantages of the RVS may be obvious, but there is at least one study (Patton et al., 2003) finding that overreliance on the RVS could lead to error propagation; where an error in the initial planning process is overlooked.

One of the latest developments is an integrated quality management system (IQM). The concept of an IQM is that the treatment beam is monitored before it reaches the patient for every irradiation. Typically an IQM consists of a large area ion chamber, divided into sections, with an integrated thermometer and barometer. The ion chamber collects charge produced by the photon beam and reports a total for each beam, control point, or segment, depending on treatment modality. In addition, the device can monitor gantry and collimator angles by the inclination of the device as measured with the integrated inclinometer. By performing a patient-specific QA check with the IQM in place, the user can establish baseline data (a checksum) by which all subsequent irradiations can be checked.

12.9 CONCLUSION

This chapter has dealt with a number of safety issues that are pertinent to modern radiation therapy. The need for a safe and accurate delivery of radiation to the patient is clear because errors can be catastrophic and irreversible. The chapter began by describing how the philosophy of quality assurance surrounding radiation oncology is shifting from merely being a prescriptive (if comprehensive) set of "rules" to a comprehensive set of checks based on an individual clinic's needs and accepted standards of practice. Furthermore, the industrial concepts of safety, as applied to radiation therapy, were explored. These industrial concepts of safety include how a safety culture can enhance the working environment and how a failure modes and effects analysis can highlight risky practices and then direct resource allocation to solve or mitigate those risks. The incorporation of industrial QA methods into clinical practice continued as three relatively underused concepts were presented listed in Section 12.5. Root cause analysis can be used to derive the ultimate cause of a catastrophic event, lean six sigma shows the advantages of streamlined process control, and class solutions offer

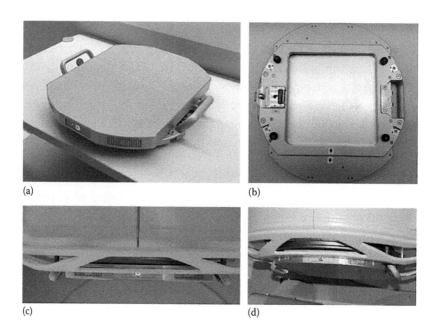

FIGURE 12.2   The Integral Quality Monitor (IQM) is a large area ion chamber (a) with a gradient of the ion chamber thickness in the axis of MLC motion. It attaches to the accessory tray holder, similar to an electron cone (b). The device has a low profile from the linac head (c and d) and connects wirelessly to a transceiver and the controlling computer.

the ability to homogenize plans so that errors or omissions are more obvious. The chapter closed with one software and one hardware addition to the radiation therapy safety armament: Incident reporting systems catalogue both incidents and near misses and can then be searched for patterns, and the integrated quality monitor can compare a planned treatment with a delivered one upstream from the patient. When coupled with portal imaging, the IQM can ensure correct dosing.

An example of an IQM device is shown in Figure 12.2.

REFERENCES

Cantone, M. C., M. Ciocca, F. Dionisi, P. Fossati, S. L. M. Krengli, S. Molinelli, R. Orecchia, M. Schwarz, I. Veronese, V. Vitolo. 2013. Application of failure mode and effects analysis to treatment planning in scanned proton beam radiotherapy. *Radiat Oncol.* 24;8:127. doi:10.1186/1748-717X-8-127.

Gawande, A. 2011. *The Checklist Manifesto: How to Get Things Right.* New York: Metropolitan Books.

Klein, E. E., J. Hanley, J. Bayouth, F. F. Yin, W. Simon, S. Dresser, C. Serago et al. 2009. Task Group 142 report: Quality assurance of medical accelerators. *Med Phys.* 36(9):4197–4212.

Kutcher, G., L. Coia, M. Gillin, W. F. Hanson, S. Leibel, R. J. Morton, J. R. Palta et al. 1994. Comprehensive QA for radiation oncology: Report of AAPM Radiation Therapy Committee Task Group 40. *Med Phys.* 21(4):581–618.

Langen K., M. N. Papanikolaou, J. Balog, R. Crilly, D. Followill, S. M. Goddu, W. Grant, G. Olivera, C. R. Ramsey, C. Shi. 2010. QA for helical tomotherapy: report of the AAPM Task Group 148. *Med Phys.* 37(9):4817–4853.

Lessard E., S.L. Kwa, B. Pickett, M. Roach, J. Pouliot. 2006. Class solution for inversely planned permanent prostate implants to mimic an experienced dosimetrist. *Med Phys.* 33(8):2773–2782.

Perks J.R., R. Jalali, V. P. Cosgrove, E. J. Adams, S. F. Sheperd, A. P. Warrington, M. Brada. 1999. Optimization of stereotactically-guided conformal treatment planning of sellar and parasellar tumors, based on normal brain dose volume histograms. *Int J Radiat Oncol Biol Phys.* 1;45(2):507–513.

Perks J.R., S. Stanic, R. L. Stern, B. Henk, M. S. Nelson, R. D. Harse, M. Mathai et al. 2012. Failure mode and effect analysis for delivery of lung stereotactic body radiation therapy. *Int J Radiat Oncol Biol Phys.* 15;83(4):1324–1329.

Patton G.A., Gaffney D.K., Moeller J.H. 2003. Facilitation of radiotherapeutic error by computerized record and verify systems. *Int J Radiat Oncol Biol Phys.* May 1;56(1):50–57.

Potters L., A. Kapur. 2012. Implementation of a "No Fly" safety culture in a multicenter radiation medicine department. *Pract Radiat Oncol.* 2(1):18–26.

Ross J. 2015. Safety culture-just what is it? *J Perianesth Nurs* 30(6):553–555.

Stelfox H. T., S. Palmisani, C. Scurlock, E. J. Orav, D. W. Bates. 2006. The "To Err is Human" report and the patient safety literature. *Qual Saf Health Care.* 15(3):174–178.

Valsangkar N.P., Eppstein A.C., Lawson R.A., Taylor A.N. 2017. Effect of lean processes on surgical wait times and efficiency in a tertiary care veterans affairs medical center. *JAMA Surg.* 152(1):42–47. doi:10.1001/jamasurg.2016.2808.

CHAPTER 13

Informatics as a Pathway for Integrating Radiation Oncology into Modern Medicine

Mark H. Phillips, Wade P. Smith, Kristi R. G. Hendrickson, and Alan M. Kalet

CONTENTS

13.1 INTRODUCTION

Biomedical informatics is a term that has no single, clear-cut definition. In general, it means the science of storing, retrieving, and using biomedical information for problem solving and decision making. Another aspect to this endeavor is knowledge representation because the choice of which data to focus on and apply is certainly linked to the overall understanding of the problem.

Informatics is often linked to computers, which is not surprising. At one end of the spectrum, information that is easily encodable is a natural match to the capabilities of computers. This has evolved hand-in-hand with our developing technology, as is witnessed in the field of genomics. Early on, our understanding was at the chromosomal level, and radiation effects were often described in terms of double-strand breaks, crossovers, and so on. As we developed the ability to identify sequences of base pairs, the amount of knowledge exploded. Both the quantity and quality of this type of data naturally led to the need for intense computerization. At the other end of the spectrum of biomedical informatics is the development of computable methods and models. For example, Benson et al. (2006) developed a model for the spread of tumor cells through the lymphatics that relied on a computable model of human anatomy (the Foundational Model of Anatomy [Rosse and Mejino, 2003]) for development of a Markov model. These and similar aspects of informatics exploit the other strength of computers, which is their ability to perform complex calculations repetitively and accurately.

The original link between biomedical informatics and radiation oncology is the presence of physicists in the latter. Computer development and use has always been integral to physics research, and as physicists took up the problems of radiation therapy, they brought with them their knowledge of computers. In addition, knowledge representation—in the form of theories and models—is another aspect of physics that provided a pathway for informatics into radiation therapy. The field of radiation biology was an early application of this penchant for formal knowledge representation. Using collected clinical data associated with radiation injury, scientists developed models of DNA damage based on their knowledge of radiation transport. Clinical trials and practice then exploited the models to determine dosing and fractionation schemes.

The ability both to handle large amounts of data and to perform accurate computations led to computers being used to calculate distributions of radiation dose and later to incorporating large image sets as well. In imaging, the most comprehensive application relevant to radiation oncology of informatics concepts and methods has been developed, namely, Digital Imaging and Communication in Medicine—Radiation Therapy (DICOM-RT) Bidgood et al. (1997) and DICOM-RT (Neumann 2002). These standards facilitate high-level representation of what an image study, an image, and a radiation therapy plan is. In addition, we have detailed methods for representing them in digital format, transferring them, and applying them to any problem of interest.

However, this early intersection of informatics and radiation oncology has not resulted in a long-term, systematic incorporation of informatics concepts and practices into the clinical practice of radiation therapy. Other medical disciplines have seized the initiative and have been much more active in this area, particularly when it comes to electronic data, clinical decision support, and the use of genomic data. A number of areas are directly relevant to the practice of radiation oncology that would be well served by a broad program to apply and develop informatics approaches. This would be an appropriate and advantageous effort that would integrate radiation oncology with other areas of medicine in the context of developments in the twenty-first century.

Abbreviations

CBCT: Cone beam computed tomography

DICOM: Digital Imaging and Communications in Medicine

DICOM-RT: DICOM in radiation therapy

DLORO: Dependency-layered ontology for radiation oncology

DVH: Dose volume histogram

EHR: Electronic health record

EMR: Electronic medical record

HL7:   Health Level 7 (international standards organization focused on clinical and administrative data)

Linac: Linear accelerator

MRI: Magnetic resonance imaging

NTCP: Normal tissue complication probability

OIS: Oncology information systems, (for example, Mosaiq [Elekta AB], Aria [Varian Medical Systems Inc.])

PACS: Picture archiving and communication system

PSA: Prostate specific antigen

RTP: Radiation treatment planning software

SQL: Structured Query Language

TCP: Tumor control probability

US: Ultrasound

XML: Extensible markup language (markup language for documents in a format readable by humans and machines)

13.2 ELECTRONIC MEDICAL DATA

"Precision medicine" and "personalized medicine" are recent and growing national initiatives. From the National Institutes of Health website, "Precision medicine is an emerging approach for disease treatment and prevention that takes into account individual variability in genes, environment, and lifestyle for each person. While some advances in precision medicine have been made, the practice is not currently in use for most diseases" (NIH, 2016a). Similarly, personalized medicine is encapsulated as follows: "NIH research is working hard to solve the puzzle of how genes and lifestyle connect to affect our lives and our health" (NIH, 2016b). What is common to both of these concepts is the need for more granular data about more aspects of normal and disease physiology.

The challenge of improving our use of medical data is twofold. First, relevant data are collected in different departments and/or institutions with independent data collection and storage mechanisms and systems. Cancer has always required a multiprong attack, and the connections between the different methods are becoming ever more complex. For example, improving the treatment of oropharyngeal cancer involves advances in genomics (identifying effects of human papilloma virus), microsurgery (reducing morbidity for certain tumors), chemotherapy (helping to determine "responsive" tumors), and radiation therapy (determining optimal dosing) (Masterson et al., 2014). Often data from one of these specialities is in a different electronic medical record (EMR), and the subsequent therapies and assessments of outcomes are not communicated or shared with all of the patient's providers.

Second, the data are in electronic formats that are difficult to access. These challenges and obstacles take multiple forms: different EMRs with different database schemas, lack of structured data, data ownership issues, and data security and privacy concerns. While these may seem to be purely administrative issues easily solved by management decisions, the barriers are high and are rooted in many aspects of the practice of medicine.

These issues are being tackled by the informatics community in a variety of ways, targeting different aspects of the problems. It should be noted that these efforts are often not coordinated; frequently, initiatives are geographically centered, for example, in the United States, Europe, and Australasia.

13.2.1 Electronic Data Storage and Interchange

The most common application of health electronic data storage and interchange is in EMRs and/or EHRs. There is no clear separation between these two, though the Centers for Medicare & Medicaid Services considers a "health record" to be more inclusive than a "medical record." These are enterprise-wide software packages that perform a multitude of roles, both administrative and clinical. On the administrative level, they serve as a legal record of the healthcare provided, which is also integral to documentation for billing purposes. On the clinical/medical level, they can handle scheduling, referrals, and medical data storage. Large commercially available software packages, for example, Epic[1] and Cerner,[2]

---

[1] Epic Systems Corporation, Madison, Wisconsin, United States.
[2] Cerner Corporation, North Kansas City, Missouri, United States.

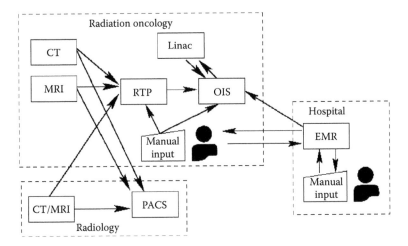

FIGURE 13.1   Flowchart of the sources and targets of data in the radiation oncology environment and the directions of data flow. (See abbreviation list.)

manage many of the activities of multiple departments in large and small medical centers. More specialized software packages are customized for a particular medical specialty, as will be discussed below for radiation oncology. A problem common to all of them is when they need to interface with other systems either within the same institution or from outside institutions and laboratories.

Figure 13.1 illustrates the sources of data in the radiation oncology environment and the data flow between them.

13.2.2 Health Level Seven

Health level seven (HL7) is a standards-developing organization "dedicated to providing a comprehensive framework and related standards for the exchange, integration, sharing, and retrieval of electronic health information." The standards have evolved since their beginning in 1987; version 2 is the most widely implemented and version 3 is the latest. The differences between the two versions reflect two different approaches to defining standards: namely, "bottom up" versus "top down." V2.x series standards have been widely implemented because low-level data elements and concepts are well-defined. This approach is very flexible, but it runs into problems when trying to communicate higher-level constructs. Detailed negotiations are necessary between the two communicating partners, and the result is unlikely to be understood outside that relationship. This also makes it difficult to support reliable conformance testing. V3 uses a reference information model (RIM) as its top-level model, from which all other aspects of the standard flow. The RIM model consists of four primary subject areas ("Entity," "Role," "Participation," "Acts") and 35 classes (http://www.hl7.org/implement/standards/product_brief.cfm?product_id=186, accessed April 20, 2017). For a given healthcare domain, an HL7 version 3 specification is based on the RIM—a common and underlying modeling framework—and includes artifacts such as: Use Case Models, Information Models, Interaction Models, Message Models, and Implementable Message Specifications (https://msdn.microsoft.com/en-us/library/ms954603.aspx, accessed April 20, 2017).

These standards define how the information is packaged and sent; for example, HL7.v3 uses XML to encode the details of the model, actions, and the data itself. Once the information is encoded properly, it is passed on to web services that use standard Internet transfer protocols, for example, TCP. In radiation oncology, we have developed our own standards (DICOM-RT) for performing much of this process. However, limiting ourselves to such a standard limits what we can learn about the medical outcomes of our treatments because that information is usually recorded outside the Radiation Oncology (RO) department. In addition, as we come to think of radiation therapy as one aspect of a sequence of therapies over time (Surveillance, Epidemiology and End Results (SEER) now considers cancer to be a *chronic* disease), we need to incorporate knowledge of these other actions into our understanding of the role of radiation. Not being able to communicate easily with the rest of the patient's EMR makes this difficult to do.

13.2.3 Oncology Information Systems

Oncology information systems (OISs) evolved from rudimentary record and verify (R&V) systems to very sophisticated systems that manage and integrate images from multiple sources, check plan parameters, and track the course of treatments. In addition, they have grown to incorporate tasks such as fractionation scheduling, chart checking, and so on. Their current evolution includes departmental scheduling, billing, and some rudimentary clinical data collection tools. Typically, they store the data in relational databases that can be queried, for example, SQL, and in text documents, and provide database management systems for interacting with these data either by structured query of the database itself or via internal data collection tools. The need for real-time communication between linac and R&V system has meant that the linac and OIS manufacturers must maintain knowledge of developments and changes in either system. These close relationships have resulted in the unintended effect that data standards have not been developed with the larger healthcare enterprise in mind, thereby limiting communication with the institutional/hospital EMRs.

13.2.4 Integrating the Healthcare Enterprise—Radiation Oncology

The mission of integrating the healthcare enterprise—radiation oncology (IHE-RO) is to advance the integration of all health information systems; radiation oncology is but one small domain. This body does not define standards but facilitates the use of existing standards, for example, HL7, DICOM, and so on. For example, for processes involving the use of DICOM-RT, IHE-RO standards give explicit instructions about which fields require which type of data. Similarly, at a higher level ("Interaction profiles"), the standards define what sort of data a system must be able to accept, for example, image sets and contours with variable z spacing. These may seem trivial or already constrained, but in fact they are neither. This has led to supposedly compatible systems failing in some instances, which can affect safe and effective treatments. The need for something like IHE-RO highlights the inevitable tension between complete compatibility achieved by defining every detail and flexibility for change and unforeseen situations in medical practice.

13.2.5 National Clinical Trial Repositories

The implementation of national clinical trials such as those run by the Radiation Therapy Oncology Group has required the development of data transfer standards. As with most other healthcare enterprises, there has been a gradual change from paper-based methods to electronic data. The standards of these trials serve two purposes: The first is to allow the data collected from multiple institutions to be compared, and the second is to reduce protocol violations in order to have a homogeneous group for analysis. The emergence of standards such as DICOM-RT has greatly reduced the overhead for the details of the radiation delivery, but the collection, transfer, and storage of other clinical variables is still idiosyncratic. In the main, these data transfers have been unidirectional—from radiation oncology departments to a central data warehouse. Data transfer out of the warehouse is on a case-by-case basis, and the data transfer process does not, in general, follow any widely accepted standard.

13.2.6 Local Solutions

Several groups have sought to establish data communication standards for the express purpose of increasing the number and diversity of patients in order to conduct studies using shared data from multiple institutions. However, the lack of a national sponsor has permitted questions of data ownership and patient privacy to become obstacles to sharing data. A group at Maastro Clinic (Netherlands) has organized a European-wide consortium in which "the data stays at home and the model goes to them" (CancerData, 2016). In this approach, the Maastro group published standards for XML representation of de-identified data. Models can be built with local data, and then the models are sent to different institutions for validation. If the model works well, then any member can use it on her or his data.

A somewhat similar approach called Oncospace was devised at Johns Hopkins University. This is a federated database at multiple institutions in which each institution instantiates an example of an SQL schema. Shared governance is used to maintain compatibility; each institution can add to the schema as long as a generalized query using the common schema will succeed. By sequestering Private Health Information (PHI) and using specialized privileges, an institution can use its data in any way it desires while sharing only that which it deems suitable. An additional goal of the Oncospace project is to increase the amount of structured data collected by means of custom clinical interfaces to the SQL database and/or a commercial OIS. This highlights two important issues. First, it shows how the collection of data can be separated from its digital format. The customized interfaces can gather any sort of data, but then the software transforms and stores it into a format that allows for generalized use, in this case, a standardized SQL schema. Second, it points out a critical issue, which is the relative paucity of structured data in radiation oncology (and indeed in larger EMRs) (Deasy et al., 2010). The power of clinical notes to transmit critical information between members of the care team means that much data is locked into document (and therefore unsearchable) formats.[3]

---

[3] Progress in natural language processing of clinical notes is impressive, but it is still plagued by many difficulties and cannot be relied on to provide a useful source for structured queries.

In addition, administrative issues and reimbursement procedures have decreased the willingness to collect structured data within the clinical workflow. Currently, too much of necessary structured data can be obtained only through a costly, time-consuming, and error-prone process of human data abstraction.

13.2.7 Summary

In the previous sections, the environment for standardized information exchange was described at the most general down to the most radiation oncology-specific levels. At the most specific level, it is clear that the field has to deal with several important issues: There is a lack of data for the type of modeling that is needed, and much of the data that exists is locked up in formats that make it difficult or impossible to extract. Addressing and creating solutions to these issues will lead to opportunities in the near future to improve our informatics capabilities and merge radiation therapy data with data from the other relevant specialties.

13.3 KNOWLEDGE REPRESENTATION

How knowledge is represented has implications far beyond the methods for storing and retrieving information. It both frames the questions that we ask and the methods that we use to study them (Rheinberger, 2010). At this time, our knowledge of the mechanisms of cancer does not often meet our needs to understand clinical decision-making factors. For a small number of cancers, there is sufficient genetic and biochemical understanding, and this has led to the development of very targeted agents. At the other end of the scale, we are still treating patients based on very broad concepts of tumor staging. Looking to better quantify understanding, physicists have been very active in developing probabilistic models of tumor control and normal tissue complications. Even one of the most basic tenets of tumor biology—unregulated growth of a single clonal cell—is no longer considered to be true. This has led to a far more sophisticated understanding of the mechanisms of tumor control, response to therapies, and recurrence. Future schemas for representation of our knowledge of any aspects of cancer treatment need to be flexible enough to adapt as our understanding expands.

13.3.1 Ontologies

One approach to representing our knowledge of a given system is through an ontology. There are numerous examples, such as Systemized Nomenclature of Medicine (SNOMED) (Snomed, 2016), National Cancer Institute Thesaurus (NCIT, 2016), Gene Ontology (GO, 2016), Radiation Oncology Ontology (ROO, 2016), and Dependency-Layered Ontology for Radiation Oncology (DLORO, 2016). The first two have the goal of introducing a standard vocabulary. At one level, this is important to facilitate information exchange between systems. However, they are more subtle than that. By using the concepts of "classes" (such as *body structure, clinical finding, event, record artifact*, etc.) and "properties" (such as *dependent of, due to, has causative agent*, etc.) the terms are linked to ideas about biology, disease, treatments, and many other aspects of medicine. The classes (and subclasses) are hierarchically related through relationships, such as "… is a …," and nonhierarchically through associative relationships, such as those listed.

Even more explicitly, the Gene Ontology "defines concepts/classes used to describe gene function, and relationships between these concepts. It classifies functions along three aspects: molecular function ..., cellular component ..., biological process ..." whose goal is to "to develop an up-to-date, comprehensive, *computational model of biological systems*, from the molecular level to larger pathways, cellular and organism-level systems" (GO, 2016).

The two radiation oncology ontologies are more explicit but differ in their classes and relationships. For example, the Radiation Oncology Ontology starts with very general classes, with more and more specific subclasses. The chain of subclasses is event → activity → behavior → occupational activity → healthcare activity → therapeutic procedure → radiation therapy → radiotherapy fraction. In the DLORO ontology, the class structure is much more direct: Prescription_parameter → Fractionation_parameter → Total_Fractions. Clearly the former allows for much more general concepts to be represented with perhaps only the addition of one or few more classes or properties, whereas the latter may be more difficult to generalize. On the other hand, the DLORO ontology was built to represent therapy treatments for the purpose of modeling the therapy process for optimization and error cases (Kalet et al., 2015). The translation to clinical software is easy in such a case. While the two ontologies are clearly related in many ways, their structures may be better suited for different types of problems. In other words, the type of problem that can be framed and addressed depends on how the knowledge is represented.

13.3.2 Mechanistic versus Probabilistic Representations

Physicists usually strive for mathematical representations of reality that involve direct relationships between cause and effect. Even statistical mechanics takes this approach because the statistics are representative of the distributions of causal mechanisms. In radiation oncology, the field of radiobiology takes this approach. Radiation damage is caused by energy transfer processes leading to chemical changes among molecules that have direct effects on death or survival. This highlights the role of stochastic processes that, again, can be described statistically.

At the level of disease phenotypes and the effects of treatment, however, our lack of knowledge dictates the type of representation that we can use. Probabilistic models are the best we can do. These representations encode the type of experiments (clinical trials) that have been performed and the results that have been measured. This type of knowledge can be represented with mathematics similar to the mechanistic, stochastic models. For example, survival curves—our most basic knowledge representation—provides the chance that a patient with a given set of characteristics will die within any particular time period after treatment. The emerging importance and understanding of the role that epigenetic factors play highlight the continuing sophistication of our models as we begin to incorporate this more detailed knowledge into our models.

The simplest probabilistic representation is embodied in tumor control and normal tissue complication probability regression models. Logistic regression on clinical trial data using a handful of variables provides our best knowledge of how dose and other variables affect outcomes. They are also the earliest examples of machine learning because the variables other than dose are selected according to their ability to best match the data without necessarily knowing why those variables are important. More advanced machine learning includes more data points, more variables, and more modeling methods (El Naqa, 2015). Models that emerge from modern machine learning methods have no mechanistic or causal structure, although with the results, conjectures can start to be made. This is certainly the hope of many genomic machine learning efforts.

A more comprehensive approach to probabilistic knowledge representations is through the use of influence diagrams (Owens et al., 1997). Built on the structure of Bayesian networks, they are directed toward knowledge representations in the context of medical decision making (Kjaerulff and Madsen, 2010). These models represent various types and sources of knowledge. One important difference compared with the above-mentioned models is that they can incorporate "degrees of belief" as well as frequentist counting data in their probabilistic representation. In these models, variables are usually chosen from a basic causal understanding, for example, dose versus response or the relationship between PSA and prostate cancer, but this is not true for all variables and/or models. One advantage of such models is that it is relatively easy to expand them as our knowledge increases; for example, a biophysical model of the role of DNA repair can replace a simple NTCP curve. In a model of the spread of disease to the lymphatics in head and neck cancers and their response to radiation, an influence diagram incorporates knowledge gained from surgical series, clinical trials of dose response, and the known accuracy of imaging (Phillips et al., 2011).

Compared to the mechanistic models in which statistical distributions represent the inhomogeneities of the population of interest, biological models use distributions to represent both inhomogeneity and uncertainty. This allows similar mathematical approaches, such as Monte Carlo techniques, to be used in numerical solutions to problems. However, it is important to keep the two concepts separate because they represent fundamentally different aspects of our knowledge.

13.3.3 Clinical Models

At the clinical end of knowledge representations, we have detailed clinical models (Goossen, 2014). At this level, the representation is intended for procedural information exchange, and there is not much room for large variations or different approaches. However, it is worth highlighting one clinical model that has been very important in the development of radiation oncology.

The DICOM paradigm can be useful to illustrate some of the important concepts. As it stands, the DICOM[4] standard prescribes how data is to be represented so that it can be transferred between two systems. Both client and server can then do with the information as they wish, for example, store it in another format, discard it, or apply an algorithm to it. If the data are to be sent to another system in the future, they will have to be repackaged in the DICOM format. This makes sense and is possible because of what image or therapy that

---

[4] The use of the abbreviation "DICOM" should also be understood to include DICOM-RT.

plan data represent. This is similar to the development of the HL7, version 3 paradigm—there is no assumption that the data storage method is universal nor are the sources or applications of the data. However, by applying knowledge representation methods to the clinical environment, we can structure the data in a way that makes sense across many different platforms.

13.4 FUTURE DIRECTIONS

The previous sections described in some detail the methods for representing knowledge and approaches for distributing and sharing that knowledge. This has been done under the assumption that such a distributed approach is necessary if we are to make substantive progress in the treatment of cancer. Several scientific and societal developments are leading us in this direction. First, our increasing knowledge of molecular, cellular, and physiological aspects of diseases and their response to therapies transcends traditional borders. This is true in the scientific realm where disciplines are no longer cut and dried, and also in the medical realm wherein a radiation oncologist or surgeon must consider the pathology report, the genomic profile, and the effects of chemotherapy. One of the truly difficult hurdles is the different vocabularies and knowledge representations that have grown up in each specialty. A quick glance through the National Cancer Institute (NCI) Thesaurus leaves no doubt about the magnitude of these challenges.

An unintended consequence of this more detailed biological knowledge is the ever-diminishing numbers of patients that can be considered part of our clinical experiments. Given the vast diversity of our genotypes and phenotypes, it is difficult to establish a cohort from which the critical "independent and identically distributed" random variables needed in clinical trials can be assured. At the same time, patients are being promised that this selectivity in therapies tailored to individual sets of variables will result in more cures and fewer complications.

For these reasons, being able to share fundamental and clinical knowledge becomes critical. Currently, many of the systems needed are only now being formulated and developed, which provides practitioners (and particularly physicists) in radiation oncology an excellent opportunity to take part and ensure that this field is not left behind. What follows are short descriptions of some opportunities in the relevant areas.

13.4.1 Genomics

At least three areas are of direct relevance between genomic data and radiation oncology: better understanding of basic radiation response mechanisms, statistical correlations with outcomes, and decision support. An example where genetic data is helpful is in breast cancer, where it has been shown that amplification or overexpression of the *ERBB2* gene is strongly associated with recurrence and poorer prognosis. On the other hand, in glioblastoma, there are many types of genetic variations and the genetic profile can vary depending on what part of the tumor is sampled. Yet at the same time, methylation of the repair enzyme O6-methylguanine-DNA methyltransferase (MGMT) DNA repair gene is associated with a better response to temozolomide.

Unfortunately, there are no universal standards for reporting the results of genomic profiles for clinical use. The amount of data to report, the format, and any links to supporting knowledge vary between institutions. In addition, the data are most often in unstructured (document) format, which makes their shared use more problematic. Complicating this data transfer problem is the growing realization that tumors often have multiple mutations that are geographically separated and that change in response to therapies.

The opportunity here is to increase efforts to gather outcome data and correlate it with the genetic information that is available. Given the fact that at any particular time, the relevant genetic information may not be understood to be important, we must develop the means to capture currently relevant data and still be prepared to retrieve data originally deemed irrelevant so that, as our understanding grows, we will be able to pair the genetic information with therapy decisions and outcome data. Such a capability requires extensive planning and coordination. As an example, the HL7 organization is working on developing standards that use existing genomic standard formats, but it is also developing a larger view so those data can be integrated more easily into clinical practice.

Decision support for clinicians in radiation oncology needs to be capable of integrating the genomic data (as illustrated above) but also to recognize and deal with its limitations. To date, our mechanistic understanding is insufficient to predict outcomes for any particular patient. Rather, this information alters the probabilities of outcomes. We must be able to retrieve the conditional probabilities and present them in a way that the decision maker can integrate this with the other factors to be considered. This is another opportunity for researchers in radiation oncology because understanding and integrating knowledge presented by computer interfaces is an active area of research in the field of biomedical informatics (AMIA, 2016).

13.4.2 Imaging

In the era of image-guided therapy and the proliferation of imaging technologies coupled with radiation treatment devices (CBCT, MRI, planar images, US, Positron Emission Tomography (PET)), radiation oncology now generates enormous numbers of images. PACS systems have developed to handle the display and storage of these images, but terabytes of pixel intensities are better classified as data than as information. The extraction of relevant information depends, of course, on the knowledge representation, as discussed above, as was its means of storage and exchange. Even the most basic of information is still stored mostly in radiologists' notes in text documents. Radiation oncologists need to develop an information structure that is adapted to our concerns. If one considers three-dimensional (3D) dose distributions as essentially images of dose, then concepts such as DVH represent a good example of knowledge extraction. Even very basic concepts such as the physical measurements of a recurrent brain tumor as imaged by MRI are important for clinical decisions regarding future therapies in light of past responses (Wen et al., 2010).

Such an effort requires a delineation of the type of information available and needed for therapy needs. Then efficient software tools have to be developed for the extraction and transformation of the raw image data into the necessary form. As an example from the cases cited above, 3D dose distributions could be correlated with longitudinal imaging studies to record the dose distribution within the region of recurrence. Whether this information is calculated as a simple statistic, such as mean dose or as a DVH, depends on what is most useful. This work will have to be carried out in cooperation with radiology departments in order to make efficient use of their PACS software.

13.4.3 Radiomics

Radiomics is an emerging field in radiation oncology in which a number of image features are extracted, and machine learning is used to discover any nonintuitive correlations between these features and clinical variables of interest. It combines elements of the problems and opportunities discussed in the previous two sections on genomics and imaging. A good review of the field is presented by Yip and Aerts (2016). This is the area in which radiation oncology can best interface with the emerging machine learning disciplines.

A study by Parmar et al. (2015) illustrates some important issues in machine learning that are not unique to radiomics or radiation oncology problems in general. Using CT and clinical datasets of non-small-cell lung cancer patients, they investigated the performance in predicting survival of 14 different feature selection methods and 12 different classification algorithms. They sought to establish which methods led to the most stable results and which were the most accurate in prediction. This study highlights the many subtleties that need to be appreciated. Clearly this is an area where work by others can be leveraged for our purposes.

13.4.4 Clinical Information

This area is most critical for success in radiation oncology and is also the most important area for integrating with the general medical environment. In the United States, laws and government regulations are pushing for more extensive and integrated use of EHRs. Other countries, such as England and the Netherlands, are also committed to similar approaches. At the beginning, these efforts are often modest, collecting rather generic data and implementing the most straightforward applications, such as improved patient education. Such a gradual approach provides many opportunities for radiation oncology to join and make substantive contributions while, at the same time, taking advantage of the work being done throughout the healthcare enterprise.

In radiation oncology, we have some historic advantages and disadvantages. Our extensive use of computers for treatment planning, imaging, and R&V systems means that we often have expertise in the collection and use of data. Recent initiatives in error detection and quality improvement demonstrate how such systems can improve clinical outcomes.

On the negative side, our collection of outcomes data is poor. While much is made of the big data side of medicine, radiation oncology suffers from "small data." We collect only a few percent of radiation therapy patients in clinical trials, and much of the data that is collected outside trials relate to short-term outcomes. The multidisciplinary nature of cancer treatments is a significant obstacle given the difficulties of merging different clinical workflows to capture relevant structured data. Institutional reluctance to share data also hampers the ability to collect meaningful amounts of information. If we are to make significant progress in the era of personalized medicine, radiation oncologists must collect data from more sources. This is definitely a nontrivial undertaking and requires significant informatics efforts in order to provide the means for clinicians to capture data in a structured format within the constraints of medical practice and reimbursement structures. Once the data are captured, then we are free to apply any of the informatics tools as described to extract as much information as we can.

13.4.5 Human–Computer Interfaces

Another area in which radiation oncology can benefit from advances in information science is in the design of interfaces for computerized systems. As mentioned above, radiation oncology was once at the forefront of the clinical use of computers. However, a brief survey of the users of any department's treatment planning or OISs is likely to illuminate the many deficiencies of these systems, many of which were designed a decade or more ago. Whether it is a system designed for physician interaction, for example, OIS or clinical decision support systems; physicist and dosimetrist use; or radiation therapists' daily tasks, poorly designed information systems can lead to fatigue, frustration, and errors (Horsky et al., 2012). In addition, we are introducing more patient–computer interfaces such as breathing control systems and quality-of-life surveys. The principles important in those systems can be quite different than those designed for professionals' use (Hartzler et al., 2016). It behooves the community to be aware of the progress that has been made in this field and to prod commercial vendors to improve their products; if the products are developed in-house, the radiation oncology community should enter into collaborations with the informatics community.

13.5 CONCLUSION

The field of medical informatics has many different aspects, some of which have been covered in this chapter. In general, the many aspects of informatics are critical to the healthy development of radiation oncology. These areas are also a great opportunity for our field to enlarge our scope and break free from the silos that seem to characterize much of our work. By embracing these challenges, we can better integrate our field with other related areas, thereby providing more comprehensive treatment approaches and regimens.

REFERENCES

Benson N., Whipple M., and Kalet I. J. 2006. A Markov model approach to predicting regional tumor spread in the lymphatic system of the head and neck. *AMIA Annu Symp Proc*: 31–35.

Bidgood W. D., Jr., Horii S. C., Prior F. W., Van Syckle D. E. 1997 Understanding and using DICOM, the dana interchange standard for biomedical imaging. *Journal of the American Medical Informatics Association*, 4(3): 199–212.

Deasy J. O., Bentzen S. M., Jackson A., Ten Haken R. K., Yorke E. D., Constine L. S., Sharma A., Marks L. B. 2010. Improving normal tissue complication probability models: The need to adopt a data-pooling culture., *Int J Radiat Oncol Biol Phys*, 73: S151–S154.

El Naqa I., Li R., Murphy M. (Eds.). 2015. *Machine Learning in Radiation Oncology*. Switzerland: Springer International Publishing.

Goossen W. T. F. 2014. Detailed clinical models: Representing knowledge, data and semantics in healthcare information. *Health Inform Res*, 20: 163–172.

Hartzler A., Weis B., Cahill C., Pratt W., Park A., Backonja U., McDonald D. W. 2016. Design and usability of interactive user profiles for online health communities. *ACM Trans Computer-Human Interaction*, 23(3): 15.

Horsky J., Schiff G. D., Johnston D., Mercincavage L., Bell D., Middleton B. 2012. Interface design principles for usable decision support: A targeted review of best practices for clinical prescribing interventions. *J Biomed Inform*, 45: 1202–1216.

http://geneontology.org/. Accessed 2 May 2, 2016.

http://www.detailedclinicalmodels.nl/dcm-en. Accessed 31 March 31, 2016.

http://www.nih.gov/about-nih/what-we-do/nih-turning-discovery-into-health/personalized-medicine. Accessed 31 March 31, 2016.

https://bioportal.bioontology.org/ontologies/DLORO?p=summary Accessed 2 May 2, 2016.

https://bioportal.bioontology.org/ontologies/NCIT. Accessed 2 May 2, 2016.

https://bioportal.bioontology.org/ontologies/ROO. Accessed 2 May 2, 2016.

https://bioportal.bioontology.org/ontologies/SNOMEDCT. Accessed 2 May 2, 2016.

https://www.amia.org/applications-informatics/clinical-informatics. Accessed 5 May 5, 2016.

https://www.cancerdata.org. Accessed 21 April 21, 2016.

https://www.nih.gov/precision-medicine-initiative-cohort-program. Accessed 31 March 31, 2016.

Kalet A. M., Gennari J. H., Ford E. C., Phillips M. H. 2015. Bayesian network models for error detection in radiotherapy plans. *Phys Med Biol*, 60: 2735–2749.

Kjaerulff U. B., Madsen A. L. 2010. *Bayesian Networks and Influence Diagrams*. New York, NY: Springer Science+Business Media LLC.

Masterson L., Moualed D., Liu Z. W., et al. 2014. De-escalation treatment protocols for human papillomavirus-associated oropharyngeal squamous cell carcinoma: A systematic review and meta-analysis of current clinical trials. *Eur J Cancer*. 50(15):2636–2648.

Neumann, M. 2002. DICOM—Current status and future developments for radiotherapy. *Zeitschrift für medizinische Physik* 12(3): 171–176.

Owens D. K., Schachter R. D., and Nease R. F. 1997. Representation and analysis of medical decision problems with influence diagrams. *Med Decis Making*, 17:241–262.

Parmar C., Grossmann P., Bussink J., Lambin P., Aerts H J Wl. 2015. Machine learning methods for quantitative radiomic biomarkers. *Sci Rep*, 5: 13087. doi:10.1038/srep13087.

Phillips M., Smith W., Parvathaneni U., Laramore G. 2011. The role of PET in the treatment of occult disease in head and neck cancer: A modeling approach. *Int J Radiat Oncol Biol Phys*, 79: 1089–1095.

Rheinberger H. J. 2010. *An Epistemology of the Concrete*. Durham NC: Duke University Press.

Rosse C., Mejino J. V. L. 2003. A reference ontology for biomedical informatics: The foundational model of anatomy. *J Biomed Inform*, 36:478–500

Wen P., Macdonald D. R., Reardon D. A., Cloughesy T. F., Sorensen A. G, et al. 2010. Updated response assessment criteria for high-grade gliomas: Response assessment in neuro-oncology working group. *J Clin Oncol*, 28: 1963–1972.

Yip S. S., Aerts H. J. 2016. Applications and limitations of radiomics. *Phys Med Biol*, 61: R150–R166.

CHAPTER 14

Big Data Applications in Radiation Oncology

John Wong, Todd McNutt, Harry Quon, and Theodore L. DeWeese

CONTENTS

14.1 INTRODUCTION

Over the past decade, advances in computer, detection, storage, and analysis technologies have led to the information explosion that underpinned the present era of "big data." Enormous amounts of the data are being made available with great speed and breadth. With new techniques and technologies to support data capture, integration, and analysis, the practice of many industries has been transformed with new insights that are revealed by big data.

In healthcare, different types of big data include electronic medical records, prescription information, medical images, laboratory results, genomic profiles, demographics, and more. (Discussions on healthcare administration data pertaining to cost, claims, reimbursement, business management, etc., are beyond the scope of this chapter.) The opportunities for improving individual and population health with big data are immense. While an earlier report from McKinsey Global Institute suggested that 90% of the providers ignored the data they generated,[1] this practice is rapidly changing. There are challenges. Big data are inevitably accompanied with noise and potential biases that can misguide its applications. There is also the daunting reality that existing policies and operation infrastructure have not been designed to reap maximal benefits from big data and thus need to be reengineered. Regardless, the potential impact of big data is such that major investments have been made by many healthcare disciplines and institutions with the goal of improving the delivery of care to their patients.[2-8] Many commercial entities, large[9] (e.g., IBM Watson Health, SAP, and Oracle) and small,[10] are also developing big data technologies to support a wide variety of initiatives in personalized medicine, data-driven decision support, drug discovery, data management, determination of comparative effectiveness, risk and waste reduction, and so on. The pace of development in healthcare applications using big data has been spectacular in the last 5 years.

14.2 BIG DATA AND ONCOLOGY

Oncology, the study and treatment of cancer, compels big data research and development. There was much enthusiasm in the early 2000s that the explosive advances in detection technologies, accompanied by the rapid increase in genomics and proteomics discoveries, would have a major impact on the development of new powerful molecular therapeutics. Anticipating the need to support the sharing of data and the conduct of clinical trials in the community, NCI, in 2004, pioneered the cancer Biomedical Informatics Grid (caBIG™) program[11] to develop data standards and software tools that broadly support the collection, management, and analysis of data from research laboratories, cancer imaging, and clinical trials. The caBIG program was led by the Center for Bioinformatics and Information Technology (CBIIT) at NCI and was supported by the core caGrid network infrastructure for secure data exchange.[12] Despite its laudable goals and perhaps naïve ambition, the caBIG program was unable to establish significant traction to address the enormous breadth and complexity of data in cancer treatment and research. An outside review found the program overly complex and ineffective, leading to the significant reduction of its budget and scope in 2011.[13] It was succeeded by the National Cancer Informatics Program in 2014.[14] Nevertheless, harnessing big data remains a key priority of NCI, which has continued to sponsor funding opportunities to support technologies and methods to generate and analyze the enormous amount of laboratory and clinical data.[15-17]

14.2.1 Advancing Clinical Research with Big Data in Oncology

The efforts of caBIG and its collaborators in the research community and industry provided invaluable insights about the many complex issues in the clinical application of big data in cancer care. The challenges range from data-related topics, such as ontology, data standards and interoperability, data storage, and import/export, to the regulatory challenges imposed by the Health Insurance Portability and Accountability Act (HIPAA),[18] to the complex cultural hurdles such as the reluctance of the provider to share practice information in the face of market competition and legal liability. These insights laid the foundation of the current developments that are beginning to make inroads in the application of big data to improve cancer care.

The problem begins with data acquisition. Significant amounts of important information such as routine and nonstandard laboratory results, dosimetry information, treatment outcomes, and so on, are stored in unstructured formats and incompatible databases. Some standardization has helped, such as Digital Imaging and Communications in Medicine (DICOM)[19] for the storage, display and exchange of medical images, Systematized Nomenclature of Medicine—Clinical Terms (SNOMED CT) for medical terminology,[20] Logical Observation Identifiers Names and Codes (LOINC) for laboratory observations,[21] NCI Thesaurus for cancer-related terminology,[22] and HL7 (Health Level 7) for healthcare information technology information,[23] and so on. Several efforts across many healthcare disciplines have been directed to the capture or extraction of case-specific structured data from patient encounters, laboratory measurements, imaging studies, and so on, so that the data are "processable" for analysis and sharing.[24,25] Unfortunately, most other data are contained in narrative reports that, even after digitization, are typically not structured for semantic content, which severely compromise the processes of query, analysis, and ultimately decision support. Natural language processing technologies are being developed for more efficient extraction of structured data, although they are not adequately robust for general use.

At present, the extraction of consistent structured data requires a fair amount of human intervention. Of particular interest and importance are the unstructured outcome and assessment texts because they drive clinical comparative effectiveness research and decision support. Nonprofit, nongovernment-based organizations such as Patient-Centered Outcomes Research Institute (PCORI),[26] via their PCOR Network, and the Commission on Cancer,[27] via the National Cancer Database as sponsored by the American College of Surgeons and the American Cancer Society, are contributing efforts to collect healthcare information that can be shared securely for research. Two notable private entities presented their initiatives at the 2015 Workshop on big data sponsored by NCI, American Society for Radiation Oncology (ASTRO), and American Association of Physicists in Medicine (AAPM).[28] One is the American Society of Clinical Oncology (ASCO), a nonprofit organization for physicians and oncology professionals that are involved in cancer care. ASCO, in collaboration with Systems, Applications, and Products (SAP), has created the CancerLinQ IT technology[29] to assemble clinical cancer information from the large data sources of participating member physicians or institutions in a powerful database that can be queried by the practitioners to help improve their decision making in the care of their cancer patients. With increasing participation from its 40,000 ASCO members, CancerLinQ has presently amassed de-identified health information from more than 1,000,000 cancer patients. The other entity is Flatiron Health, which offers a cloud-based repository of clinical and laboratory data, from more than 2,000 providers, that have been converted for structured query.[30] For example, Flatiron Health enables participating sites with the National Comprehensive Cancer Network to access data from each other. Both ASCO CancerLinQ and Flatiron Health are now well established with significant strategic partners from major device and pharmaceutical industries as well as healthcare organizations. Their technologies are demonstrative of the evolution in clinical decision support and healthcare delivery in the era of big data. Companies such as Cancer Outcomes Tracking and Analysis,[31] IBM Watson Oncology,[32] Optum,[33] NantHealth,[34] and others are also actively involved in cancer-related data initiatives.

14.3 BIG DATA AND RADIATION ONCOLOGY

For the subspecialty practice of radiation oncology, the impact of the big data initiatives in oncology has been modest. Much of the resources currently provided by CancerLinQ pertains to medical oncology, although discussions have been held between ASCO and ASTRO to collaborate on CancerLinQ for radiation oncology. Part of the challenge is that typical medical informatics is not set up to accommodate the detailed treatment plan information associated with the practice of radiation oncology. It is encouraging that a new partnership has recently been forged between Flatiron Health and Varian to engage radiation oncology with healthcare analytics.[35]

The radiation oncology community itself, on the other hand, has a long history of managing quality data for clinical research. For many decades, the gold standard of outcome research in radiation oncology has been the conduct of randomized clinical trials such as those conducted by the NCI-funded Radiation Therapy Oncology Group (RTOG).[36] RTOG was recently succeeded by the RTOG Foundation, which is part of the NRG Oncology group,[37] one of five lead protocol organizations in the new NCI-funded National Clinical Trials Network (NCTN).[38] These cooperative research groups employ formal data centers with established quality assurance (QA) standards, credentialing processes, and stringent randomization procedures to ensure data integrity and valid conclusions. Since 1968, RTOG has activated over 500 protocols and accrued more than 110,000 cancer patients from a broad spectrum of diseases. The efforts of RTOG have been instrumental in improving treatment outcome and quality of life (QOL) by advancing the practice of radiation therapy, alone or in conjunction with other therapeutic agents.

Despite the many achievements, there are also known challenges with cooperative trial research. Patient accrual is low and typically enrolls less than 3% of all patients that are under treatment. In addition to the burden of data de-identification, the centralized database incurs a large overhead in data export and import, particularly when imaging data are involved, thus lengthening the time of study completion. The collected data in a cooperative RTOG trial are inherently limited in the scope and size. As a result, data reuse other than for the specific study are rare.[39–41] After publishing the results of their studies, RTOG does not have a direct means to evaluate the impact of its research findings on the quality of practice in the community. It is thus ironic that, contrary to the anticipation that big data would greatly enhance clinical research, its sheer volume and complexity actually exacerbate the challenges of conducting the cooperative research, leading to the undesirable opposite effect of data loss.

14.3.1 National Radiation Oncology Registry

The research community in radiation oncology is well aware of the big data challenges and has explored other means to address them. Most notable was the 2011 partnership of ASTRO and the Radiation Oncology Institute (ROI), the nonprofit research foundation of the ASTRO, to develop the National Radiation Oncology Registry (NROR) for aggregating standardized population data electronically from multiple institutions about their patients, treatments, and outcomes.[42] The goals of this first-of-its-kind registry specific for radiation oncology—to address some of the shortcomings of the randomized clinical trial process by expanding the breadth and volume of clinical data and in turn help to define best practices, identify deficiencies, and support comparative effectiveness research—were most laudable.

NROR recently completed its first pilot study with prostate cancer. The experience reflected on the pros and cons of conducting research with data registry. The desire to include a larger number of participants would inherently limit the breadth and the depth of the submitted data, particularly because much of the data were entered manually. Out of the 30 initial institutions selected to participate, 14 remained for the duration of the study and provided the necessary data for the accrual target of 430 patients. The rest pulled out because of legal and/or administrative concerns, as well as issues associated with data transfer. It was recognized that manual data entry was a significant burden. More in-depth details about the delivered dosimetry were desired yet incomplete. The NROR initiative is presently under review while the community awaits the report from the pilot study.[43]

14.3.2 Big Data Initiatives in the Radiation Oncology Community

The foundation to harnessing big data is to amass and organize the data so that they can be processed efficiently and effectively. The burden of data export and import, exacerbated by their unstructured format, has made data gathering and sharing most challenging. While centralized data centers such as CancerLinQ and Flatiron Health allow efficient query and analysis, significant computational investments and manual efforts are needed to collect and convert the clinical data from the various sources into a structured format. At present, within the scope of medical oncology, both products can demonstrate to their participants the tremendous values of information from the large volume of aggregated data. It will be interesting to see how these products will evolve to accommodate the broader oncologic practices such as radiation oncology and whether their initial success will encourage the adaptation of more data standardization in the future.

It was recognized early in the big data era that there was a need to reengineer the current clinical informatics infrastructure to support clinical research and decision making. It was noted by Robert Harrington from Duke Research Institute in 2005 that "[o]ne of the greatest inefficiencies of the current model of clinical research in our country is the lack of a sustaining infrastructure (which includes shared resources, common data standards, and effective use of information technology among researchers), as well as the lack of a convenient forum to share best practices and learn from one another's mistakes and successes."[44]

The big data challenge stems from technology-driven advances; however, the solutions will not be based on technology itself. Advances in computer and detection technologies cannot keep pace with the resultant exponential increase in data. There is no better example than the practice of radiation oncology to explore new approaches to manage and harness big data in research and treatment. In addition to the already extensive amount of clinical and laboratory data that are requisite in the management of a cancer patient, the prescription of a radiation treatment is accompanied by the exquisite spatial description of the "medicine"—the three-dimensional distribution of dose, that will be delivered to different anatomic parts of the patient. In addition, treatment verification has been so ingrained in the practice that, in the present era of image guided radiation therapy (IGRT), the radiation oncologist would have a high degree of certainty that the dose was delivered correctly. Image guidance undoubtedly improves treatment quality; it has also resulted in more than an order of magnitude increase in the data volume for each patient. Ironically, in another case of data loss, the vast amount of imaging and treatment planning data generated from IGRT has not been analyzed systematically, with few exceptions, to further identify process improvements for the specific disease site or the benefits of the individual patient, such as optimizing treatment margins. Radiation oncology presents an ideal data-rich model to explore big data applications.

14.4 NEW APPROACHES FOR DATA SHARING IN RADIATION ONCOLOGY

Novel concepts of data sharing to alleviate the burden of data submission to a data center or data exchange between institutions has been considered by a few investigators in the radiation oncology community.[45,46] An effective approach is to federate the databases where the individual healthcare institutions maintain their own patient data, particularly the ones that provide long-term, multidisciplinary specialty care for the patients, such as in oncology. Patient data and information can be updated in time and recalibrated by the institution when improved methods of analysis or future discoveries become available. By creating a federated system comprised of databases from multiple, ideally autonomous institutions, data can be efficiently queried across the system so that only results of interest are extracted. The volume of data for research is significantly increased and enriched. Innovative collaborative research can be conducted using nonrandomized data that, with appropriate biostatistics methods, can help develop well-founded hypotheses for testing with randomized data. Each institution itself bears the burden of maintaining data integrity and QA, and rightfully so, to sustain its own continued advancement and quality improvement. The federated database system has been shown to be effective in various applications in life science and healthcare.[47-49]

14.4.1 The Euro-Regional Computer Assisted Theragnostics Program for Rapid Learning

Two notable efforts in radiation oncology on data sharing were started in the mid-2000s. In the Euro-regional Computer Assisted Theragnostics (EuroCAT) project,[50] Maastricht Radiation Oncology (Maastro), the host institution, developed an international computer network with advanced software to mine the clinical patient data and treatment results in the individual databases from collaborating institutions in the Netherlands, Belgium, and Germany. The goal was to set up large-scale sharing of patient information to enable rapid learning and decision support. The project formed the foundation for the development of the rapid learning system by the knowledge engineering group at Maastro.[46,50,51] Because the approach does not require the export of actual patient information, patient privacy is protected by the individual institution and simplifies the consenting process. Figure 14.1 shows the schematic of rapid learning where the coordinating, or master, center at Maastro sends queries to the participating institutions to develop prediction models based on the local data, which are combined to produce a single "master" model that can be used for generalized outcome prediction.

In an early 2009 study,[52] a prediction model for 2-year survival was developed for non-small-cell lung cancer (NSCLC) Stage I-IIIB patients treated with chemoradiotherapy to a radical radiation dose, defined as ≥ 45 Gy. Data from 326 patients from 2002 to 2006 were used for training the model. Patients could be divided into low- medium-, and high-risk groups. Validation was conducted internally at Maastro with leave-one-out analysis as well as using data from 101 patients at two outside institutions. The area under (the receiver operating characteristic [ROC]) the curve (AUC) was a respectable 0.75. In a follow-up 2014 study, the model developed in Europe was applied to 159 out of 419 eligible patients treated at 5 institutions in Australia from 2003 to 2011 according to the inclusion criteria of the earlier

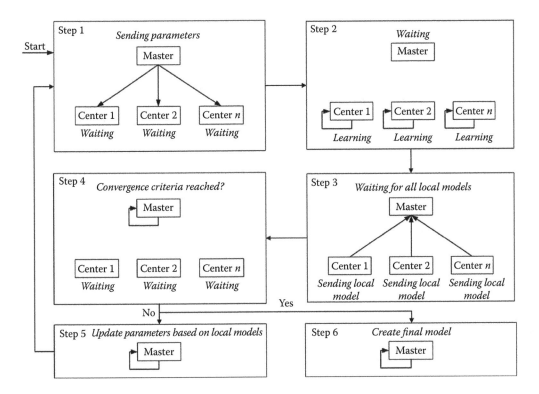

FIGURE 14.1 The schematic of rapid learning where the coordinating center at Maastro incorporates clinical data from collaborating institutes to enhance the modeling for predicting clinical outcome.

study.[53] The AUC was 0.69. The model was able to identify patients with good prognosis but could not distinguish the medium to poor prognosis groups, as with the European patients.

The results from these early studies of the EuroCAT rapid learning project were promising. They demonstrated that it would be feasible to derive information for decision support from routinely collected clinical data, and in a relatively short period compared to the traditional randomized trial studies. However, there were caveats. For both the European and Australian studies, missing data in the different databases led to significant reduction of patient data that could be analyzed. The "as is" data were not controlled and assumed to be correct. For various reasons, such as differences in practices and timing of treatment, data quality could be variable and biased. Still, a prospective comparison study shows that the model derived from rapid learning significantly outperformed the prediction made by the physicians.[54] Valuable insights were gained from the effort, even for the modest prediction end point of median survival greater than 2 years. It would be important to increase the breadth and volume of patient data to help identify bias and to extend the prediction to include toxicity outcomes that affect QOL. Incorporation of methods to ensure data accuracy and to handle missing data would be highly pertinent to data-driven rapid learning. As noted in an earlier presentation,[55] the inherently large amount of unstructured, free text data represented a major hurdle and emphasized the need to establish machine-readable data or ontologic standards.

The EuroCAT program at Maastro has made significant contributions to the development of rapid learning and decision support using big data in radiation oncology.[56] In addition, the Maastro group has also pioneered the novel application of radiomic approaches in radiation oncology where prognostic features are identified from medical images to enhance the development of prediction models.[57,58] On the other hand, its early rapid learning efforts also led to the recognition that the use of query tools customized for the data sources from different institutions would quickly become intractable and limit the number of clinical databases that can be accommodated. A new framework for data exchange and sharing with federated databases is now under consideration for major radiation oncology clinics in Europe.[59]

14.4.2 The Oncospace Program at Johns Hopkins University

The Oncospace program[45,60] is the second informatics model that forgoes the export/import of data. Oncospace is intended to be a distributed analytic database for structured information and contains many of the strategies being considered by the European community. It was modeled after the successful SkyServer "active database" program developed at Johns Hopkins University (JHU).[61] SkyServer provides Internet-based access to the public Sloan Digital Sky Survey data for astronomers and science educators. The impact of its contribution is evident in the enormous volume of queries and projects submitted by students, researchers, and the public around the world since 2001.

The Oncospace program provides a new framework that promotes data sharing between collaborating institutions to support clinical research, decision making, and quality improvement. Oncospace differs from the traditional clinical research process and infrastructure, as in RTOG, as marked by the numberings in the operation schematic shown in Figure 14.2. The concept is to (1) distribute Oncospace(s) to the collaborating clinics

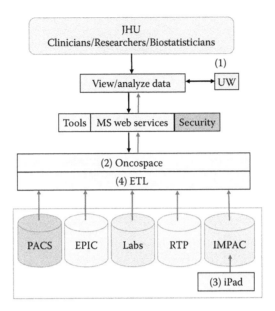

FIGURE 14.2   A schematic showing the operation of Oncospace for data collection and sharing at Johns Hopkins University (JHU). UW denotes University of Washington, a partner site that operates its own Oncospace. PACS denotes the picture archiving and communication system. EPIC is the name of the hospital information system at JHU. IMPAC is the commercial OIS system in the radiation oncology clinic at JHU. The iPAD product from Apple is the mobile technology for capturing clinical data during patient encounter.

where the adherence to (2) the Oncospace common database schema and data standards would enable more efficient and effective queries and analyses. An important and intentional feature of the Oncospace program is (3) the integration of data collection as part of the clinical workflow, such as during patient encounter, to minimize the need of repeating data extraction and entry. The final important step is (4) the implementation of efficient extract, transform, and load (ETL) processes from different requisite data sources in the radiation oncology practice, such as the commercial Oncology Information System (OIS) or the radiation treatment planning (RTP) system to populate each Oncospace. The most significant principle of Oncospace is to avoid the burden of having to export data. With the appropriate authorization and agreement, collaborators can send queries and analyses via web-based access to the distributed (2) Oncospace(s). Only results are retrieved from the queries. The Oncospace program provides a very rich forum for data mining to address many research questions and to develop a clinical decision support system (CDSS).

Figure 14.3 shows the foundation data schema of Oncospace based on an initial set of questions of interest submitted by the clinicians and research personnel. Our initial focus is on patient treatment outcomes with respect to plan quality. The data tables are arranged to support patient geometry, targets, and organs at risk (OARs) and their spatial relationships, dose distributions and dose-volume histograms (DVHs), toxicities, diagnosis and disease progression, chemotherapy and medications, laboratory values, patient histories, and demographics. Unlike an operational database where events and

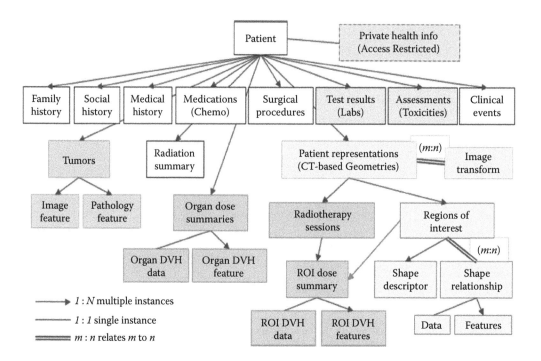

FIGURE 14.3 The data schema and table for Oncospace, where the patient is the central focus of all information. The patient is de-identified, with access to his or her private health information restricted with appropriate authorization. ROI: region of interest in the treatment plan. DVH: dose volume histogram. The mapping relationship (e.g., m:n) between data elements is also shown.

data elements are arranged in sequential order, the analytic Oncospace schema is relational in nature: The information is centered around the patient (Figure 14.3).

A key requirement of Oncospace is to protect patient privacy while allowing external queries. For each patient, personal health information (PHI) is isolated in a single table to allow the database to be anonymous when the table is removed or when access is restricted. To enhance anonymization of the patient, Oncospace stores a reference date to relate all other dates in the system for the patient. This date is typically, but not necessarily, the first day of treatment. When it is, then all dates are in days prior to, or following, the first treatment.

Another important design feature of Oncospace is to provide more complete treatment planning information that can be queried, about the individual patients than what is traditionally available in data registry. A significant amount of processing must be performed to populate the Oncospace schema. A proprietary interface with our treatment planning system (Pinnacle,[3] Philips, Madison, Wisconsin) and a DICOM RT import utility is developed for ETL of the planning information. For each patient, a series of scripts extract relevant radiation treatment information, including three-dimensional dosimetry and volume information, structures (in binary mask), and limited demographic and prescription information (medical record number, treatment site, number of beams, treatment beam geometries, prescription dose, and number of fractions). Treatment plan information is inserted into the Radiotherapy Sessions table. The novel ETL function of overlap volume histogram (OVH) describes the (intercepting) spatial relationships between an expanding or contracting planning target volume (PTV) and its surrounding critical structures.[62–64] OVHs are generated for each PTV/OAR pair using the binary masks and inserted into the Shape Relationship Data table for the patient. OVHs are used effectively to identify similar anatomic features between patients treated with radiation.

A logical human interface has been implemented with web pages to allow users to access and navigate efficiently through Oncospace with data groupings based on demographics, diagnostics, therapeutic techniques, and so on. In Oncospace, data are continuously updated as new patients, or new information about a patient, are added to the system. In addition to outcomes and dosimetry data, other interactive analysis tools are available and continue to be developed as needed. In the early phase of the Oncospace program, data were captured using the treatment assessment utilities of the OIS, Mosaiq from Elekta, at JHU. This task was tedious because the OIS is an operational database not streamlined for entry or extraction of research data. Beginning in 2011, with help from Elekta, data collection pages have been implemented on portable electronic tablets, the iPads.[65] These pages are customized for each specific disease site, allowing the providers and their assistants to capture clinical data during their encounter with the patients. The data are then sent directly via WiFi to their appropriate locations in Mosaiq. Figure 14.4

FIGURE 14.4   A sample data entry page for clinical assessment on the iPad that is captured by clinical personnel during a patient encounter. Also available is a page to capture quality-of-life (QOL) reporting by the patient. The iPad "mobile solution" has greatly improved efficiency in capturing clinical data into Mosaiq as part of clinical workflow.

shows a sample data entry page for clinical assessment by a nurse. An iPad page is also created to allow QOL reporting by the patients. The iPad "mobile solution" has greatly improved efficiency in capturing clinical data into Mosaiq as part of clinical workflow. In turn, the efficiency gain is passed on to the ETL process to populate Oncospace with data from Mosaiq.

14.5 BIG DATA RESEARCH IN RADIATION ONCOLOGY

14.5.1 Data-Driven, Knowledge-Based Treatment Planning in Radiation Oncology

While there are investigations into analyzing past clinical experience to derive decision support, a notable early formal application of big data in radiation oncology is the development of knowledge-based treatment planning. There are two similar methods, one developed at Washington University in St. Louis[66–68] and one at JHU.[62–64] Both approaches define a geometrical relationship between each critical structure and the target volume according to a distance or an OVH-shaped metric. These metrics are then used to query the database of previous treatment plans to determine what plan quality or dose constraint can be achieved. In the Washington University method, a training set of plans for a disease site from a specific clinic is used to predict a dose matrix that provides adequate sparing of the OARs. The information is used to guide the planning process for the individual patient at that clinic. The use of "training" plans with superior quality will lead to prediction of better dose matrices for sparing. This approach has recently been commercialized by Varian as the RapidPlan Knowledge-Based Planning Software.[69] In the JHU method, the OVH as determined for a specific OAR of the individual patient is used to identify those patients in Oncospace with similar OVHs treated for the same disease. The process is repeated for all OARs for the patient. The best achievable dose constraints for all the associated OARs are then used to drive the plan optimization process. With the established framework of Oncospace, the plan and OVH information are continually updated and enriched as new patients are being treated at JHU. Experience with using both knowledge-based methods for plan optimization from Varian and JHU demonstrates that superior plan quality can be achieved for the individual patient with greater efficiency and less variability than those achieved by a human planner.

14.5.2 Clinical Research with Oncospace

Beginning in 2007, Oncospace has been deployed to support clinical research and decision-making in the management of patients. Table 14.1 shows that close to 1,000,000 clinical assessments have been made of patients with head and neck (H&N), pancreas, lung, and prostate cancer treated with radiation (at the time of this writing).

TABLE 14.1   Distribution of Clinical Assessments in Johns Hopkins University Program to Date

	Head and Neck	Pancreas	Prostate	Thoracic
Toxicity	513,649	21,511	83,064	34,753
Quality of life	216,898	15,238	104,664	790

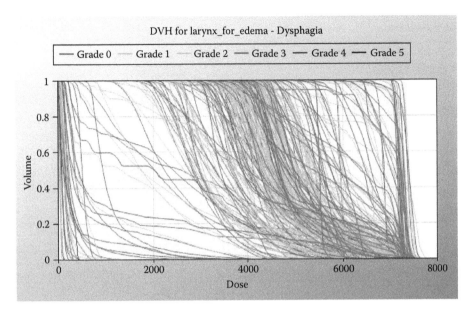

FIGURE 14.5   An Oncospace display of the dose-volume histograms (DVHs) of an organ at risk (OAR) experienced by a group of patients. The DVH is color-coded to show the toxicity grades reported for the individual patient.

For H&N cancer patients, interactive navigating tools are available for the clinicians and researchers to review and better understand the DVH and toxicity information. Figure 14.5 shows an Oncospace display of the DVHs, color-coded for increasing toxicity grade, of an OAR experienced by a group of patients. Other analyses of dosimetric information or DVHs can be conducted. The trends of various toxicities can be evaluated for the individual patient or the group of patients. Figure 14.6 shows an Oncospace display of the distribution of H&N cancer patients suffering from different grades of dermatitis and fatigue after their course of treatment. In turn, this and other information can be accessed by physicians to inform treatment planning and to inform individual patients of expected toxicity outcomes more accurately and efficiently. Thus, physician is not simply relying on his or her memory but can incorporate information from Oncospace that takes into account patient-specific variables and compares them to previously treated patients with the same or similar variables. This is a step no physician could take today because of a lack of full recall and because individual physicians typically have not treated every patient at the institution and thus would not even know the information existed.

14.5.3 Outcome Prediction with Oncospace

A major goal with knowledge gained from big data is to improve decision making in managing the care of a patient. A feasibility study of toxicity prediction, in collaboration with Toshiba Medical Systems, was retrospectively performed using the prospectively collected data in Oncospace of 391 H&N patients treated at JHU from 2007 to 2015.[70] These patients received intensity-modulated radiotherapy (IMRT) with or without concurrent

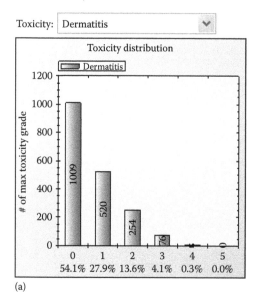
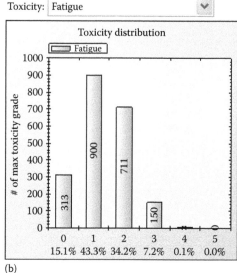

FIGURE 14.6 Oncospace displays of the distribution of head-and-neck cancer patients suffering from different grades of (a) dermatitis and (b) fatigue after their course of treatment.

or induction chemotherapy. Weight loss $\geq$ 5 kg at 3 months post-radiation therapy (RT) was chosen as the toxicity end point for prediction. A total of 3,015 variables was identified; the variables included (1) 1,777 radiotherapy and dosimetry variables, such as various target and OAR volumes, OVH, DVH, and so on; (2) 19 treatment-related variables, such as TNM stages, age, and so on; and (3) 1,219 clinical assessment variables, such as physician-assessed toxicities and patient-reported QOL that were captured longitudinally with the mobile iPAD during patient visits.

Two weight-loss prediction models were built with the classification and regression trees (CART) algorithm[71] in R 3.2.2.[72] Model 1 was developed at the planning phase of radiotherapy when dosimetric and tumor-related variables were available, as shown in Figure 14.7. The principal discriminator was tumor site (ICD-9 code). Patients with laryngeal tumors were at a lower risk of weight loss compared to patients with pharyngeal tumors, suggesting that irradiation of the propulsive swallowing structures may be more important than the coordinating role of the larynx in maintaining weight.

Model 2 provided prediction at the end of treatment (EOT) using assessment variables that were captured during the course of treatment, as shown in Figure 14.8. At EOT, the significant predicted factors were patient-reported oral intake, diagnosis, N stage, nausea, skin toxicity, pain, dose to larynx, parotid, and low dose in the PTV-larynx distance. It seems logical that the most predictive factors pertained to whether the patient had problems eating or enjoying food.

For the developed models, the AUC was 0.773 at treatment planning and 0.821 at the end of treatment, respectively. These values are respectable given the sample size of the patients that were divided into several cohorts for modeling. The results are encouraging and demonstrate that it is feasible to use data that are routinely collected during patient

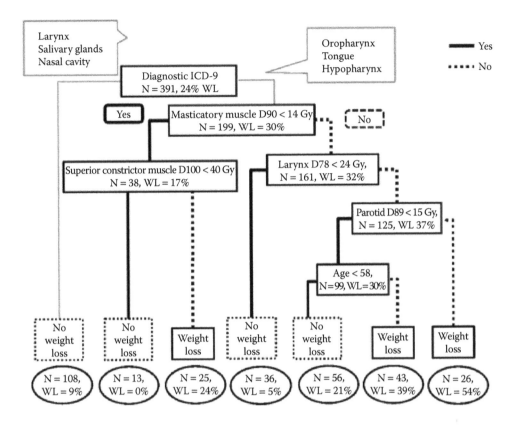

FIGURE 14.7 The outcome prediction tree during the treatment planning phase (Model 1) for 5% weight loss among head-and-neck cancer patients 3 months after treatment. Dosimetric and tumor-related variables are available. Nodes display the predicting factors and the partitioning points. The potential risk factors in the CART analysis were ICD-9 code, dose to larynx, masticatory muscle, combined parotid, age, and chemotherapy.

encounters to discover factors that can affect a treatment-related toxicity. It also seems possible that better predictions can be made if additional data other than clinical assessments, such as image data, are also included. For 87 randomly selected patients in the Model 1 prediction, combining nonimaging data, such as DVH, with imaging radiomic features, such as the shapes and textures of the contoured parotid/submandibular glands in CT images, improved the AUC of predicting xerostomia (grade $\geq$ 2) to 0.81 from the 0.70 value obtained without imaging data.[73] Similar findings were obtained for weight-loss predictions where the AUC of the predictions was 0.78 with additional radiomic features and 0.63 without.

14.5.4 Clinical Decision Support System

A highly anticipated application of a prediction model derived from big data is to provide decision support in the management of the patient care. A prototype CDSS for weight-loss prediction at time of treatment planning for H&N patients was developed at JHU in collaboration with Toshiba Medical Systems where a graphical user interface (GUI) would show

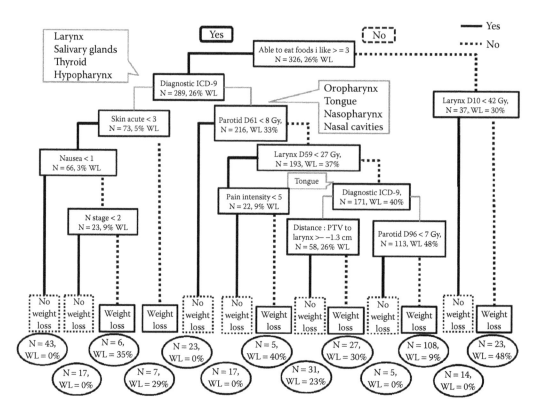

FIGURE 14.8 The prediction tree at the end of treatment (Model 2) for 5% weight loss among head-and-neck cancer patients 3 months after treatment. The assessment variables were captured during the course of treatment. The potential predictors in the CART analysis were ICD-9 code, patient reported outcome of swallow efforts, overall stage, Karnofsky performance status, pain intensity, taste disturbance, dose to larynx, parotid, inferior constrictor muscle, and masticatory muscle.

the risk of weight loss based on parameters derived from the prediction model. The utility of the CDSS in modifying the physician's treatment decision is evaluated retrospectively by a physician's ability to predict weight loss first using traditional planning information only as shown in Figure 14.9a, then with additional information from the CDSS, as shown in Figure 14.9b.

The CDSS utility study employs retrospective data from 100 patients with known weight-loss status at 3 months after treatment. Four physicians with varying degree of experience were recruited to participate in the evaluation. They consisted of a novice who does not treat H&N patients; an early-career physician; a highly experienced, community-based physician; and an expert practitioner with a focused H&N practice. Given the small number of patients from a single institution's dataset, the limited scope of the clinical information, and the simplicity of its design, it is recognized that the study would not provide quantitative information about the clinical benefits of the CDSS for weight loss. Instead, the study focused on understanding how the CDSS might be deployed, the need for additional information, and the improvement that can be made in presenting the information to the physician. Preliminary analyses show that appreciable variation in

Patient 2 is a 73-year-old male with a history of a T4 NX 1 adenoid cystic carcinoma of the right External ear canal status post resection. He denies any weight loss or dysphagia. KPS 100% and weight 67.2 kg.

He agrees to treatment with adjuvant radiation therapy of total dose: 66 Gy (2 Gy * 33, Tumor volume 90cc).

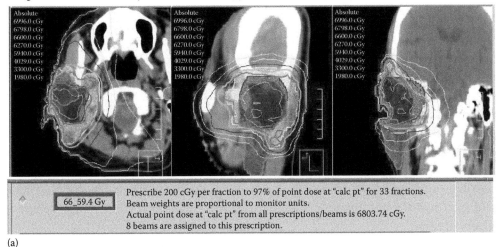

(a)

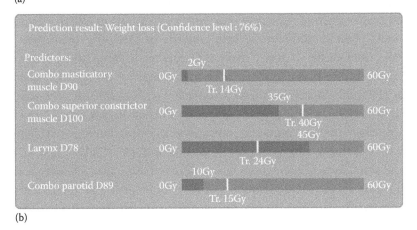

(b)

FIGURE 14.9 (a) Information content and format for a typical de-identified clinical case that is digitally presented to a physician for making a clinical prediction for weight loss. (b) Information from the CDSS in addition to those in Figure 14ba, that are presented to the physicians who are recruited to evaluate the utility of the prototype CDSS for predicting weight loss.

weight-loss prediction exists among the four physicians without the CDSS. The availability of CDSS influenced the decisions made by all physicians, particularly for the two with less clinical experience. Results of the ongoing analysis will form the foundation of a more comprehensive study of the CDSS with refined clinical information as well as the inclusion of more patients and practitioners.

At JHU, the amount of collected clinical data has grown significant since the inception of the Oncospace program in the mid-2000s. It has become very clear that the wealth of clinical data provides tremendous opportunities for advancing clinical care. As with the EuroCAT effort, the JHU's experience with outcome prediction and decision support show that routinely and efficiently captured high-quality clinical data can be used to have a significant impact on the quality of care for future patients. Efforts to develop decision support tools are now being actively pursued for other disease sites.

14.5.5 Interinstitutional Collaboration

Oncospace at JHU had been used as a research database to house common data elements of a prospective multi-institutional research trial on short-course stereotactic body radiation therapy (SBRT) of pancreatic cancer that included JHU, Stanford University, and Memorial Sloan Kettering Cancer Center[74]; although Oncospace was not distributed at these clinical sites. However, the vision of Oncospace program is to enable data sharing between institutions by supporting interinstitutional queries. Toxicity trends can vary between institutions, and these variations may be due to varying techniques in symptom management or variation of the delivered radiation dose. By having these data readily accessible, important comparisons can be made interactively to help the clinician understand interinstitutional differences in the treatment plans and possibly to improve the delivery of care.

As a feasibility study on the use of Oncospace to promote data sharing between institutions, a self-funded consortium of radiation oncology institutions in North America was formed in 2014. At present, the membership consists of JHU, the University of Washington in Seattle, the Odette Cancer Center at Sunnybrook Hospital in Toronto, and the University of Virginia; it is expected to be expanded.[75-78] Adopting the data schema developed at JHU, the consortium has to address data and software interoperability between the institutions to support the inclusion of other disease sites, data elements, and data from other sources. Syntactic interoperability (the ability to exchange data between systems) is achieved by adhering to a common set of tables and attributes in the relational databases implemented at each consortium site. The core schemas are identical, so queries can be generated and distributed without modification to all federated Oncospace's. The results can be aggregated without transformation. The architectural approach does not preclude the need for mapping and transformation of heterogeneous source data to fit the common Oncospace schema; however, it does greatly simplify the mechanics of data integration within the federated model. Unlike the globally mediated schema approach, Oncospace places that burden on the ETL layer, where the processing resources are not constrained, compared to the operational/analytical layer, where system performance for extraction will be critical. Semantic interoperability (the ability of systems to exchange data with unambiguous, shared meaning) is achieved by the use of ontologies and standardized terminologies. The consortium agrees on a common data dictionary that explicitly defines the set of concepts that populates the database and standardizes encoding. In the case of the RTP table, the consortium adopts a common structure and attributes across databases (i.e., achieving syntactic interoperability), which is populated with parameters that have concept unique identifiers (i.e., semantic interoperability), such as PTV. Together, these two components ensure that a query of treatment-planning parameters distributed to all Oncospace's will be executed successfully on each Oncospace and produce results that can be readily combined.

This common framework in effect establishes the scope of the research questions that the consortium can ask. The implementation of Oncospace at each consortium site, however, is not rigid. First, each institution owns its own data. Second, each institution is free to expand the Oncospace database to include additional data elements of interest only to that institution provided it maintains the system so that queries on any "standard" tables and variables are addressed correctly. Finally, we expect a continual evolution of the database to account for changes and advances in our understanding of the critical variables.

Data sharing between the members of the Oncospace Consortium has had varying degrees of progress, much of it is related to the availability of local technical support and the breadth of the clinical data for the selected clinical sites. The most extensive sharing is between the Oncospace at the University of Washington and JHU. The disease sites common to the two centers are H&N and prostate. To date, JHU has extracted dosimetric and anatomical information from over 1,000 H&N patients' treatment plans, and collected about 800,000 longitudinal QOL and toxicity assessments from the clinical record. About 500 H&N patient plans are stored in the University of Washington's Oncospace, with outcome data for about 100 of those patients. Approximately 180,000 assessments made on 1,600 prostate patients are stored in the Oncospace at JHU; at the University of Washington, about 50 prostate patients are entered, with no outcomes imported so far.

Various Oncospace webpages have been developed to show the relationship of dose delivered to the OARs with the assessed toxicities. Data can be combined from multiple consortium member sites to improve the correlative power between dose and outcome to specific structures. An example is shown in Figure 14.10. Figure 14.10a shows the distributions of DVHs of the larynx with various grades of dysphagia for an institution with little collected outcome data. The black curve shows the hypothetical DVH of the larynx of a patient to be treated. Figure 14.10b shows the data of Figure 14.10a combined with those of other consortium members and shows the DVHs for a larger number of patients, which can give insight about the treatment of the individual patient.

At JHU, a web server has been set up that, upon IRB approval, will allow restricted access to the Oncospace databases by the consortium members. Researchers "bring the analysis to the data" by submitting their stored analytical procedures to the JHU Oncospace data administrator, who will evaluate them for data security and install them on the shared web server if they conform to JHU data security requirements. These procedures will be able to perform analysis of the combined data from participating consortium members. For the consortium, access to the stored procedures is granted explicitly by each institution's database administrator. The output of these procedures delivered to the shared web server will comprise the analytical results of multi-institutional studies. Locally, an institution can decline participation in a given research activity by simply omitting the storage of the procedure to its database.

The Oncospace Consortium collaboration has shed significant insight on the requirements of sharing data through a federated database system. Seven key requirements have been identified and established: (1) definition of interoperability criteria to maintain software coding standards, interinstitutional system compliance, and code repository; (2) QA measures and methods to ensure and validate data integrity; (3) rigorous biostatistics

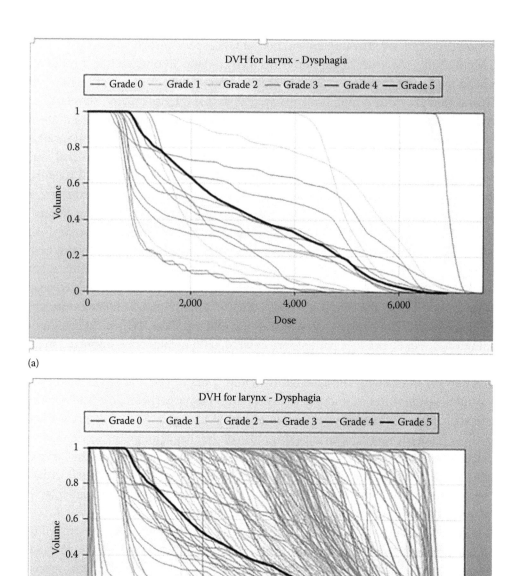

(a)

(b)

FIGURE 14.10 The black curve shows the hypothetical dose-volume histogram (DVH) of the larynx of a patient to be treated. (a) An institution with little outcome data to study can draw few conclusions about the relationship between dose to the larynx and dysphagia. (b) By combining data from member institutions, a much clearer picture about how a given dose level might affect a patient is visible.

methods for appropriate analysis of the data that are not collected under strict controlled protocol; (4) maintenance and expansion of a data dictionary to support ontology mapping and to ensure interinstitution adherence; (5) a scientific committee to evaluate the merits of the common study questions and data/analysis to be shared; (6) implementation of regulatory standards to oversee patient privacy concerns and the principles of interinstitutional sharing; and most important, (7) a governance body, comprised of leaders from within and outside the community, to provide oversight of all development.

The Oncospace program is still in the early phase of its development, a decade after its initiation. Undoubtedly, the capture of clinical assessment as structured data at the time of a patient encounter represents a major breakthrough by overcoming the most significant barrier to assess outcome data in big data research. Adherence to a common data schema and dictionary is also imperative to support data sharing for a federated system. Given the progress in its real-world implementation and albeit limited deployment, the program has demonstrated sufficient traction and worthiness to encourage expansion of the consortium, to conduct more in-depth clinical research, and to enable decision support in the community. Oncospace is a work in progress to leverage big data to improve delivery of radiation therapy.

14.6 CONCLUSION

The era of big data in healthcare is here. The immense volume of data generated from diverse sources creates exciting opportunities to transform the delivery of care. It is expected that big data can be harnessed to promote preventive care and improve health. Significant operational and regulatory barriers prevent researchers and providers from accessing, managing, integrating, and analyzing such data from widely diverse and disparate sources. Hardware and software solutions are continuously being developed to address these challenges. It is estimated that using big data can result in a substantial healthcare savings of more than $300 billion in the United States alone.[79]

The singularly most challenging obstacle to harnessing big data in healthcare is the preponderance of unstructured clinical data that are captured during patient encounters by individual providers working in their siloed environments. In oncology, several initiatives have been undertaken to address the problem. Most notable are the CancerLinQ program of ASCO and the Oncology Cloud product suite from Flatiron Health. In both cases, clinical data from participating providers or institutions are collected by various machines and manual means, converted into structured sets, and organized in a central repository to allow query by their members. Clearly, these registries with in-depth data are showing promise to improve decision support and patient management in the community.

In radiation oncology, patient dosimetry and the potential toxicities are explicitly tied to the patient's 3D anatomy. Such data contain valuable information but are difficult to incorporate in a registry. To bypass the burden of exporting and importing volumetric data, the EuroCAT program sent queries to local databases for outcome modeling and prediction. These investigations demonstrated that clinical data collected at different institutions and shared can be analyzed for deriving meaningful conclusions as an alternative to conducting formalized clinical trials. The earlier EuroCAT effort also demonstrated, however, that it would be impractical to send data queries to an increasing number of database and difficult to retrieve the detailed dosimetry delivered to the patients. A new framework of a federated database system to support data sharing has thus been proposed by the radiation oncology community in Europe.[59]

The Oncospace program at JHU for outcome research is predicated on data sharing. It is an analytic database. A critical innovation is the structured data capture at the time of a patient encounter. Important ETL tools have been developed to populate Oncospace with information from the treatment-planning system and other sources. Central to the entire effort are the definition of and adherence to a common data architecture, schema, and dictionary that can be deployed in a federation of collaboration sites. The multifaceted development of the program has taken a decade, and now a consortium of distributed Oncospaces can share data to enable decision support. At JHU, a toxicity prediction model of weight loss for its H&N patients treated with chemoradiation therapy has been incorporated into a prototype CDSS currently under development.

In summary, Oncospace and the evolving EuroCAT programs show promise in improving radiation oncology by leveraging big data. The sharing of clinical data is possibly the most transformative step in the enhancement of clinical research, distant learning, and decision support. It can also significantly improve healthcare delivery in the community. A data-sharing consortium can include a significant number of satellite community clinics and practices. The model of federated databases in radiation oncology, such as Oncospace, can ultimately be expanded to other healthcare specialties to reap the full potential of big data.

REFERENCES

1. McKinsey Global Institute. 2011. Big data: The next frontier for innovation, competition, and productivity, May 2011, Copyright @McKinsey & Company, 2011. http://www.mckinsey.com/mgi; accessed February 28, 2017.
2. Zhang X, Pérez-Stable EJ, Bourne PE, Peprah E, Duru OK, Breen N, Berrigan D, Wood F, Jackson JS, Wong DWS, Denny J. Big data science: Opportunities and challenges to address minority health and health disparities in the 21st century. *Ethn Dis.* 2017; 27(2):95–106. doi:10.18865/ed.27.2.95.
3. Levin MA, Wanderer JP, Ehrenfeld JM. Data big data and metadata in anesthesiology. *Anesth Analg.* 2015; 121(6):1661–1667. doi:10.1213/ANE.0000000000000716.
4. FitzHenry F, Resnic FS, Robbins SL, Denton J, Nookala L, Meeker D, Ohno-Machado L, Matheny ME. Creating a common data model for comparative effectiveness with the observational model outcomes partnership. *Appl Clin Inform.* 2015; 6(3):536–547. doi:10.4338/ACI-2014-12-CR-0121.
5. Yu P, Artz D, Warner J. Electronic health records (EHRs): supporting ASCO's vision of cancer care. *Am Soc Clin Oncol Educ Book.* 2014:225–231. doi:10.14694/EdBook_AM.2014.34.225.
6. Shaikh AR, Butte AJ, Schully SD, Dalton WS, Khoury MJ, Hesse BW. Collaborative biomedicine in the age of big data: The case of cancer. *J Med Internet Res.* 2014; 16(4):e101. doi:10.2196/jmir.2496.
7. Aubert CE, Schnipper JL, Williams MV, Robinson EJ, Zimlichman E, Vasilevskis EE, Kripalani S et al., Simplification of the HOSPITAL score for predicting 30-day readmissions. *BMJ Qual Saf.* 2017. doi:10.1136/bmjqs-2016-006239.

8. Bui AA, Darrell Van Horn J. NIH BD2K centers consortium. Envisioning the future of "Big Data" biomedicine. *J Biomed Inform*. 2017. doi:10.1016/j.jbi.2017.03.017.

9. https://www.healthdatamanagement.com/slideshow/the-10-biggest-big-data-companies-by-revenue; June 8, 2016; accessed February 28, 2017.

10. www.healthcareitnews.com/.../healthcare-it-startups-watch-2016-running-list-big-new, Monegain, B. December 20, 2016; accessed February 28, 2017.

11. Buetow K. Heading for the BIG Time (PDF). *The Scientist*. 2008; 22(4):60.

12. Oster S, Langella S, Hastings S, Ervin D, Madduri R, Kurc T, Siebenlist F, Covitz Om Shanbhag K, Foster I, Saltz J. caGRID 1.0: A grid enterprise architecture for cancer research. *AMIA Annu Symp Proc*. 2007; 2007:573–577.

13. Board of Scientific Advisors Ad Hoc Working Group (March 3, 2011). An Assessment of the Impact of the NCI Cancer Biomedical Informatics Grid (caBIG®) (PDF). National Cancer Institute.

14. https://cbiit.nci.nih.gov/ncip; accessed March 31, 2017.

15. NCI PAR 15-331: Advanced Development of Informatics Technologies for Cancer Research and Management (U24), August 18, 2015.

16. PAR 15-332, Companion funding opportunity to PAR-15-332: U01 Research Projects—Cooperative Agreements (U01), August 18, 2015.

17. PAR 15-334, Companion funding opportunity to 331, R21: Exploratory/Developmental Grants, August 18, 2015.

18. https://www.hhs.gov/hipaa/. Health Insurance Portability and Accountability Act, USA. Accessed March 31, 2017.

19. http://DICOM.nema.org. Digital Imaging and Communications in Medicine (DICOM). Accessed March 31, 2017.

20. www.snomed.org/snomed-ct. Systematized Nomenclature of Medicine—Clinical Terms (SNOMED CT). Accessed March 31, 2017.

21. Forrey AW, McDonald CJ, DeMoor G, Huff SM, Leavelle D, Leland D, Fiers T et al. Logical observation identifier names and codes (LOINC) database: A public use set of codes and names for electronic reporting of clinical laboratory testes results. *Clin Chem*. 1996; 42(1):81–90.

22. Sioutos N, de Coronado S, Haber MW, Hartel FW, Shaiu WL, Wright LW. NCI Thesaurus: A semantic model integrating cancer-related clinical and molecular information or cancer-related terminology. *J Biomed Inform*. 2007; 40(1):30–43.

23. http://www.hl7.org/implement/standards/, accessed March 31, 2017.

24. Varma S, Haq S, Haq MM, Raju N, Varma R. Structured data system for a breast cancer medical record. *Stud Health Technol Inform*. 2009; 143:354–357.

25. Simpson RW, Berman MA, Foulis PR, Divaris DX, Birdsong GG, Mirza J, Moldwin R, Spencer S, Srigley JR, Fitzgibbons PL. Cancer biomarkers: The role of structured data reporting. *Arch Pathol Lab Med*. 2015; 139(5):587–593.

26. http://www.pcori.org/, accessed April 30, 2017.

27. https://www.facs.org/quality-programs/cancer/ncdb, accessed April 30, 2017.

28. Benedict S. WE-H-BRB-01: Overview of the ASTRO-NIH-AAPM 2015 Workshop on exploring opportunities for radiation oncology in the era of big data. *Med Phys* 2016; 43:3842.

29. https://cancerlinq.org/, ASCO CancerLinQ | Enhance Cancer Diagnosis & Treatment, accessed April 30, 2017.

30. https://www.flatiron.com/ Flatiron Health: The New Standard for Oncology Technology, accessed April 30, 2017.

31. https://www.cotahealthcare.com Cota: Improving the lives of cancer patients everywhere, accessed April 30, 2017.

32. https://www.mskcc.org/about/innovative-collaborations/watson-oncology.

33. https://www.optum.com/,accessed April 30, 2017.

34. http://nanthealth.com/ NantHealth: Home, accessed April 30, 2017.

35. http://www.prnewswire.com/news-releases/ Varian medical systems and flatironhealth to develop next generation of cloud based oncology software. May 26, 2015, accessed April 30, 2017.
36. www.rtog.org, Radiation Therapy Oncology Group, 1818 Market Street, Suite 1600, Philadelphia, PA 19103-3604, accessed April 30, 2017.
37. https://www.rtog.org/AboutUs/NRGOncologyInformation.aspx, accessed April 30, 2017.
38. https://ctep.cancer.gov/initiativesprograms/nctn.htm, An overview of NCI's National Clinical Trials Network, accessed April 30, 2017.
39. Bradley JD, Hope A, El Naqa I, Apte A, Lindsay PE, Bosch W, Matthews J, Sause W, Graham MV, Deasy JO. A nomogram to predict radiation pneumonitis, derived from a combined analysis of RTOG 9311 and institutional data. *Int J Radiat Oncol Biol Phys*, 2007; 69(4):985–992.
40. Tucker SL, Dong L, Bosch WR, Michalski J, Winter K, Lee AK, Cheung MR, Kuban DA, Cox JD, Mohan R. Fit of a generalized Lyman normal-tissue complication probability (NTCP) model to Grade ≥ 2 late rectal toxicity data from patients treated on protocol RTOG 94-06. *Int J Radiat Oncol Biol Phys*. 2007; 69(3):S8–S9.
41. Deasy JO, Bentzen SM, Jackson A, Ten Haken RK, Yorke ED, Constine LS, Sharma A, Marks LB. Improving normal tissue complication probability models: The need to adopt a "data-pooling culture." *Int J Radiat Oncol Biol Phys*. 2010; 76: S151–S154.
42. Palta JR, Efstathiou JA, Bekelman JE, Mutic S, Bogardus CR, McNutt TR, Gabriel PE, Lawton CA, Zietman AL, Rose CM. Developing a national radiation oncology registry: From acorns to oaks. *Pract Radiat Oncol*. 2012; 2(1):10–17.
43. http://www.ascopost.com/issues/september-25-2015, Goldberg KB, NROR Radiation Oncology looks to collaboration for big data systems, accessed May 15, 2017.
44. Harrington RA, CTN best practices: Creating implementing, and sharing best practices for clinical trials networks, *Second Steering Committee Meeting* May 9 and 10, 2005, Bethesda, MD, Summary: p. 11, August 3, 2005.
45. McNutt T, Wong J, Purdy J, Valicenti R, DeWeese T. OncoSpace: A new paradigm for clincial research and decision support in radiation oncology. *Proceedings of the XVIth International Conference on the Use of Computers in Radiation Therapy*, Sonke JJ (Ed.), Amsterdam, the Netherlands, Published by Het Nederlands Kanker Instituut—Antoni van Leeuwenhoek Ziekenhuis, 2010.
46. Lambin P, Roelofs E, Reymen B, Velazquez ER, Buijsen J, Zegers CM, Carvalho S et al. Rapid learning health care in oncology—An approach towards decision support systems enabling customised radiotherapy, *Radiother Oncol*. 2013; 109(1):159–164. doi:10.1016/j.radonc.2013.07.007.
47. Doiron D, Burton P, Marcon Y, Gaye A, Wolffenbuttel BH, Perola M, Stolk RP et al. Data harmonization and federated analysis of population-based studies: The BioSHaRE project. *Emerg Themes Epidemiol*. 2013; 10(1):12. doi:10.1186/1742-7622-10-12.
48. Fleischman W, Lowry T, Shapiro J. The visit-data warehouse: Enabling novel secondary use of health information exchange data. *EGEMS* (Wash DC). 2014; 2(1):1099. doi:10.13063/2327-9214.1099.
49. Razick S, Močnik R, Thomas LF, Ryeng E, Drabløs F, Sætrom P. The eGenVar data management system—cataloguing and sharing sensitive data and metadata for the life sciences. *Database* 2014; 2014:bau027. doi:10.1093/database/bau027. Print 2014.
50. http://www.eurocat.info/, Home, accessed May 15, 2017.
51. http://www.swat4ls.org/wp-content/uploads/2014/12/Dekker-SWAT4LS-Dec-2014.pdf, accessed May 15, 2017.
52. Dehing-Oberije C, Yu S, De Ruysscher D, Meersschout S, Van Beek K, Lievens Y, Van Meerbeeck J et al. Development and external validation of prognostic model for 2-year survival of non-small-cell lung cancer patients treated with chemoradiotherapy. *Int J Radiat Oncol Biol Phys*. 2009; 74(2):355–362. doi:10.1016/j.ijrobp.2008.08.052.

53. Dekker A, Vinod S, Holloway L, Oberije C, George A, Goozee G, Delaney GP, Lambin P, Thwaites D. Rapid learning in practice: A lung cancer survival decision support system in routine patient care data. *Radiother Oncol.* 2014; 113(1):47–53. doi:10.1016/j.radonc.2014.08.013.

54. Oberije C, Nalbantov G, Dekker A, Boersma L, Borger J, Reymen B, van Baardwijk A et al. A prospective study comparing the predictions of doctors versus models for treatment outcome of lung cancer patients: A step toward individualized care and shared decision making. *Radiother Oncol.* 2014; 112(1):37–43.

55. Dekker A, Wiessler W, Nalbantov G, Bulens P, Coucke P, Dries W, Krishanpuram B, Eble M, Lambin P. euroCAT: A rapid learning health care system for Radiation Oncology, *Proceedings of the Proceedings of the 17th International Conference on the Use of Computers in Radiation Therapy*, Melbourne, Australia, May 6–9, 2013.

56. http://www.eurocat.info/community.html, accessed May 15, 2017.

57. Lambin P, Rios-Velazquez E, Leijenaar R, Carvalho S, van Stiphout RG, Granton P, Zegers CM et al. Radiomics: Extracting more information from medical images using advanced feature analysis. *Eur J Cancer.* 2012; 48(4):441–446. doi:10.1016/j.ejca.2011.11.036.

58. Parmar C, Grossmann P, Bussink J, Lambin P, Aerts HJ. Machine learning methods for quantitative radiomic biomarkers. *Sci Rep.* 2015; 5:13087. doi:10.1038/srep13087.

59. Skripcak T, Belka C, Bosch W, Brink C, Brunner T, Budach V, Büttner D et al. Creating a data exchange strategy for radiotherapy research: Towards federated databases and anonymised public datasets. *Radiother Oncol.* 2014; 113(3):303–309.

60. Chetty IJ, Martel MK, Jaffray DA, Benedict SH, Hahn SM, Berbeco R, Deye J et al. Technology for innovation in radiation oncology. *Int J Radiat Oncol Biol Phys.* 2015; 93(3):485–492.

61. http://skyserver.sdss.org/dr13/en/home.aspx, accessed May 15, 2017.

62. Wu B, Ricchetti F, Sanguineti G, Kazhdan M, Simari P, Jacques R, Taylor R, McNutt T. Data-driven approach to generating achievable dose-volume histogram objectives in intensity-modulated radiotherapy planning. *Int J Radiat Oncol Biol Phys.* 2011; 79(4):1241–1247. doi:10.1016/j.ijrobp.2010.05.026.

63. Wu B, McNutt T, Zahurak M, Simari P, Pang D, Taylor R, Sanguineti G. Fully automated simultaneous integrated boosted-intensity modulated radiation therapy treatment planning is feasible for head-and-neck cancer: A prospective clinical study. *Int J Radiat Oncol Biol Phys.* 2012; 84(5):e647–e653. doi:10.1016/j.ijrobp.2012.06.047.

64. Wang Y, Zolnay A, Incrocci L, Joosten H, McNutt T, Heijmen B, Petit S. A quality control model that uses PTV-rectal distances to predict the lowest achievable rectum dose, improves IMRT planning for patients with prostate cancer. *Radiother Oncol.* 2013; 107(3):352–357. doi:10.1016/j.radonc.2013.05.032.

65. Yang WY, Moore J, Quon H, Evans k, Sharabi A, Herman J, Hacker-Prietz A, McNutt T. Browser based platform in maintaining clinical activities—use of the iPads in head and neck clinics. *Phys. Conf Ser* 489:012095. doi:10.1088/1742-6596/489/1/012095.

66. Appenzoller LM, Michalski JM, Thorstad WL, Mutic S, Moore KL. Predicting dose-volume histograms for organs-at-risk in IMRT planning. *Med Phys.* 2012; 39(12):7446–7461. doi:10.1118/1.4761864.

67. Moore KL, Appenzoller LM, Tan J, Michalski JM, Thorstad WL, Mutic S. Clinical implementation of dose-volume histogram predictions for organs-at-risk in IMRT planning. Published under license by IOP Publishing Ltd. *Journal of Physics*, Conference Series 2014; 489:012055.

68. Shiraishi S, Moore KL. Knowledge-based prediction of three-dimensional dose distributions for external beam radiotherapy. *Med Phys.* 2016; 43(1):378. doi:10.1118/1.4938583.

69. https://www.varian.com/oncology/products/software/treatment-planning/rapidplan-knowledge-based-planning, accessed May 15, 2017.

70. Cheng Z, Nakatsugawa M, Kiess AP, Robertson SP, Moore J, Allen M, Afonso S. et al. The role of a decision tree model to predict weight loss following radiation therapy in head and neck cancer patients. *Int J Radiat Oncol.* 2015; 93(3):E335. doi:10.1016/j.ijrobp.2015.07.1401.

71. Rokach S, Maimon, O. *Data Mining with Decision Trees: Theory and Applications* 2008. Hackensack, NJ: World Scientific.
72. https://www.r-statistics.com/2015/12/r-3-2-3-is-released-with-improvements-for-windows-users-and-general-bug-fixes/, accessed May 15, 2017.
73. Nakatsugawa M, Cheng Z, Goatman K, Lee J, Robinson A, Kiess A, Choflet A et al. Radiomic analysis of salivary glands for prediction of xerostomia and weight loss. *Radiother Oncol.* 2017, submitted for publication.
74. Herman JM, Chang DT, Goodman KA, Dholakia AS, Raman SP, Hacker-Prietz A, Iacobuzio-Donahue CA et al., Phase 2 multi-institutional trial evaluating gemcitabine and stereotactic body radiotherapy for patients with locally advanced unresectable pancreatic adenocarcinoma. *Cancer* 2015; 121(7):1128–1137. doi:10.1002/cncr.29161.
75. Bowers MR, McNutt TR, Wong JW, Phillips MH, Hendrickson KRG, Song WY, Kwok P, DeWeese TL. Oncospace Consortium: A shared radiation oncology database system designed for personalized medicine and research. ASTRO Annual Meeting 2015. *International Journal of Radiation Oncology • Biology • Physics*, 2015; 93(3):E385.
76. Hendrickson KRG, Phillips MH, Fishburn MB, Evans KT, Banerian SP, Wong JW, McNutt TR, MR Bowers, Moore JA. A radiation oncology-specific multi-institutional federated database: initial implementation. NCI/ASTRO Big Data Workshop, Washington DC (2015).
77. Bowers MR, Robertson SP, Moore JA, Wong JW, Phillips MH, Hendrickson KRG, Evans KT, McNutt TR. Multi-institutional plan quality checking tools built on Oncospace: A shared radiation oncology database system. SU-F-P-35, *58th Annual Meeting of the AAPM*, Washington, DC, 2016.
78. Bowers MR, McNutt TR, Robertson SP, Moore JA, Wong JW, Phillips MH, Hendrickson KRG, Kwok P, Song WY, DeWeese TL. Oncospace Consortium: A shared radiation oncology database system designed for treatment planning support and research.*International Conference on the use of Computers in Radiation Therapy (ICCR)*, London, UK, June 2016.
79. Groves P, Kayyali B, Knott D, Van Kuiken S. The big data revolution in healthcare. 2013. http://www.mckinsey.com/industries/healthcare-systems-and-services/our-insights/the-big-data-revolution-in-us-health-care, accessed May 15, 2017.
80. Hui X, Quon H, Robertson SP, Cheng Z, Moore J, Bowers MR, Page BR et al. A risk prediction model for head and neck radiation toxicities: Novel insights to reduce the risk of head and neck radiation-induced xerostomia. *Int J Radiat Oncol Biol Phys.* 2016; 96(2S):E686. doi:10.1016/j.ijrobp.2016.06.2344.

Index

Note: Page numbers followed by f and t refer to figures and tables respectively.

Milton Keynes UK
Ingram Content Group UK Ltd.
UKHW051944071024
449327UK00026B/2156

9 780367 571542